Logical Dilemmas

Kurt Gödel at Amherst College, 1967.

Logical Dilemmas

The Life and Work of Kurt Gödel

John W. Dawson, Jr.

Department of Mathematics
Pennsylvania State University
York, Pennsylvania

A K Peters
Wellesley, Massachusetts

Editorial, Sales, and Customer Service Office

A K Peters, Ltd.
289 Linden Street
Wellesley, MA 02181

Library of Congress Cataloging-in-Publication Data

Dawson, John W. (John William), 1944—
    Logical Dilemmas : the life and work of Kurt Gödel / John W. Dawson.
        p.   cm.
    Includes bibliographical references (p.   —   ) and index.
    ISBN: 1-56881-025-3
    1. Gödel, Kurt. 2. Logicians--United States--Biography.
    3. Logicians--Austria--Biography.   I. Title.
QA29.058D39   1996
193--dc21
[B]   96-50001
        CIP

Printed in the United States of America
01 00 99 98   10 9 8 7 6 5 4 3 2

To Cheryl

*Without whose love, assistance, and forbearance
this book could not have been written.*

Contents

Preface and
Acknowledgments

EVERY BIOGRAPHER IS confronted with certain questions whose answers depend upon the nature of the particular life to be examined: What is it that sets it apart from others and makes it worth writing about? What sources are there to draw upon? To whom should one's account be addressed? What should its central focus be?

In the case of Kurt Gödel the subject was a reclusive genius whose work has generally been considered abstruse and whose life, combining elements of rationality and psychopathology, has been the subject more of rumor than of concrete factual knowledge. There is, however, no doubt that Gödel's discoveries have been of the utmost importance within mathematics, and there is growing awareness of the impact they have had on our modern world view. The problem is to make the ideas underlying his work comprehensible to nonspecialists without lapsing into oversimplification or distortion, and to reconcile his personality with his achievements.

Since a biography is not a textbook, one whose subject is a twentieth-century mathematician must of necessity be addressed to persons who possess a modicum of mathematical understanding. I have consequently presumed that readers of this volume will have some acquaintance with the large-scale structure of modern mathematics and at least a passing familiarity with some of its major figures. Those, for example, who are wholly ignorant of nineteenth century developments in analysis, when the notions of function and of real number were first made precise, cannot be expected to appreciate foundational questions that arose therefrom; and those who have never heard of Hilbert or von Neumann are unlikely to have heard of Gödel either.

I have not, however, presumed any acquaintance with modern mathematical logic, since even among mathematicians of the first rank such knowledge is often wanting. In chapter III, I have interrupted the narrative of Gödel's life to provide a précis of the development of logic up to the time of his own contributions; in chapter VI, I have reviewed in somewhat more detail the early history of set theory; and in an appendix I have included brief biographical vignettes of some of the major figures mentioned in the text.

For those well versed in logic the present work is intended as a complement to Gödel's *Collected Works*. The three volumes of that compilation so far published contain full texts of all of Gödel's publications (with parallel English translations of all German originals), as well as previously unpublished essays and lectures. Incisive commentary on the content, significance, and influence of each item is given in an accompanying introductory note, and there is an extensive bibliography of related literature. The edition is modeled on Jean van Heijenoort's deservedly lauded *From Frege to Gödel: A Source Book in Mathematical Logic, 1879–1931*, in which a selection of seminal texts leading up to Gödel's incompleteness paper is similarly presented.

In citing references I have adopted the format employed in those exemplars: Sources are indicated by a code consisting of the author's name and the date of publication (or of composition, in the case of unpublished manuscripts), both set in italics — e.g., *Gödel 1931a*. When the author is clear from context I have sometimes omitted the name, and when more than one work by an author appeared during the same year I have used letter suffixes to distinguish among them.

For the most part, information concerning sources has been put into numbered endnotes, cited within brackets, that are collected together at the back of the volume. Explanatory footnotes, on the other hand, appear at the bottom of the page.

◇ ◇ ◇

Because Gödel's circle of acquaintances was so restricted, accounts of his life must rely primarily on documentary sources rather than interviews. The most important collection of such sources is Gödel's own *Nachlaß* (literary remains, designated herein by the letters GN), housed at the Firestone Library of Princeton University. It appears that Gödel retained almost every scrap of paper that crossed his desk, including library request slips, luggage tags, crank correspondence, and letters from autograph seekers and mathematical amateurs, so his *Nachlaß* is perhaps best described as a scholarly midden. Within it, for those willing to sieve through the mass of accessory material, lies embedded a rich trove of information.

Outside the *Nachlaß* four other major sources chronicle particular phases of Gödel's life. For his childhood there is his brother Rudolf's memoir of their mother (*R. Gödel 1987*). For his years in Vienna and his semester at Notre Dame the recollections of Karl Menger (*1981*) are particularly valuable. For the period 1945–66 Gödel's letters to his mother, preserved at the Wiener Stadt- und Landesbibliothek and designated here by the letters FC (family correspondence), yield otherwise unobtainable insights into the human side of his character. And for the decades 1947–77 the diaries of his friend Oskar Morgenstern (cited as OMD) are indispensible.

In the years since Gödel's death several overviews of his life have appeared in print, including the obituary memoirs by Georg Kreisel (*1980*), Curt Christian (*1980*), and Stephen C. Kleene (*1987b*); my own sketch *1984b*; and the fine essay by Solomon Feferman (*1986*) in Volume I of Gödel's *Collected Works.* Two books by the late Hao Wang, *Reflections on Kurt Gödel* and *A Logical Journey: From Gödel to Philosophy*, also contain chapters devoted to Gödel's life. Neither of them, however, is a full-scale biography.

The present work is the first study of Gödel's life and work to draw upon all of the sources listed above. It grew out of my experiences as the cataloger of Gödel's *Nachlaß* and as a co-editor of his *Collected Works.* I am deeply grateful to the Institute for Advanced Study, Princeton, and its former director, Dr. Harry Woolf, for the invitation to undertake the cataloging, for stipendiary support during the two years that task required and for permission to quote from Gödel's writings. Supplemental support for travel and photocopying expenses was provided by the Campus Advisory Board at Penn State York. I am indebted, too, to my colleagues on the editorial board of the *Collected Works*, with whom I have had many illuminating conversations and from whom I have learned much. I wish especially to thank Solomon and Anita Feferman, whose loving support and friendship, wise counsel and constructive criticism helped to sustain and enhance my efforts. Both they and Charles Parsons took the time to read through my draft manuscript and to offer detailed suggestions for improvement.

Archivists at several institutions have cordially assisted my endeavors. At Princeton's Firestone Library Alice V. Clark, Margarethe Fitzell, Marcella Fitzpatrick, and Ann van Arsdale were especially helpful. At the Institute for Advanced Study Elliott Shore, Mark Darby, Ruth Evans, and Momota Ganguli provided ongoing assistance during and after my cataloging of Gödel's papers. At the Hillman Library of the University of Pittsburgh Richard Nollan and W. Gerald Heverly answered my queries concerning the papers of Rudolf

Carnap and Carl G. Hempel, transcribed passages from Carnap's shorthand diaries and helped to arrange permission for me to quote from them. Wendy Schlereth of the University of Notre Dame Archives kindly searched Gödel's personnel files there and was instrumental in obtaining permission for me to see and quote from letters in the administrative files of former University president John F. O'Hara. I am grateful also to Linda McCurdy and William Erwin, Jr. of the Perkins Library of Duke University for assistance during a week spent there and for permission to copy extracts from the diaries of Oskar Morgenstern.

Several individuals helped me to obtain copies of photographs and documents from European repositories. Foremost among them are Eckehart Köhler and Werner DePauli-Schimanovich of the University of Vienna; Blazena Švandová of the Kurt Gödel Society in Brno, Czech Republic; Beat Glaus of the ETH-Bibliothek in Zürich; and Wolfgang Kerber of the Zentralbibliothek für Physik in Wien.

I am grateful as well to the many associates of Kurt and Adele Gödel who consented to be interviewed about them or who shared their recollections with me in correspondence: Dr. Franz Alt; Professors Paul Benacerraf, Gustav Bergmann, and Herbert G. Bohnert; Dr. George Brown; Professors Freeman Dyson and Paul Erdös; Mrs. Adeline Federici; Professor Herbert Feigl; Mrs. Elizabeth Glinka; Dr. med. Rudolf Gödel; Dr. Herman Goldstine; Professor and Mrs. Carl Kaysen; Professor John Kemeny; Harry Klepetař; Professors Georg Kreisel, Saunders Mac Lane, Karl Menger, and Deane Montgomery; Mrs. Dorothy Morgenstern Thomas; Mrs. Louise Morse; Professor Otto Neugebauer; Mrs. Dorothy Paris; Professors Atle Selberg and Olga Taussky-Todd; Miss Carolyn Underwood; and Professors Hao Wang and Morton White.

For bringing to my attention documents of which I would otherwise have remained unaware I wish to thank Mrs. Phyllis Post Goodman, Gerhard Heise, and Professors Peter Suber and Christian Thiel. For information concerning Gödel's medications I am grateful to Eugene and Lucille Sire. And for escorting me to sites associated with Gödel in and around Brno I am indebted to Professor Jiří Horejš.

In addition to those already mentioned, acknowledgment is due the following organizations and individuals for granting permission to use the materials indicated:

The Graphische Sammlung Albertina, Vienna, and Heinrich Moser, for the color reproduction of the painting "Homage to the American dancer Loïe Fuller" by Koloman Moser.

The ETH Bibliothek, Zürich, and Dr. med. Ludwig Bernays, for quotations from the letters of Professor Paul Bernays.

Bibliopolis, Naples, and its director Francesco del Franco, for quotations from the book *Gödel Remembered.*

The Carnap Collection Committee of the University of Pittsburgh Library, for quotations from the diaries of Rudolf Carnap. Quoted by permission of the University of Pittsburgh. All rights reserved.

The late Professor Alonzo Church, for quotations from his letters to Gödel and to me.

Harcourt Brace Jovanovich, Inc., and Faber and Faber Ltd., for the excerpt from "Burnt Norton" in *Four Quartets,* copyright 1943 by T.S. Eliot, renewed 1971 by Esme Valerie Eliot. Reprinted by permission of Harcourt Brace Jovanovich, Inc.

The Rare Books and Manuscripts Division of the New York Public Library and the Astor, Lenox, and Tilden Foundations, for quotations from the records of the Emergency Committee in Aid of Displaced Foreign Scholars.

Frau Regula Lips-Finsler, for quotations from the letters of her uncle, Professor Paul Finsler.

Professor Carl G. Hempel, for extracts from his interview of 17 March 1982 with Richard Nollan.

Wolfgang Ritschka of Galerie Metropol, Inc., for the reproductions of photographs from the book *Sanatorium Purkersdorf.*

Leon M. Despres, trustee of the Karl Menger Trust, for quotations from the writings of Professor Karl Menger.

Dorothy Morgenstern Thomas, for quotations from the diaries of her late husband, Professor Oskar Morgenstern.

Oxford University Press, for quotations of commentary from the volumes of Kurt Gödel's *Collected Works.*

Phyllis Post Goodman, for quotations from the letters of her father, Professor Emil L. Post.

The Copyright Permissions Committee of the Bertrand Russell Archives, McMaster University, Hamilton, Ontario, for the quotation from Russell's letter of 1 April 1963 to Professor Leon Henkin.

The University of Vienna, for quotations of materials from the personal file of Kurt Gödel, in the *Dekanats-Bestand* of the philosophical faculty.

The Zentralbibliothek für Physik in Wien and Professor Walter Thirring, for the quotation from Professor Hans Thirring's letter to Gödel of 27 June 1972.

Professor Marina von Neumann Whitman, for quotations from the papers of her father, Professor John von Neumann.

The Universitätsbibliothek, Freiburg im Breisgau, for quotations from the correspondence between Gödel and Ernst Zermelo.

Klaus Peters and his staff at AK Peters, Ltd., have been most cordial, helpful and efficient in preparing this text for publication. Special thanks are due Joni Hopkins McDonald, Erin Miles, Iris Kramer-Alcorn, and an anonymous copy-editor for their careful proofreading and correction of my manuscript and their creative contributions to the design and production of this book. Their efforts were further enhanced by Alexandra Benis, who coordinated and expedited communications among us.

I owe the greatest debt of all to my wife, Cheryl, who has borne with me through all the stresses of authorship and who, in a rash moment, volunteered to learn to read Gödel's Gabelsberger shorthand. Without the window thus opened on his thoughts, this account would be much the poorer.

John W. Dawson, Jr.
York, Pennsylvania
May 1996

I
Der Herr Warum
(1906–1924)

> The child, who supposes that there are reasons for every-
> thing, asks Why ... [not only when] a reason exists, but
> also ... in cases where the phenomenon is fortuitous but
> ... the child sees a hidden cause.
>
> —Jean Piaget and Bäbel Inhelder, *The Origin of the Idea of
> Chance in Children*

KURT GÖDEL WAS an exceptionally inquisitive child. By the time he was four years old his parents and older brother had begun to call him "der Herr Warum" (Mr. Why), and in an early family portrait he stares at the camera with an earnestly questioning gaze (Figure 1).

Herr Warum's queries were typical of those that children are wont to ask. They were sometimes embarrassing — as when he once asked an elderly visitor why her nose was so long — but more often were just the kinds of questions that adults regard as *not having* answers. What set Kurt Gödel apart was not that he asked such questions but that he never stopped asking them. Throughout his life, he refused to accept the notion of fortuitous events. A few years before his death he declared, "Every chaos is merely a wrong appearance," and in an undated memorandum found among his papers after his death he listed fourteen principles he considered fundamental. First among them was "Die Welt ist vernünftig" (The world is rational) [1].

To persist in asking "unanswerable" questions can quickly lead to social isolation, for the questioner is more likely to be reckoned a crank than a genius. Indeed, to seek rationality in all things is, from a modern point of view, a profoundly *irrational* act. It is not just that causal determinism is in opposition to the contemporary scientific *Zeitgeist*. It is that there are seemingly insuperable obstacles to the rational explanation of much human behavior.

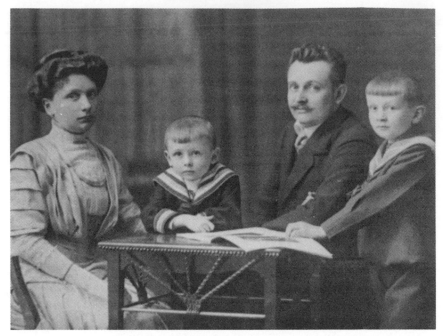

FIGURE 1. The Gödel family, ca. 1910: Marianne, Kurt, father Rudolf, son Rudolf.

Consequently, an archrationalist who is determined, as Gödel was, to find hidden causes to account for such behavior is likely to become distrustful of human motives.

At the same time, one who is convinced of the underlying orderliness of the world is likely to be attracted to mathematics. As Gödel himself once wrote, "In the world of mathematics everything is well poised and in perfect order." But Gödel went further, asking, "Shouldn't the same be assumed (expected) for the world of reality, contrary to appearances?" [2].

Gödel's choice of profession, his Platonism, his mental troubles, and much else about him may thus be ascribed to a sort of arrested development. He was a genius, but he was also, in many respects, a man/child. Otto Neugebauer, the eminent historian of ancient mathematics whose association with Gödel spanned nearly fifty years, described him as a precocious youth who became old before his time; and Deane Montgomery, another of Gödel's colleagues, noted that like a child, Gödel always needed to be looked after [3]. Despite his prodigious intellect, he often exhibited a childlike naiveté. His tastes remained unsophisticated, and his well-being depended on the efforts of those who were willing to shield him from the outside world, to tolerate his sometimes bizarre behavior, and to see that he received treatment in times of physical and mental disability.

Gödel's childhood may thus be expected to provide insight into the development of his personality, so long as the question "Why?" is not pressed too persistently.

◇ ◇ ◇

The word "Gödel" is not to be found in modern German dictionaries. In Weigand's *Deutsches Wörterbuch* of 1909, however, "Godel" appears as a main entry, said to be a variant of "Gote"; and under that entry, in turn, is the remark "auch dim. *Gödel*." The meaning given for all three terms is "Pate" (godparent).

It is unclear when the word "Gödel" became a surname, but Kurt Gödel's paternal forebears can be traced back at least four generations, to a Carl Gödel, who died in 1840. About him, his wife, and five children few details survived in family memory. It was recalled only that the family came from Bohemia and Moravia — then part of the Austro-Hungarian monarchy — and that several of its members lived in Brno (or Brünn, as it is called even today by Germans and Austrians), where they were employed in the leather industry as merchants, bookbinders, or, in one case, as a factory owner. In general they were not successful in their trades, so the family's resources gradually became diminished [4].

Carl's son Josef had five children of his own, including Kurt Gödel's grandfather, also named Josef. The latter lived with his wife, Luise (Aloisia), in Vienna, where he, too, was employed in the leather industry. His son Rudolf August, who would become Kurt's father, was born in Brno in 1874 [5]. Shortly thereafter the younger Josef died (allegedly a suicide), causing his wife to entrust the care of young Rudolf to two of Josef's siblings, Anna and August, who reared the boy in Brno as their foster child.

Rudolf got on well with his aunt and uncle, but he did not do well in grammar school. Accordingly, at about age twelve he was sent to a weaver's school — a wise decision, as it turned out, for Brno was the center of the Austro-Hungarian textile industry, and the boy's interest in and talent for the clothmaking trade soon became apparent. Indeed, after completing his studies "with distinction ... [he] immediately obtained a position in the ... renowned textile factory of Friedrich Redlich, [where] he worked ... until his death," rising swiftly within the firm to become its director and, in the end, a partner. Along the way, however, he became estranged from his mother, in part because she "made too many pecuniary demands of him" during the years before he became fully established [6].

Rudolf's interest in the textile industry may also have been fostered by the Gödels' friendship with the Handschuhs, a family of seven that shared the apartment building at 9 Bäckergasse (now Pekařská) where the Gödels lived [7]. Gustav Handschuh (literally "hand shoe," the German word for glove) was a weaver who had emigrated from the Rhineland. By dint of hard personal effort, he had slowly worked his way up to the position of purchasing agent in the prosperous Schöller firm. He was active in public life, and one of his three daughters, Marianne, was close to Rudolf in age.

There was ample opportunity for the two families to become acquainted. The apartment house was constructed "in the Biedermeier style, with open galleries on which the neighbors would meet in the evenings" to chat or play music together [8]. So it seems reasonable to suppose that the Gödels might have consulted with Herr Handschuh about their foster son's schooling. In any case, he certainly influenced young Rudolf's life later on, for in due course he became Rudolf's father-in-law.

◇ ◇ ◇

The wedding of Rudolf Gödel and Marianne Handschuh took place in Brno on 22 April 1901. In the view of their elder son, Rudolf (born 7 February 1902), the marriage "was not a 'love match'" though "it was certainly built on affection and sympathy." Rather, it was an enduring bond forged out of mutual needs and respect: Marianne "was ... impressed by [Rudolf's] energetic capable nature," while "he, being of a more serious and ponderous disposition, ... found pleasure in her cheerful friendly nature" [9]. As such, the union conformed to prevailing social norms; in particular, it was typical of the times that Marianne, who had attended a French *lycée* in Brno, was far more cultured and better educated than her husband.

Following their marriage, the couple moved to an apartment at 15 Heinrich Gomperzgasse. Soon after their elder son's birth, however, they moved back to 5 Bäckergasse, next door to their parents (Figure 2). And there, on 28 April 1906, their second and last child, Kurt Friedrich, came into the world.

On 14 May, aged sixteen days, the baby was baptized in the German Lutheran congregation in Brno, with Friedrich Redlich, his father's employer, serving as godfather — Gödel's own "Gödel," so to speak. (It was from Redlich, presumably, that the child's middle name was taken.) The ceremony,

Figure 2. Gödel's birthplace, 5 Pekařská Street, Brno, 1993. (Brno Archives)

however, seems to have been *pro forma*, as the parents were not churchgoers. Rudolf August was nominally Old Catholic, while Marianne was Protestant, having been reared in a household where "enlightened piety," including regular Sunday church attendance, prevailed. As an adult, however, she seems not to have practiced her faith. The children were reared as freethinkers, and neither of the sons subsequently belonged to any religious congregation. By his own admission, Rudolf remained a somewhat regretful agnostic throughout his long life [10]. Kurt, however, did eventually become a believer: In an unsent reply to a questionnaire sent to him in 1975 by the sociologist Burke D. Grandjean [11] he described his belief as "theistic rather than pantheistic, following Leibniz rather than Spinoza"; and in a letter to his mother written in 1961, he declared that there is much more that is rational in religion than is generally believed, notwithstanding that from earliest youth we are led to the contrary view through books, experience, and bad religious instruction in the schools [12].

The Grandjean questionnaire is one of several sources that shed light on details of Gödel's childhood and youth. Others include his school records, which he carefully preserved, family photographs, recollections of his brother, and occasional references to childhood events in his postwar correspondence with his mother. The picture that emerges from all those sources is that of an earnestly serious, bright, and inquisitive child who was sensitive, often withdrawn or preoccupied, and who, already at an early age, exhibited certain signs of emotional instability.[1]

Both the Gödel boys were strongly attached to their "liebe Mama." Indeed, at age four and five, little Kurt reportedly cried inconsolably whenever his mother left the house [14]. And though she oversaw the running of the household and frequently entertained guests, Marianne found ample time to read, sing, and play the piano for her children. She was, it seems, a model homemaker and hostess.

With their father, "personal contact was perhaps a little less warm," if only because the children saw much less of him. A man "of thoroughly practical disposition," Rudolf August devoted a good deal of his attention to his business. Still, he was remembered as a "good father who fulfilled many [of his sons'] wishes" and "provided plentifully" for their education [15]. Through his efforts the family became wealthy enough to employ several servants, including a governess for the children.

---

[1] In a letter to Hao Wang of 29 April 1985 [13], Rudolf Gödel stated that at about age five his brother suffered a "leichte Angst Neurose" (mild anxiety neurosis).

Thus the Gödel boys grew up in privileged circumstances, within the community of Sudeten Germans then dominant in Brno society. The brothers had few playmates but apparently got along well together, with only infrequent conflicts. As younger children they played "mostly quiet games" with blocks, toy trains, and such [16]. They also accompanied their parents on excursions to the Moravian countryside and to spas such as Aflenz and Marienbad.

The memory of such occasions remained fresh in Gödel's mind decades later. He recalled visits to Achensee and, especially, Mayrhofen, where he loved to play in the sand piles [17]. With equal nostalgia, both he and his brother remembered their childhood excitement when, a few weeks before Christmas, they were allowed to select some of their forthcoming gifts from the catalog of a Viennese toy store. Christmas itself was celebrated at home, in the company of many of their relatives, and was preceded by a visit to the home of their maternal grandparents, where "the Christmas tree was ... fixed to the ceiling of the room ... [in a way that allowed it to] turn freely on its axis" [18]. It seems, however, that such gatherings did not include their father's relatives, apart from Aunt Anna and Uncle August.

As the boys grew older the Gödel family began to feel the need for larger quarters. Rudolf August therefore began to look for a suitable building site. He found one only two blocks away, at the base of a steep hill just west of the historical town center.

Called the Špilberk, the hill was then, as it is now, Brno's dominating landmark. It is crowned by a brooding stone edifice of the same name, built as a castle in the late thirteenth century and converted to a fortress in the 1740s. The ramparts and outer fortifications were destroyed on Napoleon's orders early in the nineteenth century following his victory at nearby Austerlitz, but the building retained its forbidding character, becoming first a Habsburg and later a Nazi prison. Nevertheless, the destruction of its defensive walls opened the town to residential and industrial expansion, notably that of the developing textile industry, and obviated the need to keep the slopes of the Špilberk free of vegetation.

Trees were planted there about 1870, so that by the first decades of the twentieth century the hill had become an oasis of greenery. As views of its summit were increasingly obscured, the ominous aspect of the Špilberk was diminished and the region at the base of its southern slope became a site for residential development. By around 1910 the area had become quite a fashionable neighborhood, and it was there, at 8A Spilberggasse (now Pellicova, renamed in honor of Silvio Pellico, an Italian poet who was imprisoned in

FIGURE 3. The Gödel villa, Pellicova 8a, Brno, 1983. (J. Dawson)

the Špilberk fortress), that Rudolf August built a three-story gabled villa, a residence commensurate with his family's growing prosperity (Figure 3).

The Gödels moved into the new home in 1913, when Kurt was seven and his brother eleven. The boys and their parents occupied the ground floor, which consisted of five rooms and a large hall, the latter containing a nook decorated in the prevailing *Jugendstil*, with furniture and upholstery designed by the famous Wiener Werkstätte. Aunt Anna lived on the floor above, while the top floor was intended as an apartment for Marianne's unmarried sister Pauline. Before she was to move in, however, a falling out occurred, and though a reconciliation was later effected it was not until World War II that Aunt Pauline finally shared the house with her sister [19].

The lot behind the house extended part way up the slope of the Špilberk, and there the Gödels planted a large garden with many fruit trees. The spacious yard provided ample room for the children to play with the family's two dogs, a Doberman and a small terrier, and the location afforded fine views of the city and the surrounding countryside. With the aid of a small telescope the children could examine the spires of the nearby Cathedral of Saints Peter

FIGURE 4. The Evangelische Privat-Volks- und Bürgerschule, Brno, 1993. (Brno Archives)

and Paul or look south across the Moravian plains to the distant chalk cliffs beyond [20].

But play was no longer the boys' sole occupation, for the previous year Kurt had followed his brother off to school. He was first enrolled on 16 September 1912, at the *Evangelische Privat-Volks- und Bürgerschule*, located on Elisabethstrasse (now Opletalova) just a few blocks away (Figure 4).

The school's name — an awkward compound even by German standards — reflects the evolution and reform of the Austrian school system that had occurred during the previous century. The school was, first of all, private and Protestant (*Evangelische*). Such schools had been permitted ever since the Toleration Charter of 1781, but in practice the Catholic Church had retained near-monopolistic control of education in the Austro-Hungarian monarchy until 1861, when an imperial *Patent* reaffirmed the rights of Protestants to organize their own schools and to determine the extent and kind of religious instruction to be offered therein.

The phrase *Volks- und Bürgerschule* referred to the fusion of two types of schools that had been created under the common school law of 1869. That statute mandated eight years' schooling for all pupils, to be provided by "a five-year elementary school (*Volksschule*) and a three-year higher elementary or grammar school (*Bürgerschule*)" [21]. Kurt Gödel, however, attended his combined Volks- und Bürgerschule for only four years before going on to a *Realgymnasium*, another hybrid type of school created at about the same time.[2]

That the curriculum Gödel encountered was, in fact, a balanced one is attested by the report cards he saved. Those from his primary school show that he received instruction in religion, reading, writing (in the old script, then still in use), German grammar, arithmetic, history, geography, natural history, drawing, singing, and physical education (*Turnen* — literally, gymnastics, but more accurately, calisthenics or floor exercise). A few of Gödel's workbooks from his very first school year have also survived, and from them the extent of drill work can be gauged (Figure 5).

Throughout his primary school career Gödel received the highest marks in all his subjects. He was, however, rather frequently absent (23.5 days in 1913–14, 35.5 in 1914–15 and 16.5 in 1915–16) — though always excusably so — and he was also excused from participation in physical education during all of 1915–16, evidently because of a health problem. Rudolf Gödel later confirmed that at about the age of eight his brother had contracted rheumatic fever, a serious illness that sometimes causes cardiac damage.

---

[2]Traditionally there had been a rigid distinction between *Gymnasien*, which offered a classical education to children from the upper classes, and the vocationally oriented *Realschulen*, where children of the lower middle class might go to learn a trade. Training in the sciences was largely confined to the latter, especially after 1819, when "the curriculum of the Gymnasien was altered to put more emphasis on Latin and Greek, to reduce the time spent on history and geography, and to eliminate natural history, geometry, and physics altogether" [22].

Following the Revolution of 1848 both the Gymnasien and the Realschulen were reorganized, so that the former became "eight-year school[s] receiving pupils of ten or eleven years of age, emphasizing ancient languages, and embodying the idea that the culture of the ancients [was] a model for all time," whereas the Realschulen became "six-year institution[s] ... emphasiz[ing] mathematics, mechanics and mechanical engineering, architecture and technical chemistry, and includ[ing] a modern language as an optional course" [23].

Subsequently, many of the practical courses that had been offered at the Realschulen were dropped in favor of greater emphasis on the sciences and modern languages. Separate vocational schools were then established, as well as the six-year preparatory Realgymnasien, whose "purpose was to provide training past the first four years of Volksschule that would enable the pupil to proceed into either the last four years of a Gymnasium or of a Realschule" [24]. The fundamental "two cultures" dichotomy persisted, however, until 1908, when, in response to increasing criticism, the Realgymnasien were converted to eight-year institutions whose curricula attempted to bridge the cultural gap. It was that structure that was in place during the years the Gödel children attended school.

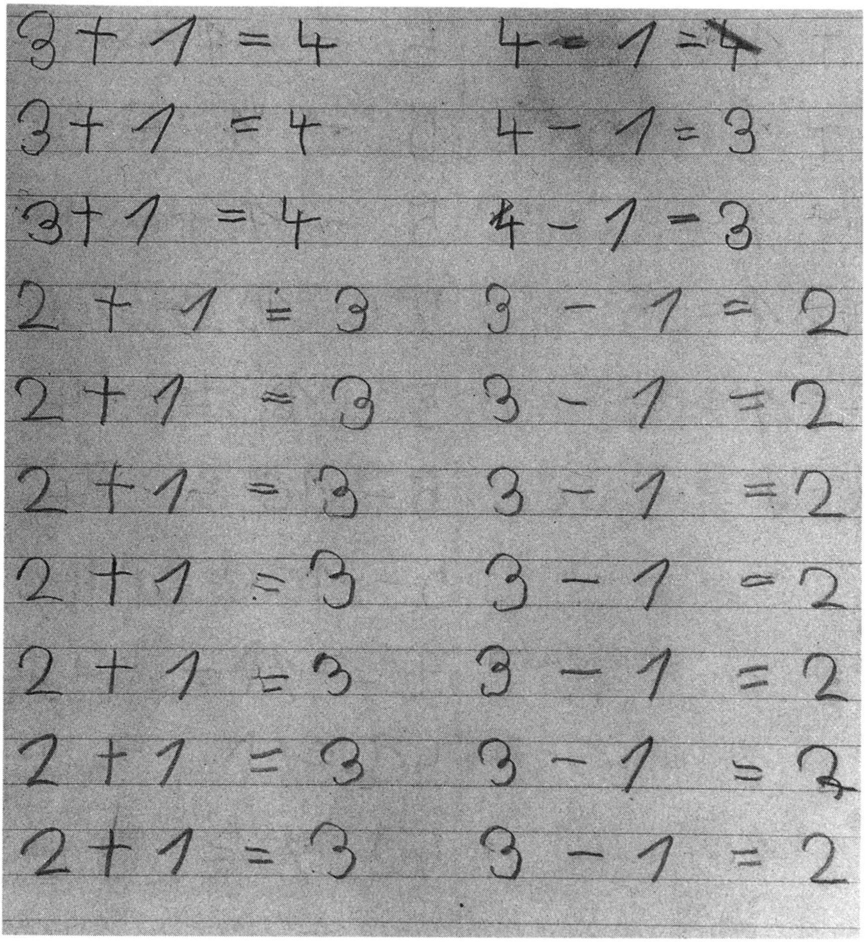

FIGURE 5. A page from Gödel's first arithmetic workbook, 1912-1913.

In Gödel's case the disease seems to have left no lasting physical effects. Nevertheless, the affliction became a turning point in his life, for, as might have been expected of *der Herr Warum*, through reading about the disease he learned for himself about its possible side effects and came to believe, despite his doctors' reassurances, that his heart had been affected. That unshakable conviction, so his brother believed, was the source of the hypochondria that in later years was to become such an integral part of Kurt Gödel's life.

In the meantime, the outbreak of World War I overshadowed everything else. For the Gödel family the effects of the war were delayed and indirect. Brno was far from the battlefields, the boys were too young for service, and their father, then forty, was not called up. But like many other ethnic Germans within the Austro-Hungarian monarchy, the Gödels cultivated their national heritage. Father Rudolf invested much of his wealth in German War Loans, and consequently, with Germany's defeat, he lost a substantial part of his financial resources. Nevertheless, he was not forced to sell the villa, and he was able to continue to pay to send his sons to school. After the war Allied assistance enabled the lands within the newly founded Republic of Czechoslovakia to achieve rapid economic recovery, so within a few years the Gödels were able to resume a high standard of living [25].

What effect, then, did the Great War have on the Gödel children? Some six decades later Kurt Gödel was to write, somewhat uncertainly, that his family had been "not much affected" by the war and the subsequent inflation [26]. On the other hand, his brother noted that the war did affect the boys' lives in one respect, by arousing their interest in games of strategy such as chess. (Within a few years, in fact, young Kurt had apparently become quite a good chess player. One of his classmates at the Realgymnasium testified that only one other student there, a chess master, could defeat him [27].) In any case the war did not interrupt the progress of the boys' education — though, insofar as it depleted the ranks of their teachers, it may have affected the quality of instruction they received. On 5 July 1916 ten-year-old Kurt graduated from the Evangelische Schule, and the following fall he enrolled at the K.-K.[3] *Staatsrealgymnasium mit deutscher Unterrichtssprache*, located on the same street as his father's factory (Strassengasse, now Hybešova) (Figure 6).

Like the primary school the Gödels attended, the Realgymnasium enrolled few students whose native tongue was not German. It was, however, a public school (though it charged tuition and admitted students by competitive examination), so that in other respects the student body was markedly different from that at the Evangelische Schule. In particular, Gödel was one of a tiny minority

---

[3] *Kaiserliche-Königliche* (imperial-royal), that is, pertaining to the Austrian crownlands; to be distinguished both from *königliche* alone (pertaining to the kingdom of Hungary) and from *kaiserliche und königliche* (jointly administered by Austria and Hungary). Following the *Ausgleich* of 1867, these pretentious epithets were applied ubiquitously to public institutions throughout Austria-Hungary [28]. In his novel *Der Mann ohne Eigenschaften* Robert Musil satirized the practice by dubbing imperial Austria "Kakania" — a name ostensibly referring to the German pronunciation of "K.K." but suggesting as well colloquial German *Kaka* (baby poop) [29].

FIGURE 6. Historical postcard view (n.d.) of the Gymnasium that Gödel attended. In the distance is the smokestack of the Redlich textile factory. (Brno Archives)

of Protestant students there: 55 percent of his classmates were Catholic and 40 percent Jewish. Of the 444 students enrolled in the school in the fall of 1916, 92 were in Gödel's entering class. Four years later, however, his class would number just 36 (out of a school population of 336), including the only girl (of 41 altogether) enrolled beyond the third-year level. The attrition seems largely to have occurred at the end of the preceding year, as students completed their eight years of mandatory school attendance [30].

These statistics highlight the social and educational milieu in which Kurt Gödel spent his formative years. How relevant that milieu was to his own social or intellectual development is an interesting question, however, since the introversion and isolation that were to become such fundamental aspects of his personality were by then already becoming manifest. As his concern over his health intensified, for example, his interest and participation in physical activities markedly decreased. Once again, a pattern of frequent excused absences appears in his school reports, and during the school year 1917–18 he was again exempted from physical education.[4] He turned away from swimming and calisthenics, activities he had previously enjoyed, and preferred to stay home and read rather than accompany the family on weekend jaunts to the Moravian countryside — behavior that became a source of alienation from his father [31].

At school he remained aloof from most of his classmates. One of the few there who came to know him well was Harry Klepetař, who shared a bench with him throughout their eight years at the Realgymnasium. Klepetař recalled that "from the very beginning ... Gödel kept more or less to himself and devoted most of his time to his studies. He had only two close friends. One was Adolf Hochwald [the chess player mentioned earlier] ... [and the other] was myself." Klepetař went on to say that "Gödel's interests were manyfold" and that "his interest in mathematics and physics [had already] manifested itself ... at the age of 10" [32].

According to Rudolf Gödel, his brother "took very different degrees of interest in the various subjects" he was taught, preferring mathematics and languages to literature and history [33]. No such preference is evident from his grades, for only once did he ever receive less than the highest mark in a

---

[4]The latter may have been due to an appendectomy, mentioned without reference to date, but with reference to a schoolmate who underwent a similar operation, in Gödel's letter to his mother of 14 February 1962 (FC 181).

subject (mathematics![5]); but foreign languages do appear prominently among his list of courses. Latin and French were required subjects at the Realgymnasium (Gödel studied the former for eight years and the latter for six), and English was one of the two courses that Gödel chose as electives (Figure 7).

He did not, however, elect to study Czech — an omission duly noted later by Czech authorities. Klepetař recalled that Gödel was the only one of his fellow students he never heard speak a word of Czech,[6] and that, especially after October 1918 when the Czechoslovak Republic declared its independence, "Gödel considered himself always Austrian and an exile in Czechoslovakia" [35].

Actually, Germans constituted almost a quarter of the population of the republic. In 1910 some 3,747,000 Germans lived in Bohemia, Moravia, and Silesia, mostly in "solid blocks" near the borders of Germany and Austria, with whose cultures they identified. Unlike the Czechs, who were "for the most part country folk," the Germans lived primarily in industrialized cities. They tended to be better educated than the Czechs, whom they considered lower class, and many of them, including the Gödels, employed Czechs as household servants [36]. Gödel's attitude toward the Czech language might therefore be taken as symptomatic of a more general prejudice against Slavs; and in fact, following the expulsion of Germans from Czechoslovakia at the end of World War II and the later influx of Czech refugees to Vienna, Marianne Gödel does seem to have exhibited such an attitude. But in his letters to her of 27 June and 31 July 1954 [37], Kurt himself declared that except for the ultranationalists among them he did not find Slavs "unsympathisch."

Gödel's interest in foreign languages — at least their formal aspects — continued beyond his school years. His *Nachlaß* contains notebooks on Italian, Dutch, and Greek in addition to the languages already mentioned, and his personal library included various foreign-language dictionaries and grammars. But little in the way of scientific or literary works in languages other than French, German, or English are to be found there, and those appear to be the only languages in which he attained spoken fluency.

---

[5] In his biographical sketch of Alfred Tarski, one of the few logicians whose stature might be compared with that of Gödel, Steve Givant has noted a similar irony: "He [Tarski] was an exceptional student in high school, [but] logic was the one subject in which he did *not* receive the top grade ... " [34].

[6] Apparently he did learn some Czech in the course of his residence in Brno, for one of the cashiers in the dining hall at the Institute for Advanced Study remarked that he had once spoken "Slavisch" to her.

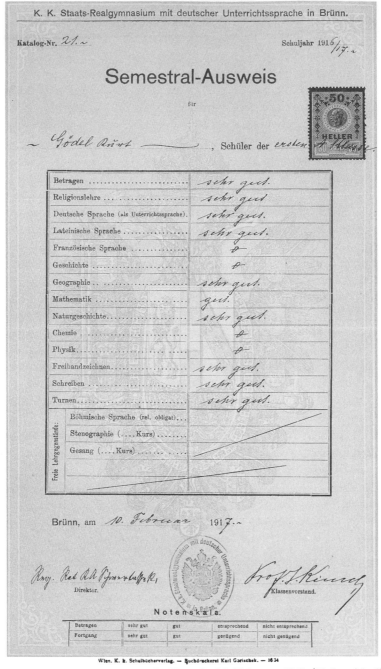

FIGURE 7. Semester report for Kurt Gödel, February 1917. (H. Landshoff)

In the sciences, the curriculum at the Realgymnasium included eight years of mathematics, six each of physics and natural history, and five of chemistry. For some of those courses Gödel's notebooks have been preserved, and from them some idea of the syllabi can be inferred. That for third-year physics, for example, included information on units of measurement, elements of basic astronomy, and the theory of forced harmonic oscillation (in particular the motion of coupled pendulums, including details on techniques for solving second-order linear differential equations). Mathematics instruction covered algebra, plane and solid descriptive geometry,[7] trigonometry, and perhaps some calculus, though it is difficult to say how much since that material is scattered among several notebooks and may have been filled in at other times or have resulted from Gödel's own independent study. In the natural history courses the subject matter varied from year to year. The more advanced topics included zoology (taxonomy and morphology), human physiology, and mineralogy.

Other required courses at the Realgymnasium included German, history, and geography (eight years each); religion (five years); free-hand drawing (four years); and, during the last two years, introduction to philosophy. A composition book from Gödel's fifth class year (1920–21) contains essays and exercises on a wide range of topics, including a descriptive report on "Metals in the service of mankind," translations and interpretations of passages from the *Niebelungenlied* and from the writings of Walter von der Vogelweide, a survey of major German poets up to the twelfth century, and an analysis of Emanuel Geibel's poem "The Death of Tiberius" [38].

Apart from the evidence afforded by the notebooks, there are sharply divergent accounts of the quality of the education that the Realgymnasien provided. In *The World of Yesterday*, Stefan Zweig's autobiographical account of Viennese life at the end of the previous century, Zweig speaks of "the invisible barrier of 'authority'" that stood between teacher and pupil, "the total lack of spiritual and intellectual relationship[s]" between them, and the "unfeeling and soulless method of ... education" that "wearied, hindered, and suppressed" students' interests, energies, and aspirations [39]. But reforms had taken place in the meantime, and while acknowledging that there was less interaction between students and instructors than in American schools, Gödel's schoolmate Klepetař described the Staatsrealgymnasium in Brno as "one of the best schools in the Austrian monarchy and later in Czechoslovakia."

---

[7]The geometry notebooks contain little in the way of Euclidean deductions. They are devoted rather to elaborate perspective drawings, including polyhedra and projections of cylinders and cones.

Certainly the faculty/student ratio appears to have been quite favorable, and many of the faculty were highly qualified: Of the twenty-one full-time faculty employed by the Realgymnasium during all or part of Gödel's years there, eleven held the doctorate. Among the latter, Klepetař mentioned in particular Dr. Georg Burggraf, the teacher of mathematics and physics, who, he said, was his own, Gödel's, "and probably everybody's favorite." Rudolf Gödel, however, declared that it was *especially* in mathematics that he and his brother "had a professor who was hardly suited to interesting us in his discipline" [40].

Gödel himself spoke disdainfully of Brno's Realgymnasium in a letter to his mother of 11 September 1960. Commenting there on a book she had sent him about the history of Brno, he noted that it failed to mention the Realgymnasium at all, "probably [because] its past is worthy of little praise, or is even discreditable" — a circumstance which, he said, "wouldn't surprise me at all, considering the conditions [there] at the time I attended it" [41].

I have found no reference in Gödel's writings to Georg Burggraf. On the Grandjean questionnaire, in response to the question "Are there any influences you would single out as especially important in the development of your interests (for example, a certain schoolteacher, a particular book or author, parents' interests, etc.)?," Gödel mentioned only the introductory calculus text in the "Göschen" series. And in letters to his mother he attributed the awakening of his interest in mathematics and science not to his courses at school but to an excursion the family had made to Marienbad in 1921, when he was fourteen [42]. There, he recalled, they had read and discussed Houston Stewart Chamberlain's biography of Goethe, and it was Chamberlain's account of Goethe's color theory and his conflict with Newton that, in hindsight, Gödel felt had led indirectly to his own choice of profession.[8]

There was, however, one subject among those Gödel studied at the Realgymnasium that did contribute significantly to his life of scholarship and that confronts all who would study his work: The shorthand that he selected as the second of his two electives. In today's world, especially in America, shorthand is considered relevant only to business and legal curricula, so such a choice of elective by one inclined toward the sciences would be exceptional. But in that time and place it was by no means unusual: Many European scientists of the period used shorthand for note-taking, and some even for correspondence.

---

[8] "So spinnen sich durch's Leben merkwürdige Fäden, die man erst entdeckt, wenn man älter wird." (Thus during one's life are curious threads spun, which we discover only as we grow older.)

FIGURE 8. School portrait of Kurt Gödel, ca. 1922.

It saved both time and space and was regarded as a useful skill for those going on to study at the universities.[9]

Gödel graduated from the Realgymnasium on 19 June 1924. Its *Jahresbericht* for 1923–24 is not among those preserved in his *Nachlaß*, but in that for 1922–23 his name is one of only four among his class of twenty-two to be printed in boldface type, signifying his status as a superior student. A school portrait of him from about that time (Figure 8) confirms the image of a studious young man confident of his future, an image enhanced by the eyeglasses, whose thick dark frames were to remain an unchanging feature of his countenance from then on. (An optometric prescription dated 13 May 1925 shows that by then he had already become distinctly myopic.)

---

[9] Shorthand was not employed for purposes of concealment. Indeed, because it was widely taught, it would have been quite inappropriate for such a purpose. Unfortunately, however, two competing systems of German shorthand, Gabelsberger and Stolze-Schrey, were in widespread use at that time. They merged into the modern *Einheitskurzschrift* a few years after Gödel learned the Gabelsberger system, so scholars today must contend not only with the obsolesence of the earlier system, but with the confusing similarity — in appearance, but not in meaning — between its symbols and those of the script that replaced it. (The old longhand script that Gödel studied at the beginning of his school career does not pose a problem, as it became obsolete before he left the Realgymnasium and he never subsequently employed it.)

II
Intellectual Maturation
(1924–1929)

> I owe a great deal to the Circle of Vienna. But it is solely
> the introduction to the problems and their literature.
>
> – Kurt Gödel to Herbert G. Bohnert, 1974

IN BROAD TERMS Kurt Gödel's life may be divided into three periods, corresponding both to his place of residence and to the changing nature of his intellectual interests: First, his childhood in Brno; second, his years in Vienna as student and *Dozent*, when he obtained his greatest mathematical results; and finally, the years following his emigration to America, when he directed his attention primarily to physics and philosophy. The transitions between periods were abrupt in each case, marked by sharp contrasts between differing intellectual, social, economic, and political milieus.

Gödel embarked on the second phase of his life in the autumn of 1924, when he matriculated at the University of Vienna. No longer an Austrian "exile" living among a Czech majority within a provincial industrial town, he was suddenly the opposite: A citizen by fiat of the new Czechoslovakian state, residing in the overcrowded metropolis that only shortly before had been the political and cultural center of the Austro-Hungarian empire.

By then, of course, Vienna was the impoverished capital of a small country having little influence in world affairs. Austria's runaway inflation had finally been slowed, so the economic outlook was considerably brighter than it had been four years earlier when Gödel's brother had arrived at the university to begin his medical studies. But shortages of housing, food, coal, and other commodities remained severe, and German and Austrian universities did not have dormitories. Rudolf was thus quite fortunate to find an apartment near the university (at Florianigasse 42/16) large enough to provide separate rooms for him and his brother.

21

In most respects, the political and economic boundary that separated the Gödel sons from their parents had relatively little effect upon their lives, for despite the passports that were a legacy of the Great War, travel between Austria and Czechoslovakia was not yet seriously restricted. The brothers were able to return home when necessary, especially during the summers, and on his frequent business trips to Vienna their father often brought them food and other staples from Brno, where commodities were cheaper. Kurt and Rudolf also had relatives in Vienna, with whom they perforce had reason and opportunity to become better acquainted.

The partitioning of the empire did mean that the brothers' prospects for making a career back in Moravia were no longer favorable. Their attitudes toward that circumstance, however, differed markedly. In Rudolf's view, "we had to get used to the idea of remaining in Austria" [43]. For Kurt, though, Vienna's cultural institutions had considerable allure. Prior to taking up residence there he had known the city more by reputation than by experience. He had visited the metropolis on occasion and had heard his brother's reports of activities there. Nevertheless, he had had little contact with Vienna's intellectual life "except through the newspaper *Neue Freie Presse*"[44].

The city's intellectual heritage, its architectural monuments, and its cultural institutions had remained largely intact after the war, despite the privations and the loss of empire. The university, in particular, had retained its eminence. And the very intensity of the problems the city then faced had forced its leaders to undertake a vigorous and long overdue program of civic renewal — most notably the ambitious building project begun in 1923, which, within five years, would result in the construction of some 30,000 dwelling units.

For Kurt and Rudolf the university was, of course, the center of their lives. Like other Austrian universities, that in Vienna was based on the German model, an institution whose structure had evolved from medieval ancestors. Of the four faculties, three (theology, law, and medicine) were devoted to professional training, while the philosophical faculty emphasized the spirit of free inquiry and research in the liberal arts and sciences. (Engineers, teachers, and musicians attended separate *Hochschulen*.) Because administrative control of the universities was exercised by the state, "neither a chapel nor a stadium" had any place in them; but, until the Nazis destroyed them, student fraternities devoted to drinking and fighting were prominent and obnoxious fixtures of German university life [45].

The character of the science and mathematics teaching that Gödel encountered at Vienna has been described by his illustrious colleague, Hermann

Weyl. According to Weyl, "Three forms of instruction were commonly employed" in the German universities: "lectures to large groups, practical exercises in the classroom or laboratory ..., and seminars for training in research" [46]. Core subjects were generally taught by professors, with assistants available outside of class to answer student questions and to help with the grading of papers. Sections were few and might be very large. At Vienna, for example, Professor Furtwängler's course in number theory — one that had a great impact on Gödel — attracted such large audiences (three to four hundred students) that it became necessary to issue alternate-day seating passes [47]! On the other hand, highly specialized research seminars were often taught by *Privatdozenten*, who had the right to charge modest fees for their courses but who, unlike the salaried professors, were otherwise unpaid.

The universities charged no tuition and mandated no required curricula. Students were free to enroll in whatever courses they saw fit, with several weeks to shop around (*hospitieren*) and attend lectures before having to commit themselves or pay any fees. In courses for which students did enroll no examinations or grades were given, and, at least at the University of Vienna, there was reportedly no distinction between undergraduate and graduate students [48].

Nevertheless, "This seemingly boundless freedom of the student[s]," so different from what they had experienced in the secondary schools, was "practically abridged by the necessity of passing a state examination at the end of [their] university career[s]," an examination covering specific fields rather than particular courses [49]. Before taking the examination candidates were required to have taken university courses for a definite period (usually three or four years), though not necessarily all at the same university. Indeed, one of Gödel's contemporaries, the noted philosopher of science Carl Hempel, recalled that "it was quite customary for students to study at various universities," staying at each one perhaps "a term or a year or several years" [50]. Hempel himself, for example, began at the University of Göttingen and moved successively to Heidelberg, Berlin, and Vienna before finally returning to Berlin for his doctorate. That Gödel chose to remain at the University of Vienna throughout his years as a university student was somewhat exceptional; and after receiving his doctorate he stayed on to become a *Dozent*, leaving finally only as a result of political and economic circumstances. As Weyl noted, "What attracts [a student] to a special university is often the fame of a great teacher or scientist under whom he wants to study" [51]. In Gödel's case, though, Vienna's proximity to Brno and the opportunity for him to live with his brother during his first few years away from home may have been the decisive considerations. Given the university's reputation, there was no particular reason for him to travel farther afield, but the particular eminence of Vienna's mathematical faculty — consisting at that time of *ordentliche Professoren* Wilhelm Wirtinger, Philipp Furtwängler, and Hans Hahn, *ausserordentlicher Professor* Alfred Tauber, and *Dozenten* Ernst Blaschke, Josef Lense, Eduard Helly, Leopold Vietoris, and Lothar Schrutka — almost certainly was *not* a factor, since Gödel enrolled at the university intending to study physics, as he in fact did for the first one or two years. Only later did he switch to mathematics, as a result of the impression made on him by those lectures of Furtwängler (the most wonderful he ever heard, he later said).

Just which courses Gödel took at the university is difficult to ascertain. Because of the peripatetic nature of their studies, students at European universities were issued a *Meldungsbuch*, a sort of academic passport in which all the courses for which they had enrolled were recorded. But Gödel's *Meldungsbuch* has apparently not survived, and, in contrast to all that is preserved from his earlier school years, his *Nachlaß* contains relatively few documents that can be definitively associated with courses he took as a university student. Documents at the University of Vienna record what courses were taught there during Gödel's student years, but they give no indication of which ones he attended. To reconstruct his course of study it is therefore necessary to collate references from a variety of sources. Even so, however, the record is incomplete.

Among the notebooks in Gödel's *Nachlaß* that can be dated is one containing extensive notes for a course on the history of European philosophy that was taught by Professor Heinrich Gomperz during the winter term of 1925. The syllabus covered the span from the pre-Socratics to the Reformation. A second notebook, dated 1926, contains notes on a sequel, also taught by Gomperz, that surveyed philosophers from Bacon to Schopenhauer, including Descartes, Leibniz, Spinoza, Hobbes, Locke, Rousseau, Kant, and Hegel, as well as many lesser figures. Both notebooks also contain miscellaneous notes on mathematics, but they do not appear to be for courses that Gödel was then taking.[1] Furtwängler's "Einführung in die Zahlentheorie" was presum-

---

[1] It is difficult to say, since Rudolf Gödel attested that his brother had, through independent study, already "mastered University Mathematics by his final Gymnasium years," much "to the astonishment of his teachers and fellow pupils" [52]. In addition, Gödel had a habit of filling the spaces left over in his notebooks, sometimes with notes on quite unrelated material taken at much later dates. In many cases he turned the notebooks over and began writing from back to front, so their contents read from both directions.

ably one of those, since he taught it only every third year, in rotation with algebra and differential and integral calculus, and is listed as having done so in 1925–26 [53].

Although physics was then the primary focus of Gödel's attention, of the several notebooks in his *Nachlaß* devoted to that subject there is only one, for a course on the kinetic theory of matter given by Professor Kottler in the summer of 1926, that unquestionably dates from his student days [54]. There is, however, another important source that sheds considerable light on his studies: The library request slips that he dutifully saved. From them it is apparent that during the years 1924–27 he consulted many physics texts, as well as Bernhard Riemann's writings on partial differential equations and their applications to physical problems. In addition, he read extensively among the mathematical classics, including Euclid's *Elements*, Euler's *Introductio in Analysin Infinitorum*, Lagrange's *Mécanique Analytique*, and Dirichlet's *Vorlesungen über Zahlentheorie*. During his first year at the University he also delved into Kant's *Metaphysiche Anfangsgründe der Naturwissenschaft* [55].

From a memoir by Olga Taussky, a fellow student of Gödel's who later became an eminent mathematician in her own right, it is known that Gödel also participated in a weekly seminar that the philosopher Moritz Schlick directed during the academic year 1925–26 [56]. Devoted to Bertrand Russell's *Introduction to Mathematical Philosophy*, that seminar may have been Gödel's first acquaintance with Russell's writings. Hans Hahn had directed an extracurricular seminar on *Principia Mathematica* the year before, but Gödel apparently did not take part in it [57].

It was probably sometime in 1925 or 1926 that Gödel first met Hahn, perhaps in the latter's course on set theory. No record of Gödel's enrollment in any of Hahn's lecture courses has been preserved, but Hahn must nevertheless be considered one of Gödel's principal mentors (in Gödel's own judgment, second only to Furtwängler in his overall influence on him). A scholar of exceptional breadth, Hahn made notable contributions to the calculus of variations, set theory, set-theoretic geometry, the theory of real functions, and Fourier integrals, in addition to the work in functional analysis — particularly the Hahn-Banach extension theorem — for which he is now best known [58]. He published several books and monographs, including a classic two-volume text on the theory of real functions, and was also highly revered as a teacher — one who, as Gödel approvingly recalled, explained everything to the last detail. The diversity of his interests was reflected in the varied areas of specialization of his doctoral students, among whom, in addition to Gödel, Karl Menger and Witold Hurewicz were especially distinguished.

That a logician should have written his dissertation under Hahn's direction might seem surprising. But by the early 1920s, shortly before Gödel's arrival in Vienna, Hahn's interests had shifted toward the philosophy and foundations of mathematics. Though he proved no results in logic, he gave a number of courses and seminars on logical topics, wrote several essays on mathematical philosophy [59], and was instrumental in bringing Schlick to Vienna in 1922 to occupy the chair in Philosophy of the Inductive Sciences once held by Hahn's teacher, Ernst Mach. Subsequently Hahn was one of a small group of scholars, inspired by Mach's positivistic philosophy, who gathered for discussion once a week in an old Vienna coffeehouse.[2]

Though Schlick was the acknowledged leader of that group — one of many informal circles centered around the leading intellectual figures in Vienna — it was reportedly Hahn who directed its members' attention toward logic [60]. By 1924 the discussions had begun to attract a larger following, so, at the urging of his students Friedrich Waismann and Herbert Feigl, Schlick agreed to establish a more formal colloquium. Like the original circle it met on Thursday evenings, but the site was moved to the *Hinterhaus* of the university's mathematical institute (now the meteorological institute).

Such were the beginnings of what later came to be known as "The Vienna Circle" (*Der Wiener Kreis*, so named from the title of a manifesto published in 1929 under the signatures of Rudolf Carnap, Otto Neurath, and Hahn). Admission to its sessions was by invitation only, so it was presumably Hahn or Schlick (or perhaps both) who invited Gödel to come. He first did so in 1926, when the Circle was engaged in a second reading of Wittgenstein's *Tractatus* [61]. From then until 1928 he attended regularly (only seldom thereafter), but in later years he was at pains to stress that from the very beginning of his participation he was *not* in sympathy with the Circle's views. He disagreed in particular with the idea — promulgated especially by Carnap — that mathematics should be regarded as "syntax of language." He shied away from controversy, however, and so held back from open criticism of the Circle's tenets. As was often his habit in such formal gatherings, he was content most of the time to listen to what others had to say, only occasionally interjecting incisive comments.

---

[2]Others in the group included Richard von Mises, Otto Neurath, and Neurath's wife, Olga (Hahn's sister, who was herself a logician). Philipp Frank, then in Prague, also attended on occasions when he visited Vienna.

It may be that Gödel found the expression of contrary views a sort of stimulus that helped him to formulate his own ideas more clearly; indeed, he once intimated that his later friendship with Einstein was based more on their differences of opinion than on their points of agreement [62]. But the Circle's primary impact on him was in introducing him to new literature and in acquainting him with colleagues with whom he could discuss issues of common interest.

Among the latter, the most important at the time were Carnap and Menger.[3] Carnap, a student of Gottlob Frege, had impressed Hahn as one who "would carry out in detail what was presented merely as a program in some of Russell's epistemological writings" [63]. Largely through Hahn's influence, he was invited to Vienna in 1926 as a *Privatdozent* in philosophy, and on arrival he was immediately invited to join the Circle. He began to attend the sessions at about the same time as Gödel, whose abilities he quickly came to appreciate. The two engaged in a lively exchange of ideas, much of it in private discussions outside the Circle, and Gödel also attended one of Carnap's lecture courses (probably that on the philosophical foundations of arithmetic, which met two hours each week during the winter semester of the academic year 1928–29), which he cited on the Grandjean questionnaire as having been especially influential with respect to his own papers of 1930 and 1931 [64].

Menger on the other hand, as already noted, was one of Hahn's own students. After receiving his doctorate in 1924 he was awarded a Rockefeller fellowship to Amsterdam in the summer of 1925. There he continued the fundamental work in dimension theory that he had begun in his dissertation. The following year he received his *Habilitation* and was appointed *Privatdozent*, but he soon came into conflict with L. E. J. Brouwer — the beginning of a long period of highly publicized strife between the two, largely over questions of priority regarding results in dimension theory. At Hahn's invitation Menger returned to Vienna in the fall of 1927, where he served as a *Privatdozent* for only a few months before being appointed *ausserordentlicher Professor* [65].

During that autumn term Menger taught a one-semester course in dimension theory in which Gödel was one of the enrollees. Menger also joined the Circle then, and the following year he established a mathematical colloquium of his own, whose proceedings were later published as the journal *Ergebnisse eines mathematischen Kolloquiums*. In October of 1929 Menger invited Gödel to

---

[3] It was in the Circle, too, that Gödel first met the economist Oskar Morgenstern, who would eventually become one of his closest friends. But it appears that their association was rather casual until they became reacquainted as emigrés in Princeton.

participate in the colloquium, and thereafter Gödel played a most active role. He assisted Menger in the editing of seven of the eight volumes of the journal, to which he also contributed numerous short articles and remarks [66].

Gödel also developed a close friendship with Herbert Feigl and Marcel Natkin, two other student members of the Circle. Feigl later recalled that they "met frequently for walks through the parks of Vienna, and of course in cafés had endless discussions about logical, mathematical, epistemological and philosophy-of-science issues — sometimes deep into . . . the night" [67]. Their association was shortlived, for in 1930 Feigl emigrated to America, there to become a prominent philosopher of science, while Natkin left academia for a career in business soon after receiving his doctorate. (He settled in Paris, where he became a well-known photographer.) But despite their separation, the three friends continued to correspond, and when Natkin visited the United States in 1957 they met in Princeton for a nostalgic reunion.

The Vienna Circle comprised a wide spectrum of personalities. Its members differed not only in their backgrounds, interests, and philosophical views, but also in their reactions to each other. Feigl, for example, described Schlick as a warm, kindly man who was "extremely calm and unassuming in his self-effacing modesty," [68] whereas Carl Hempel thought Schlick "an aristocratic personality" who, though "never dogmatic or authoritarian," was "conservative and somewhat distant" [69].

Schlick was also, in Feigl's view, "an extremely lucid thinker and writer" who was "well informed in [both] the history of philosophy and the history of science." Hahn, on the other hand, was less familiar with traditional philosophy. According to Menger, his favorite author was Hume. Like Gödel, he was a great admirer of the works of Leibniz. Unlike Gödel, however, he greatly disliked Kant [70].

Carnap and Neurath were another contrasting pair. They were close friends who, insofar as both were "somewhat utopian social reformers," had "a great deal in common" [71]. Yet in many other respects they were polar opposites: Carnap was "introverted, cerebral, and thoroughly systematic," while Neurath was "a lively, witty, extroverted man [of] boundless energy" [72].

As might be expected within such a diverse group of intellectuals, there were disagreements. Most were friendly, but a few were more divisive. Hahn, for example, "took strong exception to [Waismann's] view" (which echoed that of Wittgenstein) that "one could not speak about language." On that issue, "Schlick sided with Waismann," whereas Gödel and Menger, "though reticent in most debates in the Circle, strongly supported Hahn," as did Neurath [73].

The pamphlet *Wissenschaftliche Weltauffassung: Der Wiener Kreis,* written mainly by Neurath, also provoked dissension. Intended as a tribute to Schlick, its character as a manifesto for a movement in fact displeased him. It estranged Menger even more (he "asked Neurath to list [him] henceforth only among those *close* to the Circle") and was among the factors that further alienated Gödel from the group [74].

Still another source of disagreement was Carnap's and Hahn's interest in parapsychological phenomena — an interest shared also by Gödel, as his letters and some of his private papers reveal. According to Hempel (*1981*), it was Wittgenstein's discovery of a book on a parapsychological topic in Carnap's library that precipitated the final break between those two; and in his "Intellectual Autobiography" Carnap recalled that Neurath was among those who "reproached Hahn because he ... took [an] active part in séances in an attempt to introduce stricter scientific methods of experimentation" — behavior that Neurath felt "served chiefly to strengthen super-naturalism," but that Carnap and Hahn defended on the ground that scientists should have "the right to examine objectively ... all processes or alleged processes, without regard for the question of [how] other people [might] use or misuse the results" [75].

Reportedly, Hahn's interest in the subject arose in part from "an influx of mediums" that Vienna experienced during the years immediately following World War I. In particular, an attempt by two of Hahn's university colleagues (Professors Stefan Meyer and Karl Przibram of the physics faculty) to ridicule the mediums by staging a fake séance led to "great indignation in the intellectual community" and prompted "a group including besides Schlick and Hahn the eminent ... physician Julius Wagner-Jauregg, the physicist Hans Thirring and a number of others ... [to form] a committee for the serious investigation of mediums" and their claims [76].

The committee was short-lived, and "by 1927 apart from nonscientists only Hahn and Thirring were left." Nonetheless the affair is of interest because of the close parallel between Hahn's attitudes toward such things and those later expressed by Gödel.

Hahn, for example, "pointed out that many mediumistic revelations are *so* trivial" that they are not only far below the level of the individuals alleged to be speaking through the mediums but are "in fact ... definitely below the medium's own level"; and far from suggesting fakery, to him this indicated that "in many cases one [was] dealing with a *genuine* phenomenon *of some kind*" [77].

Such unorthodox reasoning — especially the attempt to find hidden causes underlying events for which there appear to be mundane explanations — has a distinctly Gödelian twist to it. It bears comparison especially with a statement that Gödel made in one of his letters to his mother: Her antipathy toward occult phenomena was, he said, "quite justified, insofar as it is difficult to disentangle genuine phenomena from the mix of fraud, gullibility, and stupidity." But, he went on, "the result (and the sense) of the fraud is ... not that it *simulates*, but that it *masks*, the genuine phenomena" [78].

No sources list Gödel as a member of the investigatory committee that Hahn set up. Among Gödel's papers there is, however, one memorandum that appears to be a shorthand record of a séance [79]. The overall significance of the item is hard to judge, but it is clear that Gödel's interest in the parapsychological went beyond mere open-mindedness. His library slips from the University of Vienna, for example, include two for a book by Alfred Lehrmann entitled *Aberglaube und Zauberei* (Superstition and Sorcery), and in another of his letters to his mother he wrote that investigators at "a local university with great strength in the sciences" (Princeton, presumably) had established that "every person" possesses the ability to predict numbers that will turn up in games of chance, though most of us have that ability "only to a quite meager degree" [80]. He counted his wife among those who possessed the ability to an exceptional degree, a fact he claimed to have verified "incontestably" in some two hundred trials. He believed also in the possibility of telepathy, and late in his life he remarked to Oskar Morgenstern that in several hundred years it would seem incomprehensible that twentieth-century investigators had discovered the elementary physical particles and the forces that hold them together but had failed even to consider the possibility ("and high probability") that there might exist elementary *psychic* factors [81].

Whatever one may think of such beliefs in extrasensory perception, it is important to note how well they accord with Gödel's belief that mind is distinct from matter, and with his later advocacy of mathematical Platonism. In an oft-quoted passage he averred that "despite their remoteness from sense experience, we do have something like a perception also of the objects of set theory," and he saw no reason "why we should have less confidence in th[at] kind of perception, i.e., in mathematical intuition, than in sense perception"; nor did he believe that such intuitions must be regarded as "something purely subjective" just because "they cannot be associated with actions of certain things on our sense organs" [82].

◊   ◊   ◊

Because he was among scholars whose interests and abilities accorded more closely with his own, Gödel had a larger circle of friends in Vienna than in Brno. Still, though his penetrating intellect was quickly appreciated and highly respected, he remained a very private person. Recollections of him by colleagues are remarkably consistent:

> He was a slim, unusually quiet young man. . . . [In the Circle] I never heard him take the floor. . . . . He indicated interest solely by slight motions of the head — in agreement, skeptically or in disagreement. . . . His expression (oral as well as written) was always of the greatest precision and at the same time of exceeding brevity. In nonmathematical conversation he was very withdrawn. (*Menger 1981*, pp. 1–2)

> [He] was well trained in all branches of mathematics and you could talk to him about any problem and receive an excellent response. If you had a particular problem in mind he would start by writing it down in symbols. He spoke slowly and very calmly and his mind was very clear. . . .
> It became slowly obvious . . . that he was incredibly talented. His help was much in demand . . . [and] he offered [it] whenever it was needed. . . . But he was very silent. I have the impression that he enjoyed lively people, but [did] not [like] to contribute . . . to nonmathematical conversations. (*Taussky-Todd 1987*, pp. 33 and 36)

> [His] great abilities were quickly appreciated . . . . [He] was a very unassuming, diligent worker, but his was clearly the mind of a genius of the very first order. (*Feigl 1969*, p. 640)

Evidently Gödel devoted himself intently to his studies. Yet he was not asocial. As noted earlier, for example, he spent a good deal of time in the coffeehouses that were then so central to Viennese intellectual and cultural life. Their singular appeal was described by Stefan Zweig:

> The Viennese coffeehouse is . . . comparable to [no] other in the world. . . . It is a sort of democratic club to which admission costs the small price of a cup of coffee. Upon payment of this mite, every guest can sit for hours on end, discuss, write, play cards, receive his mail, and, above all, can go through an unlimited number of newspapers and magazines [83].

Then as now, different cafés catered to different clienteles. Literati, for example, congregated at establishments such as the *Herrenhof*, while members of the Schlick Circle frequented cafés like the *Reichsrat, Schattentor,* and *Arkaden.* (For special occasions within the Gödel family preferred pubs included the cafés *Meißel und Schaden, Elisabeth,* and *Coq d'Or.*) The appeal of a particular café might be based on such factors as the eminence of its patrons, its ambience, or the selection of periodicals it offered. For mathematicians it was reportedly the color of the marble table tops (black or white) that was the overriding consideration: White tops were preferred because equations could be written on them [84].

While the cafés provided a backdrop for serious discussions, there were other, lighter diversions as well. The Gödels were especially fond of the theater, and when Vienna's artistic life revived during the latter years of the twenties they often attended performances together at the Burgtheater, the Volks- or Staatsoper, or, especially, the newly opened Theater in der Josefstadt, where they held box seats. (Marianne's taste for Lieder, and for symphonic and chamber music, was not shared by her sons; but Kurt did retain a liking for opera, and especially operetta, throughout his life.)

The family also traveled together, especially in the summers. Indeed, in 1926, when the Chrysler Corporation opened a factory in Czechoslovakia, Rudolf August was among the first in the country to purchase one of their cars. In it the family journeyed to various alpine resorts and spas, both in the vicinity of Brno (Plansko, Adamstal, and Vranau) and farther afield (Marienbad, Mariazell, and the Salzkammergut). The Gödels employed a chauffeur, but on such occasions the sons often preferred to do the driving themselves [85].

Apart from such excursions, Kurt and Rudolf mostly went their separate ways. Though they continued to live harmoniously together, neither seems to have had much contact with the other's friends. In part, of course, that circumstance reflected the great disparity in their intellectual interests and abilities. But Rudolf was also much less concerned with another, quite different object of his brother's attention: Women.

There are conflicting accounts of just how Gödel's interest in the opposite sex was first awakened. According to Kreisel, "he came upon his first romantic interest without much waste of time: she was the daughter of family friends who were frequent visitors ... [and] was regarded as an eccentric beauty. Because she was ten years older his parents objected strongly and successfully" [86]. In a letter to his mother, however, Gödel himself declared that it was during an excursion to the Zillertal that he experienced his first love — a girl whose name he vaguely recalled as Marie [87].

Whichever story is correct, by the time Gödel arrived at the university there was no longer any doubt that he "had a liking for members of the opposite sex, and ... made no secret" of the fact [88]. He exhibited a preference for older women, but also, according to his brother, simply enjoyed looking at pretty girls. Rudolf recalled in particular a small family-run restaurant near their Florianigasse apartment where the twenty-year-old daughter served as waitress. She was unusually attractive and, Rudolf believed, was probably the reason Kurt often went there [89].

◇ ◇ ◇

In April of 1927, for unknown reasons, Gödel moved from the apartment on Florianigasse to one at Frankgasse 10/11. He remained at that address only until 20 July — five days after a mob of angry demonstrators burned the Palace of Justice and was fired upon by the police, resulting in the deaths of 89 persons [90]. The riots may have had nothing to do with Gödel's departure, but in any case he did not reestablish residence in the city until 6 October, when he and his brother moved into an apartment at Währingerstrasse 33/22. So whether or not he was present at the time of the massacre, he was away during the tense days that immediately followed.

The building on Währingerstrasse was a particularly convenient abode for Gödel, since it also housed the *Café Josephinum*, another gathering place for members of the Schlick Circle [91]. Yet there, too, the brothers remained for less than a year: In early July of 1928 they rented an apartment at Langegasse 72/14, one whose three large rooms were spacious enough to provide accommodation for their parents when they came to visit [92]. Unfortunately, though, it was to serve that purpose for only a few months; for on 23 February 1929, just five days before his fifty-fifth birthday, Rudolf August died, suddenly and unexpectedly, of a prostatic abcess.

For Marianne the loss was devastating. Left isolated in Brno, she became increasingly lonely and distraught, to the point that her sons grew seriously alarmed about her health. They decided they dared not leave her alone, so the villa was rented out and the three of them, together with Aunt Anna, moved into another large apartment, at Josefstädterstrasse 43/12a, opposite the theater they had so often patronized [93].

At the time of that move the brothers had lived in the Langegasse apartment for only about sixteen months. Nevertheless, it was there that Gödel wrote his doctoral dissertation, there that he became a citizen of Austria,[4] and, most important, there that he met the woman who would later become his wife.

---

[4]His application was initiated sometime after 11 July 1928, the date a lawyer in Vienna wrote to Rudolf August to advise him what fees the process would entail. Gödel was released from Czechoslovakian citizenship on 26 February 1929, just three days after his father's death, but was not granted Austrian *Landschaft* until 6 June.

Her maiden name was Adele Thusnelda Porkert. Born 4 November 1899 in Vienna, she was the eldest of the three daughters of Josef Porkert, a portrait photographer who lived with his wife, Hildegarde, at Langegasse 65, diagonally across the street from the Gödels' apartment. At the time Kurt and Adele first met she was already married to a man named Nimbursky, about whom little is now recalled save that he, too, was a photographer. That marriage was both unhappy and short-lived, however, so the way soon became clear for Adele to become the object of Gödel's attentions.

Their courtship was a most protracted one, since Gödel's parents strongly disapproved of the match. In their eyes Adele had many faults: Not only was she a divorcée, older than their son by more than six years, but she was Catholic, she came from a lower-class family, her face was disfigured by a port wine stain, and, worst of all, she was a *dancer*, employed, according to several accounts [94], at a Viennese nightclub called *Der Nachtfalter* (Figure 9).[5]

Adele herself claimed to have been a ballet dancer — though, if so, she must just have been an extra in a corps (perhaps at the Volksoper). But whether she danced in a cabaret or a ballet troupe, the stigma was equally great. As Stefan Zweig remarked in *The World of Yesterday* with reference to the period before World War I, "A ballet dancer ... was available for any man at any hour in Vienna for two hundred crowns" [95]. To marry someone with such associations could destroy even a well-established career, as illustrated by the case of Hans Makart, the most lionized of nineteenth-century Viennese painters. According to William Johnston, "In 1881, he undid his popularity by marrying a ballet dancer; three years later he died of venereal disease" [96]. No wonder, then, that Rudolf and Marianne, imbued with the mores of the past generation, opposed their son's interest in Miss Porkert.

---

[5] "The Moth". The German term, much more than its English gloss, suggests a shadowy creature of the night. The underlying connotation is captured perfectly in this illustration, a reproduction of Koloman Moser's contemporary print of the dancer Loïe Fuller.

Figure 9: Koloman Moser, "Homage to the American Dancer, Loïe Fuller," shown portraying a nocturnal moth. (Graphische Sammlung Albertina, Vienna)

III
Excursus
A Capsule History of the Development of Logic to 1928

THE DIRECTION GÖDEL'S life's work would take was significantly influenced by four events that occurred in 1928: The appearance of the book *Grundzüge der theoretischen Logik,* by David Hilbert and Wilhelm Ackermann; Hilbert's address "Probleme der Grundlegung der Mathematik" (Problems in laying the foundations of mathematics) at the International Congress of Mathematicians, held that September in Bologna; and a pair of lectures that the iconoclastic Dutch mathematician L. E. J. Brouwer delivered in Vienna.

To understand the significance of those events requires some acquaintance with the basic concepts and prior history of mathematical logic. The present chapter provides a primer for those lacking such knowledge.

◇　◇　◇

The systematic study of valid modes of inference was initiated by Aristotle, whose *Prior Analytics*–part of the posthumous anthology known as the *Organon*–was the basis for the study of logic for more than two thousand years. Aristotelian logic was devoted to the classification and analysis of *syllogistic* forms of argument, to which Aristotle believed that "all proof, properly so called" could be reduced [97].

A syllogism, as the reader may recall, consists of three statements, the major and minor *premises* and the *conclusion,* each of which is of one of four forms: *universal affirmative* ("Every X is Y"), *universal negative* ("No X is Y"), *particular affirmative* ("Some X is Y") or *particular negative* ("Some X is not Y"). The premises are linked by a common "middle term" (X or Y, as the case may be)

that is absent from the conclusion. Not all combinations of the four statement
forms yield valid deductions, however. Aristotle recognized fourteen patterns
that do, which he divided into three groups, or *figures*, according to the relative
positions of the middle term in their premises (before the copula in both, after
it in both, or after it in the first and before it in the second).

Although Aristotle's basic scheme was elaborated by later logicians throughout the Middle Ages and beyond, their efforts are of little relevance to the
mathematical logic of today. It is rather the differences between ancient and
modern logic that are of concern.

In the first place, Aristotelian logic was *not symbolic*. Apart from "term
variables" such as X and Y, syllogisms were expressed in words rather than
abstract symbols — a fact that may explain why it took so long for syllogistic
arguments to be interpreted as relations among classes [98]. Second, within
the statements of a syllogism *no* logical *connectives* other than negation occur,
although the *conjunction* of the premises is understood to *imply* the conclusion.
The *quantifiers* "some" and "every" do appear prominently in both premises
and conclusion, but their use can be avoided if implication is taken as an
additional connective. For example, the universal statement "Every man is
an animal" can be rendered as "Being a man implies being an animal," and
the particular statement "Some men are evil" as the negation of "Being a man
implies not being evil." That is so because syllogisms contain only *monadic
predicates* — those that express properties of individual entities, such as "is
a man" or "is evil" — so *nested* quantifiers never occur [99]. In contrast, in
statements that contain *polyadic predicates* — those that express relations *between*
entities — the order of the quantifiers affects the meaning. "Every integer is
less than some prime," for example, is a theorem of Euclid, while "Some
prime is greater than every integer" is simply false.

Consequently, though Aristotle's interest in logic seems to have arisen out
of concern for rigor in geometric demonstrations, his syllogistic was fundamentally inadequate for analyzing many quite simple deductions that occur
both in reasoning about inequalities among natural numbers and in geometric
arguments that were known in his own time.

Mathematics, however, was not the only stimulus to the development of
logic. An equally important current within ancient Greek thought, especially
among the Megarians and Stoics, was the tradition of dialectical disputation.
It was out of that tradition that the classical paradoxes arose, including the
Antinomy of the Liar, attributed in its original form ("All Cretans are liars")
to Epimenides of Crete. And it was that aspect of logic that was responsi-

ble for the subject's survival during the Middle Ages. Unlike mathematics, which languished in the West for centuries after the fall of the ancient Greek civilization, the Aristotelian and Stoic traditions in logic were kept alive by scholastic philosophers, who sought to apply logical principles in their theological debates. Of particular importance to Gödel's own later work was Anselm's attempt to deduce the existence of God from the assumption of His perfection — the so-called ontological argument.

With the coming of the Renaissance, the fortunes of the two disciplines were reversed: Mathematics revived and prospered, while logic — precisely *because* it had become the province of philosophers rather than mathematicians — was little affected by the great advances in scientific thought that occurred during the sixteenth and seventeenth centuries.

The one figure from the latter period whose logical ideas merit special mention is Gottfried Leibniz. In contrast both to his predecessors and to such later figures as Kant (who declared in the second edition of his *Critique of Pure Reason* that logic had neither "had to retrace" nor "been able to advance" a single step since Aristotle, and so was "to all appearances complete and perfect"), Leibniz did not think that all reasoning could be reduced to syllogistic forms. Instead, he envisioned the creation of an artificial language (*lingua characterica*) within which reasoning about any concepts whatever could be carried out mechanically by means of certain precisely specified rules of inference. By appeal to such a *calculus ratiocinator* disputes of every sort could, he thought, be resolved through simple calculation.

Leibniz made little progress toward realizing his utopian vision. Nevertheless, his foresight was remarkable. The notion of an abstract symbolic algebra, the idea that reasoning might be mechanized, and above all the suggestion that logic was not merely a part *of* mathematics, but that mathematical procedures could also be applied *to* logic — all those insights foreshadowed important later developments; and though few of Leibniz's technical contributions to logic were to prove of lasting significance, one of them, the idea of denoting primitive concepts by prime numbers and logical combinations of them by composite integers, was later exploited by Gödel as a key device in the proof of his incompleteness theorem. (Leibniz certainly did not anticipate Gödel's sophisticated coding technique. But Gödel was a great admirer and student of Leibniz's work, so it may be that the idea of using such a device was suggested to him by his reading of Leibniz during the years 1926–28.)

Real progress in developing a logical calculus was first achieved by George Boole in his *Mathematical Analysis of Logic*, published in 1847. Boole recognized the formal analogies between the use of the arithmetical operations "plus" and "times" and the logical connectives "or" and "and", and he showed how Aristotle's four basic statement forms could be expressed as equations within a symbolic algebra, which contained, in addition to the variables and operation symbols, the constant symbols 0 and 1. By taking 1 to represent truth and 0 falsity the proposition "Every X is Y" — construed as "Whatever *thing* is X is also Y" — can, for example, be rendered by the equation $x(1 - y) = 0$, where the variables x and y are understood to take the value 1 if the "thing" in question is X or Y, respectively, and the value 0 otherwise.

Boole took certain equations as axioms, but because he intended his system to admit of various interpretations he did not fully axiomatize it. In particular, he did not require that the variables take *only* the values 0 or 1; he emphasized, for example, that his algebra could be interpreted probabilistically by allowing the variables to take values in the interval $[0, 1]$, or as equations among classes if 0 were taken to denote the empty class and 1 the universe of discourse.[1]

In effect, Boole introduced what is known today as *propositional* logic — the formal logic of connectives. Statements involving nested quantifiers remained outside Boole's domain of analysis because his system was still restricted to monadic predicates. That limitation was later overcome, however, by Charles Sanders Peirce and Ernst Schröder, who built on Boole's ideas to develop a far-reaching "calculus of relatives" based on a similar algebraic perspective.

Peirce, in particular, created a symbolism that included both quantifiers and polyadic relation symbols. Its advantages were immediately recognized by Schröder, who adopted it for his own *Vorlesung über die Algebra der Logik* (1890–95), a three-volume compendium that served as the standard text on logic for some twenty years. Schröder's treatise was eventually displaced by Bertrand Russell and Alfred North Whitehead's *Principia Mathematica*, but Peirce's notation was retained in modified form, to become the basis for the symbolism used today.

Peirce also established a number of important theorems of quantification theory. He noted, for example, that every formula of his system was logically equivalent to one in *prenex form* (wherein all quantifiers stand at the front, before a Boolean expression made up of a disjunction of conjunctions of relations), and he devised a proof procedure that was a precursor of what are now called natural deduction systems [100]. But Peirce gave no axioms for his system, and his formalism differed from those of today in two important re-

---

[1] A century later Boolean algebras with infinitely many "truth values" would be found to have important applications within logic and set theory. But those applications depended on ideas that had yet to be formulated at the time of Boole's work.

spects: It did not employ function symbols such as $+$ or $\cdot$ for the formation of terms (pronominal expressions, like $1 + x$ or $x \cdot y$, built up from constants and variables), and the quantifiers — for which, following Boole, Peirce used the symbols $\sum$ and $\prod$ — were interpreted as disjunctions or conjunctions taken over the totality of elements of a given domain of interpretation. In contrast, the interpretations of modern systems are not fixed in advance. They are required only to satisfy the axioms of the system. So, for example, in the context of the natural numbers the Peircean formula $\sum_x P_x$ would represent the assertion "the property P holds of 0 or of 1 or ...," whereas the modern formula $\exists x P(x)$ would assert that "the property P holds of *some* object in the domain of interpretation," which might be any collection of objects satisfying the formal axioms for the natural numbers.[2]

Peirce's ideas were quite influential in their day, but in the end they were overshadowed by the work of the German mathematician and philosopher Gottlob Frege. In 1879, six years before the appearance of Peirce's article on the algebra of logic (*1885*), Frege published a small book entitled *Begriffsschrift* (Ideography), in which, like Peirce, he recognized that "polyadic predication, negation, the conditional, and the quantifier [were] the bases of logic." In the spirit of Leibniz, Frege gave precise meaning to the notion of a formal system, in which mathematical demonstrations were to be "carried out ... by means of explicitly formulated syntactic rules" [101]; he introduced the truth-functional interpretation of the propositional calculus; and, most important, he "analyzed propositions into function(s) and argument(s) instead of subject and predicate ... [and gave] a logical definition of the notion of mathematical sequence" [102].

Regrettably, Frege's logical works, including his ambitious *Grundgesetze der Arithmetik* (The Basic Laws of Arithmetic, published in two volumes in 1893 and 1903), were couched in a cumbersome two-dimensional notation that deterred most readers. Recognition of their importance thus came only belatedly, largely through the efforts of Bertrand Russell, who translated Frege's ideas into a Peircian style of notation and brought them to the attention of the mathematical community.

---

[2]Some readers may protest that the natural numbers are the *only* objects that satisfy Peano's axioms (discussed further below). That is true of second-order versions of those axioms, in which quantification over *properties* of objects is allowed, but not of *first-order* formalizations, in which the quantifiers range only over the objects themselves.

In doing so, however, Russell also discovered an inconsistency in Frege's work — the paradox that now bears Russell's name. As he explained to Frege in a now-famous letter that Frege received just as the second volume of his *Grundgesetze* was about to be published, by allowing an arbitrary predicate to serve as argument of another Frege had unwittingly made possible the construction of a self-referential antinomy; for the predicate "to be a predicate that cannot be predicated of itself" can neither be nor not be predicated of itself.

◇ ◇ ◇

Russell's antinomy was one of a spate of paradoxes that appeared around the turn of the new century. Some, like that of Richard (1905), concerned the notion of definability within a language. Others, like that of Burali-Forti (1897), made reference to set-theoretic notions such as the totality of all ordinals or all sets. And some that seemed not to refer to such notions could be recast in ways that did. Russell, for example, noted that his own antinomy could be expressed in terms of the self-referential totality of all sets that were not members of themselves (in symbols, $\{x : x \notin x\}$).

The paradoxes were both a reflection of and a stimulus to the concern for greater rigor in mathematical demonstrations, which had arisen half a century earlier in response to issues involving the definition and representation of functions. Particularly vexing were questions concerning the convergence of trigonometric series, provoked by Joseph Fourier's claim that "any" function could be represented as the limit of such a series.

Fourier's assertion, made in his 1822 book *Théorie Analytique de la Chaleur*, was widely disbelieved at first and was eventually shown to be untrue. Nevertheless, investigations revealed that the class of functions that were definable as limits of trigonometric series was much larger than had been thought, and some of those functions — particularly Karl Weierstrass's example of one that was continuous but nowhere differentiable — seemed decidedly pathological.

Weierstrass presented his example two years after Georg Cantor, then twenty-five years old, had proved an important result on the uniqueness of Fourier representations. In his paper *1870* Cantor showed that if a Fourier series converges to zero throughout the interval $(-\pi, \pi)$ then its coefficients must all equal zero. Hence for series that converge everywhere uniqueness is guaranteed. In subsequent papers Cantor went on to investigate what happens if there are exceptional points, and in the course of that research he developed ideas that were to germinate a few years later into what became his theory of transfinite ordinals.

Specifically, given a set P of real numbers Cantor defined its (first) derived set to consist of all the limit points of P. The notion of derived set can be iterated, and Cantor showed that if a Fourier series converges to zero everywhere in $(-\pi, \pi)$ except for a set P whose n-th derived set $P^{(n)}$ is empty for some finite n, then its coefficients are again uniquely determined. In the same paper (*1872*) he also developed a rigorous theory of irrational numbers, based on the notion of a Cauchy sequence of rationals and the axiom that to each such sequence there corresponds a unique point of the real line. (Richard Dedekind's book *Stetigkeit und irrationale Zahlen* [Continuity and Irrational Numbers], in which the irrationals were developed in terms of the notion of "cut" and the corresponding continuity axiom, appeared later that same year.)

At the time, Cantor chose not to raise the question whether the notion of derived set might be iterated beyond the finite. Privately, however, he had observed that there were sets P all of whose finite derivatives were nonempty, and for such sets he had considered the intersection of all the $P^{(n)}$'s, together with *its* sequence of derivatives. Eight years later he employed the symbols $\infty, \infty + 1, \ldots$ as indices for those transfinite derivatives, and in 1883, in his landmark book *Grundlagen einer allgemeinen Mannigfaltigkeitslehre* (Foundations of a General Theory of Manifolds), he took the bold step of conceiving such indices as transfinite ordinal *numbers* [103].

The notion of transfinite *cardinal* number evolved during the same period, as an outgrowth of Cantor's discovery that the real numbers could not be placed in one-to-one correspondence with the integers. As his correspondence with Dedekind makes clear, that discovery took place in December 1873, at a time when his interests were turning "away from trigonometric series to an analysis of [the notion of] continuity" [104].

Both Cantor and Dedekind were aware that the real numbers were, in some vague sense, far more numerous than the rationals. But the precise notion of *denumerability* needed to make that intuition precise was lacking, and though Cantor established the uncountability of the reals in his paper *1874*, he gave precedence there to the theorem that the *algebraic* numbers *could* be enumerated by the integers — a result due apparently to Dedekind (to whom, however, Cantor failed to give credit [105]). He employed the non-enumerability of the reals primarily as a tool for proving that every interval contains infinitely many transcendental numbers (a result already established by Liouville, who gave explicit examples of such numbers), and his original proof of non-enumerability involved the construction of a nested sequence of intervals, very much in the spirit of his earlier studies of point sets. Modern proofs, by contrast, employ the diagonal method, whereby, given any sequence $r_1, r_2, r_3 \ldots$ of reals, one constructs a decimal $.d_1 d_2 d_3 \ldots$ whose n-th digit d_n differs from that of r_n. That technique, fundamental to many arguments in analysis and recursion theory, is also due to Cantor, but came only much later.

Between the years 1878 and 1897 Cantor published a series of papers in which the concepts of transfinite set theory gradually took shape. In his 1878 paper "Ein Beitrag zur Mannigfaltigkeitslehre" (A contribution to the theory of manifolds) he first made the notion of one-to-one correspondence explicit, and it was there, too, that he first asserted that every infinite set of real numbers was equinumerous either with the set $\mathbb{N}$ of all integers or the set $\mathbb{R}$ of all real numbers — a weak form of what would later be called the Continuum Hypothesis (CH).

Cantor thought at first he had proved the assertion. Soon, though, he became aware of difficulties, and so turned to a consideration of special cases. From his letters to the Swedish mathematician Gösta Mittag-Leffler it is clear that by 1882 he had conceived the notions of well-ordering and of classes of ordinal numbers and had proved that the corresponding property held for what he would later call the second number-class (the class of ordinals having only countably many predecessors): Every infinite subset of it was equinumerous either with the integers or with the whole class [106]. To establish his earlier claim it would therefore suffice to show that the real numbers were equinumerous with the class of all countable ordinals (the Continuum Hypothesis in its full-fledged form).

Two years later, in an attempt to prove the CH, Cantor succeeded in showing both that every *perfect* subset of $\mathbb{R}$ was equinumerous with $\mathbb{R}$ and that his original conjecture held for all *closed* subsets of $\mathbb{R}$; but he failed to establish the CH itself, and even briefly thought he had refuted it.

Between 1885 and 1890 Cantor refrained from publishing his work in mathematical journals, having been stung by the harsh criticisms of Leopold Kronecker and by the hostile reaction his theories had provoked within the German and French mathematical communities. The response had been so negative, in fact, that Mittag-Leffler advised him to withhold publication of further papers on transfinite set theory until he had succeeded in applying his methods to obtain "new and very positive results," such as a definite determination of the power of the continuum [107].

The theorem that every set has cardinality strictly less than its power set was such a result, since it implied that there is no upper bound to the size of infinite magnitudes. Its quasi-paradoxical proof — based on consideration of the set $\{x \in A : x \notin f(x)\}$, where f is a putative one-to-one mapping of the elements of a set A onto all of its subsets — relied on the diagonal method, which Cantor presented together with the theorem in his paper *1891*.

Finally, in 1895, as part of his general theory of cardinal and ordinal arithmetic, Cantor employed the aleph notation that has since become standard. Given that notation, together with the fact that $\mathbb{R}$, regarded as the set of all binary numerical expansions, is equinumerous with the power set of $\mathbb{N}$, he could then express the CH succinctly by the equation $2^{\aleph_0} = \aleph_1$. He did not, however, go on to assert that $2^{\aleph_\alpha} = \aleph_{\alpha+1}$ for all ordinals α. That conjecture, the so-called Generalized Continuum Hypothesis (GCH), was first formulated in 1908 by Felix Hausdorff.

It was during the period of Cantor's estrangement from the mathematical community that Schröder's treatise on logic appeared, as well as Dedekind's book *Was sind und was sollen die Zahlen?* (What Are Numbers, and What Should They Be?). In the latter, published in 1888, Dedekind proved the fundamental theorem justifying definitions by recursion, gave the recursion equations for addition and multiplication, and stated four axioms characterizing what he called "simply infinite systems" (structures generated inductively from an element 1 lying inside the domain but outside the range of a given one-to-one mapping of a set N into itself). Those axioms, together with the principle of complete induction that Dedekind had derived as a theorem, were then reformulated the following year by Giuseppe Peano in his book *Arithmetices Principia, Nova Methodo Exposita* (Principles of Arithmetic, Explained by a New Method).

A few months before [108], Peano had introduced many of the symbols that have since become standard in set theory: $\in$ for the membership relation, $\cup$ and $\cap$ for union and intersection (and also for the propositional connectives "inclusive or" and "and"), and the colon notation for set abstraction (which, however, he soon abandoned).[3] In his 1889 book he expressed the axioms for arithmetic in terms of $\in$, the constant symbol 1, the symbol $=$ (used between terms to denote numerical equality and between equations to denote equivalence) and the successor function $+1$. He also employed an inverted C (rendered below by the modern symbol $\supset$ into which it later evolved) to

---

[3] In later papers Peano also introduced $\ni$ for "such that" and an inverted ι for "the sole member of" [109]. The symbol $\exists$ is his as well, though he used it not for existential quantification, as did Russell and later logicians, but rather to express the subtly different assertion that a class is nonempty. The symbol $\forall$ for universal quantification, however, is due not to Peano but to Gerhard Gentzen, who introduced it in his dissertation in 1934.

denote both deductive entailment and class containment. The first four of his five axioms involved only numerical terms:

1. $1 \in N$

2. $a \in N \supset a + 1 \in N$

3. $a, b \in N \supset a = b . = . a + 1 = b + 1$

4. $a \in N \supset a + 1 \neq 1$

The induction axiom, however, made reference to the class K of all classes, whose antinomial character Peano failed to recognize:

5. $k \in K . \therefore 1 \in k . \therefore x \in N . x \in k :\supset_x x + 1 \in k :: \supset . N \supset k.$

In ordinary language this axiom states that for all classes k, if 1 is in k and if, for all numbers x, x in k entails $x + 1$ in k, then all numbers are in k. So expressed, reference to the problematic class K is replaced by quantification over all *sets* of numbers; the corresponding axiomatization is then second-order.

Since Peano's axiomatization preceded the discovery of the logical and set-theoretic paradoxes, it did not arise in response to them. Nor did it constitute a true *formalization* of number theory since Peano did not specify any rules of deduction. His proofs were carried out informally, as a list of formulas representing a sequence of successive inferences. It was left to the reader to justify the passage from one formula to the next.

The reason for that "grave defect," as one prominent historian has described it [110], is presumably that Peano did not aim, as Frege did, to reduce arithmetic to logic. He sought rather to characterize the natural numbers up to isomorphism, as his axioms (and Dedekind's) in fact did.

His concern for doing so very likely arose from developments in geometry, especially Eugenio Beltrami's construction of a model for Lobachevskian geometry and Moritz Pasch's axiomatization of projective geometry, events that had taken place in 1868 and 1882, respectively; for in the same year that his *Arithmetices Principia* appeared Peano also published a small pamphlet entitled *I principii di geometria logicamente esposti*, in which he built upon Pasch's work to give a set of axioms for the "geometry of position" (in modern terms, axioms of incidence and order). He returned to the subject again in 1894, and at that time focused attention on the metatheoretical question of the independence of his axioms, a feature that he demonstrated in a few instances by means of models. His work was overshadowed, however, by the definitive axiomatization of elementary geometry that Hilbert gave five years later in his book *Grundlagen der Geometrie* [111].

Meanwhile, in volume I of his *Grundgesetze* (1893) Frege presented a rigorous second-order formalization of arithmetic, in which he derived Peano's postulates as consequences of a single axiom (Hume's Principle, according to which "The number of *F*s is the same as the number of *G*s just in case the *F*s can be correlated one-one with the *G*s"). After Russell demonstrated the inconsistency of the logical system that underlay the *Grundgesetze*, that contribution was all but forgotten; but on re-examination it has turned out that the fragment of the system in which Frege carried out that derivation *is* consistent and the derivation itself is correct [112].

◇   ◇   ◇

Such was the situation in logic at the beginning of the new century, just before the emergence of the paradoxes and the ensuing "crisis in foundations." It was a time for optimism, and the Second International Congress of Mathematicians, held in Paris in 1900, was an occasion both for celebrating the gains that had been made and for calling attention to questions that remained unanswered.

It was there that Hilbert delivered his famous address on the problems of mathematics, in which he reflected on past accomplishments and offered a tentative list of ten "particular ... problems, drawn from various branches of mathematics, from the discussion of which an advancement of science m[ight] be expected." Every such "definite mathematical problem," he believed, "must necessarily be susceptible of an exact settlement, either in the form of an actual answer ... or by [a] proof of the impossibility of its solution" by given means; and, in principle at least, it should always "be possible to establish the correctness" of a proposed solution "by means of a finite number of steps based upon a finite number of hypotheses which are implied in the statement of the problem and which must always be exactly formulated" [113].

At the head of his list Hilbert placed problems concerning "the principles of analysis and geometry," fields in which he deemed "the arithmetical formulation of the concept of the continuum ... and the discovery of non-Euclidean geometry" to have been "the most suggestive and notable achievements" of the preceding century. He cited Cantor's work in particular and, as problem one, called for a proof of the Continuum Hypothesis, as well as a direct (and preferably constructive) proof of Cantor's other "very remarkable" conjecture that the reals could be well-ordered.

Hilbert was aware of the paradoxes engendered by inconsistent multitudes such as "the system of *all* cardinal numbers or of *all* Cantor's alephs," but he was not troubled by them. In his view such inconsistent multitudes simply did not exist as mathematical entities, whereas he was convinced that a consistent axiomatic description of the continuum *could* be given. A first step toward obtaining such a characterization would be to prove that the arithmetical axioms were noncontradictory — the task he proposed as the second of his problems.

At the time, Hilbert seems to have expected that the second problem would prove more tractable than the first, which had, he noted, already withstood "the most strenuous efforts." In a sense he was right, for the consistency question was settled long before the Continuum Hypothesis (though hardly in the way Hilbert had anticipated). But while the CH itself continued to defy all attempts at proof or disproof, only four years later Ernst Zermelo established the well-orderability of the reals — and indeed, of any set whatever (the Well-Ordering Principle, first conjectured by Cantor in 1883) — albeit without showing how particular well-orderings could be defined.

To do so Zermelo introduced his controversial Axiom of Choice, according to which, given any collection $\{X_i\}$ of nonempty sets indexed by a set I, there is a function f with domain I that selects one element $f(i)$ from each X_i; and that axiom, together with the paradoxes, served to precipitate an uproar over foundations that lasted more than three decades.

At issue in the sometimes bitter disputes was the relation of mathematics to logic, as well as fundamental questions of methodology, such as how quantifiers were to be construed, to what extent, if at all, nonconstructive methods were justified, and whether there were important connections or distinctions to be made between syntactic and semantic notions.

Partisans of three principal philosophical positions took part in the debate. Logicists, like Frege and Russell, saw logic as a universal system, within which all of mathematics could be derived. For them the validity of a mathematical statement rested ultimately on its logical *meaning*, so that mathematical truths possessed a tautological character. Formalists, such as Hilbert, sought to justify mathematical theories by means of formal systems, within which symbols were manipulated according to precisely specified syntactic rules.[4]

---

[4] The term "formalist", sometimes reserved for those who believe that mathematics is a "game" played with symbols that are inherently meaningless, is used here in the broader sense indicated. Hilbert did *not* deny the meaningfulness of mathematical statements — far from it — but sought rather to secure the validity of mathematical knowledge without appeal to semantic considerations.

Their criterion for mathematical truth — and indeed for the very existence of mathematical entities — was that of the *consistency* of the underlying systems, since they presumed that any consistent, complete theory would be *categorical*, that is, would (up to isomorphism) characterize a unique domain of objects. And finally constructivists,[5] including Hermann Weyl and Henri Poincaré, restricted mathematics to the study of concrete operations on finite or *potentially* (but not actually) infinite structures; completed infinite totalities, together with *impredicative definitions* that made implicit reference to such, were rejected, as were indirect proofs based on the Law of Excluded Middle. Most radical among the constructivists were the intuitionists, led by the erstwhile topologist L. E. J. Brouwer, a crusader who went so far as to renounce his earlier mathematical work. Brouwer opposed the idea of formalization and developed an alternative mathematics in which some of the theorems (such as the nonexistence of discontinuous functions) contradicted fundamental results of classical analysis.[6] Out of the rancor, and spawned in part by it, there arose several important logical developments. First among these was Zermelo's axiomatization of set theory (*1908a*), discussed more fully in chapter VI [114]. That was followed two years later by the first volume of *Principia Mathematica*, in which Russell and Whitehead showed how, via the theory of types, much of arithmetic could be developed by logicist means. And then Hilbert, goaded by the criticisms of Brouwer and Weyl, entered the lists with a bold proposal to secure the foundations of mathematics once and for all.

Hilbert introduced his program, which he called *proof theory*, in his paper *1923*, and developed it further in an address that he delivered to the Westphalian Mathematical Society on 4 June 1925. In the published version of the address (*1926*) Hilbert acknowledged that "the situation [then prevailing] ... with respect to the paradoxes" was "intolerable," but he expressed his conviction that there was nevertheless "a completely satisfactory way of escaping the paradoxes without committing treason" against the spirit of mathematics. That way lay in "completely clarifying *the nature of the infinite*," a notion that he granted was "nowhere ... realized" in the physical universe and was "not admissible as a foundation" for rational thought, but that had proved useful

---

[5] The term is employed here to encompass a wide spectrum of views, including strict finitism, intuitionism, and predicativism. Poincaré adhered to the latter standpoint, while Weyl began as a predicativist but later embraced a form of intuitionism.

[6] The theorems of intuitionistic logic and arithmetic, on the other hand, form a proper *subset* of the statements provable within the corresponding classical systems. Formal axiomatizations for those theories were given by Brouwer's student Arend Heyting in 1930.

as a theoretical construct in a manner analogous to the introduction of ideal elements in algebra and geometry [115].

Hilbert proposed to replace references to the actual infinite by reasoning of a purely *finitary* character, a notion he illustrated by example but never defined precisely. Such reasoning was to be applied not to the mathematical entities themselves (numbers, sets, functions, and so forth) but to the *symbols* of a formal language in which the concepts had been axiomatized. The formulas of the language were to be regarded as mere sequences of those symbols. The aim, however, "was not to *divest* infinitistic formulas of meaning, but rather to *invest* them with meaning by reference to finitistic [notions] ... [whose] meaning ... [was] unproblematic" [116]. The "single but absolutely necessary" condition for the success of the enterprise was a finitary proof of the consistency of the axioms — or, rather, a series of such proofs for stronger and stronger axiomatic systems. For example, once the consistency of arithmetic was established (as Hilbert mistakenly thought Ackermann had already done) that of analysis was to be reduced to it, again by finitary means. And so on. The logical framework within which Hilbert expected his program to be carried out included variables for numbers, functions, sets of numbers, countable ordinals, and more. Within that system, however, Hilbert distinguished a subsystem that he called the "restricted functional calculus," in which bound function variables were not allowed. It comprised statements (*well-formed formulas*) formed by applying logical connectives and the universal and existential quantifiers to *atomic formulas*, composed in turn of polyadic relation symbols whose arguments were terms built up from function symbols applied to variables and constants denoting *individual* objects. (Languages with such syntax are now called *first-order*, a term coined by Peirce [117].) The axioms and rules of inference of the restricted functional calculus were those of pure logic (pertaining just to the use of connectives and quantifiers). First-order *mathematical* theories were then obtained by adjoining *proper* axioms, such as Peano's, that described particular mathematical entities.

In his proof theory Hilbert expressed quantification by means of a logical choice function, which, given a relation A holding of some object(s), selected a particular such object, ϵ_A. The existential statement $\exists x A(x)$ could then be expressed simply as $A(\epsilon_A)$, and the universal statement $\forall x A(x)$, that is, $\neg \exists x (\neg A(x))$, as $\neg(\neg A(\epsilon_{\neg A}))$.

The use of choice functions to express quantification was by no means unique to Hilbert. Indeed, one commentator has declared that "The connection between quantifiers and choice functions or, more precisely, between quantifier-dependence and choice functions, is the heart of how classical logicians in the twenties viewed the nature of quantification" [118]. In particular, the Norwegian logician Thoralf Skolem had employed choice functions to eliminate existential quantifers. By introducing new function symbols f and g, for example, an expression of the form $\forall x \forall y \forall z \exists u \exists v \Phi(x, y, z, u, v)$ could be replaced by one of the form $\forall x \forall y \forall z \Phi(x, y, z, f(x, y, z,), g(x, y, z))$. In 1920 Skolem used that device to extend a theorem of Leopold Löwenheim, who (in effect) had proved that any first-order formula that is satisfiable at all is satisfiable in a countable model [119]. Skolem showed that the same was true for denumerable sets of such formulas, and in a later paper (*1923b*) he applied that result to obtain the so-called Skolem Paradox: That first-order set theory, in which the existence of uncountable sets is provable, itself has a model whose domain is denumerable.[7]

The Löwenheim-Skolem Theorem is one fruit of the attention logicians of that era paid to questions concerning the satisfiability or validity of quantificational formulas. The central question, known as the *Entscheidungsproblem*, asked whether there was an effective way to determine the satisfiability of any such formula. But a change of perspective had begun to take place, first in the study of propositional logic. In his doctoral dissertation of 1920, Emil Post introduced the method of truth tables as a means of determining which propositional formulas were formally *derivable* within the system of *Principia Mathematica*. The method is viewed today as a procedure for determining which propositional formulas are *tautologies* (universally valid statements), but there is some question whether Post so conceived it. His primary concern appears to have been syntactic rather than semantic, as evidenced by his proof that propositional logic is syntactically complete: If any unprovable formula be added to the propositional axioms, inconsistency results. [120]

Independently of Post, and somewhat earlier, Paul Bernays had also described a decision procedure for propositional logic and likewise proved syntactic completeness [121]. In contrast to Post, however, Bernays spoke of "universal validity" and noted that propositional logic was also complete in the sense that every universally valid formula was derivable.

In the fall of 1917 Hilbert invited Bernays to become his assistant at Göttingen and to serve as "official note-taker" for a course he was preparing entitled "Principles of Mathematics and Logic." Bernays accepted, and the notes that

---

[7]Skolem saw the paradox as establishing the relativity of set-theoretic notions (cf. *Moore 1988*). Later commentators, however, have seen it rather as a mark of the *weakness* of first-order logic; see, for example, *Wang 1974*, p. 154.

resulted from that collaboration became the basis for Hilbert and Ackermann's joint book (*1928*) — the first of the four events mentioned at the beginning of this chapter [122].

In their book Hilbert and Ackermann singled out first-order logic as a tractable object for study and posed the question of its semantic completeness as an open problem. In his Bologna address Hilbert raised the question of syntactic completeness as well, both for first-order logic and for formalized arithmetic [123]. Brouwer, however, took a very different view. For an intuitionist there was no reason to expect of every formula that either it or its negation should admit of a constructive proof, and in his Vienna lectures, entitled "Wissenschaft, Mathematik, und Sprache" and "Über die Struktur des Kontinuums,"[8] he went even further: In the first of them he drew a sharp distinction between "consistent" theories and "correct" ones — an idea that seems to have suggested to Gödel that even within classical mathematics formally undecidable statements might exist [124].

---

[8]That is, "Science, mathematics, and language" and "On the structure of the continuum," delivered in Vienna on 10 and 14 March 1928, respectively.

IV
Moment of Impact

(1929–1931)

> One of themselves, even a prophet of their own, said, The
> Cretans are always liars . . . . This witness is true.
>
> – Titus 1:12–13

THOUGH THERE IS no way to pinpoint when Gödel first began work on his dissertation, internal evidence, together with the many library request slips he saved, some of his later statements, and Carnap's memoranda of conversations he had with him, shows that it was sometime in 1928 or early 1929.

During that period Gödel requested books both from the University of Vienna library and the library of the *Technische Hochschule* in Brno. A comparison of the request slips from those two institutions reveals a marked difference in subject matter. Those from Brno are for standard works on number theory (Dirichlet), function theory (Bieberbach), and differential geometry (Blaschke and Clebsch), while those from Vienna include, in addition to the books of Blaschke and Bieberbach (again), many works in logic and philosophy: Schröder's *Vorlesung über die Algebra der Logik*, Frege's *Grundlagen der Arithmetik*, Heinrich Behmann's *Mathematik und Logik*, Schlick's *Naturphilosophie*, Leibniz's *Philosophische Schriften*, and the *Proceedings of the Fifth Scandinavian Mathematical Congress* (Helsingfors, 1922), in which one of Skolem's papers appeared [125]. Interestingly, Hilbert and Ackermann's *Grundzüge der theoretischen Logik* is *not* among the texts requested; but, as noted earlier, it was at this time that Gödel purchased a copy of Whitehead and Russell's *Principia Mathematica*. The request slips for the books by Schröder and Frege date from the following October, which suggests that Gödel sought them in connection with the course by Carnap mentioned in chapter II. The shift in Gödel's mathematical interests

53

away from more classical mathematical fields toward logic and foundations thus seems to have occurred between the summer and fall of 1928.

It is hard to say what drew Gödel's attention to the completeness question that became the subject of his dissertation. As noted in the previous chapter, the corresponding questions for the *propositional* calculus had been settled independently by Emil Post in his *1921* and by Paul Bernays in his *1926*. But Gödel had not yet become acquainted with either of those individuals, nor, apparently, with their work, since neither of their papers is mentioned in the dissertation itself [126]. In footnote c of the dissertation Gödel does refer to a related completeness result in an unpublished manuscript by Carnap, so it may have been he who first made Gödel aware of the issue; but it is just as plausible to suppose that Gödel spoke with Carnap about the matter sometime after he began work on his dissertation. Alternatively, one might presume that it was Hahn, in his role as Gödel's dissertation supervisor, who suggested the topic. Yet Gödel told Hao Wang that he had already completed the dissertation before ever showing it to Hahn [127]. In footnote 1 of the published version he did, however, explicitly thank Hahn "for several valuable suggestions that were of help ... in writing this paper" [128].

In the absence of hard evidence to the contrary it seems best to suppose that Gödel came upon his dissertation topic on his own, as a result of his reading or attendance at lectures. If so, then an examination of the text of the dissertation might be expected to yield further clues as to its conceptual genesis.

Of particular significance in that regard are Gödel's introductory remarks, which were *omitted* from the published paper. In the first sentence Gödel cites both *Principia Mathematica* and *Hilbert and Ackermann 1928* and states that his goal is to prove the completeness of the axiom systems given in those sources for the restricted functional calculus. In the next sentence he states explicitly that by "completeness" he means that "every valid formula expressible in the restricted functional calculus ... can be derived from the axioms by means of a finite sequence of formal inferences," which, he goes on to say, is "easily ... seen to be equivalent" to the assertion that "Every consistent axiom system [formalized within that restricted calculus] ... has a realization" — that is, a model — and, as well, to the statement that "Every logical expression is either satisfiable or refutable," the form in which he actually proves the result. The importance of the result, he points out, is that it justifies the "usual method of proving consistency" — that of exhibiting such a realization — and so provides "a guarantee that in every case the method leads to its goal." For those who might fail to perceive the *need* for such justification, he noted further that "Brouwer ... [in opposition to Hilbert] has emphatically stressed that from the consistency of a system we cannot conclude without further ado that a model can be constructed." Indeed (to Gödel at least), to think that "the existence of ... notions introduced through ... axiom[s]" is guaranteed "outright" by their consistency "*manifestly presupposes* ... that every mathematical problem is solvable" (emphasis added) — a presupposition that, as noted in chapter III, was so manifest to Hilbert and his followers as to be indisputable [129].

Having raised the specter of unsolvability, Gödel was careful to observe that what was "at issue ... [was] only unsolvability by certain *precisely stated formal* means of inference"; and then, cautiously and almost offhandedly, he remarked that such reflections were "intended only to properly illuminate the difficulties ... connected with ... a definition of the notion of existence, without any definitive assertion being made about its possibility or impossibility."

These comments strongly suggest that by 6 July 1929, when his dissertation was formally approved by Professors Hahn and Furtwängler, Gödel had already begun to think along the lines of his incompleteness discovery. They also suggest that he was spurred to consider the question of completeness both by Hilbert and Ackermann's mention of it as an unsolved problem and by Brouwer's distinction between "correct" and "consistent" theories. At the same time, however, the remarks show that Gödel disagreed both with Hilbert's naively optimistic belief in the limitless efficacy of formal methods and with Brouwer's contempt for the very idea of formalization. Above all, the more speculative of those passages (and especially their omission — if done at Gödel's own behest — from the manuscript he submitted to *Monatshefte für Mathematik und Physik*) exemplify that combination of conviction and caution that Solomon Feferman has rightly recognized as a salient characteristic of Gödel's personality [130].

Before discussing further the content of Gödel's dissertation it should be noted that though Brouwer's role in stimulating Gödel's thought seems beyond doubt, how Gödel became aware of Brouwer's work remains uncertain. One might assume that he attended Brouwer's Vienna lectures, but the evidence on that point is oddly equivocal. In a private memorandum of 23 December 1929 [131] Carnap noted that he had spoken with Gödel that day about the "inexhaustibility" (*Unerschöpflichkeit*) of mathematics and that Gödel's thoughts on the matter had been "stimulated by" (*durch ... angeregt worden*) Brouwer's lecture. Yet after Brouwer's death in 1966 Gödel declined to write an obituary of him for the American Philosophical Society, on the grounds that he was "entirely unqualified" for the task, having "seen Brouwer only on one occasion, in 1953, when he came to Princeton for a brief visit" [132]. Had Brouwer's ideas not provided such a stimulus to Gödel's thoughts, one might think that with the passage of nearly forty years Gödel had simply forgotten about his attendance at Brouwer's lecture. It seems unlikely, though, that he could have forgotten such a seminal event. So it may be that he learned of the lecture's content at second hand.

In any case, Gödel was certainly well acquainted with the tenets of intuitionism. Toward the end of his introductory remarks he defended the "essential use" he had made of the Law of Excluded Middle, arguing that "From the intuitionistic point of view, the entire problem [of a completeness proof] would be a different one," one that would entail "the solution of the decision problem for mathematical logic, while in what follows only a transformation of that problem, namely its reduction to the question [of] which formulas are provable, is intended." He also observed that "it was not the controversy regarding the foundations of mathematics that caused the problem treated here to surface" and that the problem of the completeness of the logical axioms and rules could meaningfully have been posed within "naive" mathematics, even had its contentual *correctness* never been called into question.

Although well warranted historically, the latter comments, too, were omitted from Gödel's published paper *1930a*. There Gödel claimed that once Whitehead and Russell had endeavored to construct logic and mathematics on the basis of formal deductions from "certain evident ... axioms ... by means of ... precisely formulated principles of inference," the question "at once" arose whether the system of axioms and inference rules was complete. Yet in fact, it was some eighteen years after the publication of *Principia Mathematica* before Hilbert and Ackermann posed the problem of its completeness — for so long as the logicist conception of logic as a *universal* language held sway, no metalinguistic questions of any kind could be regarded as meaningful [133]. That the question of completeness seemed to Gödel to present itself so immediately indicates how advanced his own understanding of fundamental logical issues was.

The text of his dissertation (*1929*) already exhibits the concise clarity that was to become a hallmark of Gödel's writings. Following his introductory remarks, Gödel describes the details of the formalism to be employed and makes precise the terminology he will use. He takes particular care to distinguish semantic from syntactic notions, as of course he must. (But as others had not; see below for a discussion of Skolem's "near miss.") He also lists several basic theorems of logic that are essential to his proofs, especially the completeness of the propositional calculus, the provable equivalence of every first-order formula to one in any of several different prenex normal forms, and the fact that every formula of the form

$$(\forall x_1)(\forall x_2)\dots(\forall x_n)F(x_1,x_2,\dots,x_n)\wedge(\exists x_1)(\exists x_2)\dots(\exists x_n)G(x_1,x_2,\dots,x_n)$$

$$\to (\exists x_1)(\exists x_2)\dots(\exists x_n)[F(x_1,x_2,\dots,x_n)\wedge G(x_1,x_2,\dots,x_n)]$$

is provable.

The proof itself proceeds by reduction and induction. First, Gödel shows that it suffices to establish the satisfiability or refutability of prenex formulas in which all the variables are bound and in which the prefix (if any) begins with a block of universal quantifiers and ends with a block of existential quantifiers. After defining the *degree* of such a formula to be the number of distinct blocks of universal quantifiers in its prefix, he shows that if all formulas of degree k are either refutable or satisfiable, so are those of degree $k + 1$. For that he makes use of a device of Skolem, the so-called Skolem normal form for satisfiability.

The brunt of his efforts is then directed to showing that every formula F of degree 1 is either refutable or satisfiable. Toward that end he defines a sequence A_n of quantifier-free formulas (whose structures depend on that of F) and shows that for each n the formula $F \to (E_n)A_n$ is provable, where (E_n) denotes a block of existential quantifiers binding all the variables of A_n. It is in that proof that he uses the last of the basic theorems mentioned above, and there, most important, that his procedure differs crucially from that employed earlier by Löwenheim and Skolem.

This is not the place to enter more fully into the details of Gödel's proofs, nor to examine in detail the complex relationship between Gödel's work and that of Skolem (*1923a*) or of Jacques Herbrand (*1930* and *1931*); for that, the reader should consult the texts of *Gödel 1929* and *1930a* in Volume I of Gödel's *Collected Works* (*1986–*), together with the incisive commentary on them by Jean van Heijenoort and Burton Dreben. Suffice it here to summarize their conclusions: By the Law of Excluded Middle — applied *outside* the formal system — each of the quantifier-free formulas A_n either is, or is not, truth-functionally satisfiable. In the former case, Skolem and Gödel both concluded that the original formula F is satisfiable in the domain of natural numbers, whereas Herbrand, though noting "that such a conclusion could be drawn, [nevertheless] abstain[ed] from doing so" because the argument involved "non-finitistic notion[s] that he consider[ed] to be alien to metamathematical investigations." In the second case, Skolem concluded only that F

would not be satisfiable, whereas Gödel and Herbrand showed that F would in fact be formally refutable [134].

One may well ask why Skolem did not also draw that conclusion; and to that query Gödel himself offered an answer. In a letter to Hao Wang of 7 December 1967 he acknowledged that

> The completeness theorem is indeed an almost trivial consequence of [*Skolem 1923a*]. However, the fact is that, at that time, nobody (including Skolem himself) drew this conclusion (neither from [*Skolem 1923a*] nor, as I did, from similar considerations of his own). . . . This blindness . . . of logicians is indeed surprising. But I think the explanation is not hard to find. It lies in a widespread lack, at that time, of the required epistemological attitude toward metamathematics and toward non-finitary reasoning. . . . The aforementioned easy inference from [*Skolem 1923a*] is definitely non-finitary, and so is any other completeness proof for the predicate calculus.[1] Therefore these things escaped notice or were disregarded [135].

In addition to the completeness result, Gödel answered a further question posed by Hilbert and Ackermann: He demonstrated the independence of the axioms he employed. He also went beyond their questions in two directions, in that, having established his basic result, he extended it both to languages incorporating the equality symbol and to countably infinite *sets* of sentences.

As already noted, the text of Gödel's published paper differed from that of the dissertation itself in that the speculative philosophical remarks at the beginning were deleted and many more citations to the literature were included. More important, however, in the published paper the completeness theorem was obtained as a consequence of a new result, known today as the (countable) compactness theorem: A countably infinite set of first-order formulas is satisfiable if and only if each finite subset of them is satisfiable.

The importance of the latter result lay unrecognized for many years afterward — in large part because of its purely *semantic* character [136]. In that regard, it is significant too that even the notion of truth in a structure, central to the very *definition* of satisfiability or validity, is nowhere analyzed in either Gödel's dissertation or his published revision of it. Rather, in a crossed-out passage of an unsent reply to a graduate student's later inquiry, Gödel asserted that at the time of his completeness and incompleteness papers, "a concept of

---

[1] The nonfinitary step in Gödel's own proof involved what amounted to an application of König's infinity lemma (that every binary tree of unbounded height possesses an infinitely long branch).

objective mathematical truth ... was viewed with [the] greatest suspicion and [was] widely rejected as meaningless" [137]. Such sentiments might be construed merely as further evidence of Gödel's deep-seated caution (not to say paranoia), but they are confirmed as well by the writings of others. For example, in discussing Tarski's epochal paper on the concept of truth in formalized languages [138] Carnap commented that when he invited Tarski to speak on the subject at the September 1935 International Congress for Scientific Philosophy, "Tarski was very skeptical. He thought that most philosophers, even those working in modern logic, would be not only indifferent, but hostile to the explication of the concept of truth." And in fact, "At the Congress it became clear from the reactions to the papers delivered by Tarski and myself that Tarski's skeptical predictions had been right. ... There was vehement opposition even on the side of our philosophical friends." Carnap went on to declare that "it may be difficult for younger readers to imagine how strong the skepticism and active resistance was in the beginning" [139]. In the end, Tarski was willing to confront such resistance; Gödel was not. Nevertheless, Gödel recognized early on the essential distinction between provability and truth. In particular, independent of Tarski, he too discovered the formal undefinability of the latter notion.

$$\diamond \quad \diamond \quad \diamond$$

There is no way to ascertain how much Gödel's work on his dissertation was interrupted by his father's death. He seems to have remained quite unaffected, though, by the growing political turmoil within and around Vienna. He did discuss economic and political matters with some of his friends, but the content of those discussions has for the most part not been recorded.

Whatever his views were of the sinister events then taking place, he cannot have remained unaware of what was going on, for already by 1926 "the forces of reactionary violence had entrenched themselves ... at the University. Socialist leaders," for example, "were prevented ... from delivering speeches in the University buildings," and because "the autonomy of the University ... forbade the police to enter th[ose] buildings ..., at more or less regular intervals Socialist and Jewish students were dragged out of classrooms and severely beaten" [140]. Gödel himself, of course, was not Jewish; but his adviser Hahn was, and so were many of the other professors with whom Gödel had studied. So in Nazi eyes he was tainted. Moreover, despite his family's pride in their German heritage, there is no indication that he had any sympathy for the Pan-German nationalists or the Austrian clerical-fascists. Indeed, according to a memorandum of Carnap, Gödel at that time was "for" socialism and was reading Lenin and Trotsky [141].

Although by 1929 political developments in Austria were becoming increasingly ominous, *Anschluß* was still nine years away. The economic crisis was much more imminent, but it, too, would not fully grip Austria for some months. (The Austrian Credit-Anstalt collapsed on 12 May 1931.) In the meantime, by Rudolf Gödel's account, the brothers spent their inheritances rather freely, "in order to be able to live well" [142]. For Rudolf that meant much traveling about, as well as visits to museums and attendance at public lectures — a lifestyle that probably did not seem improvident to him, both because he was accustomed to enjoying his family's wealth and because by then, following his graduation from the university and a long trip to North Africa (a graduation present from his father), he had obtained a position at the Wenkebach clinic.

For Kurt, however, the prospect of earning a regular income was still a long way off. Although he was granted his Ph.D. on 30 February 1930, possession of such a degree by no means enabled him to embark on an academic career. For that one needed to obtain the *Habilitation*, which entailed, among other requirements, the writing of yet another major paper (the *Habilitationsschrift*). Only then could one hope to begin teaching as an unpaid *Dozent*, while awaiting the still uncertain offer of a permanent appointment. At the beginning of 1930, Gödel thus faced two tasks: that of making his dissertation results known to a wider audience, and that of finding a suitable problem to serve as the topic for his *Habilitationsschrift*.

The revision of Gödel's dissertation was received by the editors of *Monatshefte für Mathematik und Physik* on 22 October 1929, but it was nearly a year before it appeared in print, for a receipt in Gödel's *Nachlaß* shows that he was shipped offprints of the paper only on 19 September 1930. In the meantime he had spoken about his results on at least two occasions: before Menger's colloquium on 14 May [143], and to an international audience at the second Conference on the Epistemology of the Exact Sciences, held in Königsberg, East Prussia, on 6 September. On 28 November he also lectured on his completeness theorem at the meeting of the Vienna Mathematical Society.

The completeness theorem was clearly a significant accomplishment since it gave the answer to a question posed in a prominent textbook. The answer, however, was the expected one, and as already noted, the method of proof was similar to methods that Löwenheim and Skolem had employed. (The compactness theorem was actually the greater discovery — certainly less expected and, in the long run, arguably more important in its consequences.) For the *Habilitation* it was desirable to find a problem whose solution would attract greater attention; and for someone as confident of his abilities as Gödel was, an obvious source for such a question was the list of problems that Hilbert had put forward in 1900.

◇   ◇   ◇

The second of those problems, that of giving a finitary consistency proof for the axioms of analysis, was seen by Hilbert as the first step in his "bootstrapping" program for securing the foundations of mathematics. It is not known when Gödel first turned his attention to the problem, but by the fall of 1930 he had found a solution that was profoundly unexpected.

By his own account, he set out not to destroy Hilbert's program but to advance it. In the same letter draft in which he spoke of the "prejudice" then prevailing against the concept of "objective mathematical truth," [144] he explained that

> The occasion for comparing truth and demonstrability was an attempt to give a relative model-theoretic consistency proof of analysis in arithmetic. Th[at] leads almost by necessity to such a comparison. For an arithmetical model of analysis is nothing else but an arithmetical $\in$-relation satisfying the comprehension axiom:
>
> $$(\exists n)(x)[x \in n \equiv \phi(x)].$$
>
> Now, if in the latter "$\phi(x)$" is replaced by "$\phi(x)$ is provable," such an $\in$-relation can easily be defined. Hence, if truth were equivalent to provability, we would have reached our goal. However (and this is the decisive point) it follows from the **correct** solution of the semantic paradoxes that "truth" of the propositions of a language *cannot be expressed* in the same language, while provability (being an arithmetical relation) *can.* Hence true $\neq$ provable.

In this passage the term "analysis" means second-order arithmetic, in which variables for sets of natural numbers are allowed. Apparently Gödel's initial idea was to interpret such variables, within structures for *first*-order number theory, as ranging over the *definable* subsets of the natural numbers. Assuming that the underlying formal language is countable, there will be only countably many such definable subsets, which can therefore be indexed by a numerical variable ("n" in the displayed formula above). The notion of set-theoretic *membership* will thereby be replaced by that of the *truth* of a general arithmetical formula — and there's the rub.

In view of the suspicions Gödel had expressed in the introduction to his dissertation concerning the possible existence of formally undecidable statements within arithmetic, one may wonder why he should have attempted to obtain a *positive* solution to the consistency question. Such a view, however, conflates two distinct problems: that of showing that arithmetic is *incomplete*, in the sense that there *are* arithmetical statements that are formally undecidable, and that of showing that its *consistency* is not only *expressible* within the theory itself, but that it is a particular *example* of such an undecidable statement. That is precisely the distinction between Gödel's first and second incompleteness theorems, which he obtained in that order, with a short but significant lapse of time in between.

Given Gödel's association with the Vienna Circle, it is not at all surprising that he should have been alert to the problem of speaking about a theory within the language of that theory itself, for the question of the extent to which one could meaningfully speak about language within language was central to Wittgenstein's philosophy (a prime stimulant to the development of the Circle's ideas) and also to Carnap's. Moreover, in February of 1930, just when Gödel is likely to have been confronting such issues, Alfred Tarski came to Vienna to give a series of lectures to Menger's colloquium, "one on mathematics and two on logic, the latter being thrown open to the Schlick Circle" as well [145]. In his "Intellectual Autobiography" Carnap recalled the topic of the joint lectures to have been "the metamathematics of the propositional calculus." More important, he recalled that "We also discussed privately many problems" of common interest:

> Of special interest to me was his emphasis that certain concepts used in logical investigations, e.g., the consistency of axioms, the provability of a theorem in a deductive system, and the like, are to be expressed not in the language of the axioms (later called the object language), but in the metamathematical language (later called the metalanguage). . . . My talks with Tarski were fruitful for my further studies of the problem of speaking about language, a problem which I had often discussed, especially with Gödel. Out of these problems and talks grew my theory of logical syntax [146].

Gödel, too, had a private discussion with Tarski at that time. Menger recalled that "After the lectures, Gödel asked me to arrange an appointment with Tarski, as he wanted to report to the visitor about the contents of his dissertation" — in furtherance, that is, of the first of the two tasks mentioned earlier. And in fact, "Tarski showed great interest in the result" [147].

Gödel may also have spoken with Tarski about some of the topics that Carnap mentioned. But if so, he was not influenced by Tarski's views in the way that Carnap was. On the contrary, it was precisely Gödel's recognition that such notions as "the consistency of the axioms" and "the provability of a statement within the system" *can* be expressed (indirectly) in the language of arithmetic that was the key to his proofs of the incompleteness theorems.

Much has been written about that idea, the so-called arithmetization of syntax. The essence of it is that not only the symbols of the language, but finite sequences of such symbols and, in turn, finite sequences of *them*, can be assigned numerical code numbers in an effective, one-to-one fashion. Among the sequences of symbols are those that constitute well-formed formulas, some of which are taken as axioms; and the properties of the sequences that make them recognizable as such translate into recognizable arithmetical properties of their numerical codes. Similarly, among the sequences of formulas are those that represent proofs — to wit, those in which the first formula is an axiom and subsequent formulas are either axioms or are obtained from earlier formulas in the sequence through the application of a finite number of recognizable rules; and that property, too, can be shown to correspond to a recognizable property of the codes.

In this précis of the method of "Gödel numbering" the term "recognizable" has been employed as a gloss for the precise mathematical concept "primitive recursive." Without entering into the technical definition of the latter, it may be understood — still informally, but accurately — as expressing the idea of being *testable by means of a predictably terminating algorithm*. Thus, given a natural number, it can be tested successively to determine whether or not it is the code number of a formula or a sequence of formulas. If the latter is the case, it can then also be determined whether it is the code number of a proof, and if so, of what. Each of those tests requires only finitely many steps, and a bound to the number of steps can be determined in advance.

Crucially, however, the modal notion "provable" is *not* so testable. Given a purported proof of a formula, it is possible to determine whether it really is such by means of a predictably terminating sequence of tests; but given only a purported theorem, there is no effective way of guessing *how* to prove it. One can only examine proofs, one at a time, to see whether the final formula of one of them is the formula in question. If so, then that formula is indeed a theorem. But one cannot predict in advance when (or if) such a proof will turn up. If the given formula is *not* provable, then the search for a proof will never terminate.

The relevance of these facts to the proof of the *second* incompleteness theorem is that to assert the consistency of a theory is to assert that a particular formula within it (of the form $A \wedge \neg A$, say) is unprovable. But if the property of being unprovable is not, in general, finitely testable, then the hope of proving the consistency of arithmetic within that theory (much less in a weaker one) must rest on the particular *form* of the consistency statement.[2]

Hilbert's presumption, and Gödel's too, was that arithmetic *is* consistent. The question was only the *means* necessary to prove that. The possibility that Hilbert (but not Gödel) failed to entertain was that such means might go beyond what was available within arithmetic itself. (If a theory is inconsistent, then *every* formula within it will be formally provable; so as soon as a proof of an absurdity appears, the inconsistency will be demonstrated. Historically, that is just how inconsistencies have always been discovered. The trouble is that, lacking a proof of consistency, formal or otherwise, one cannot know in advance when or if that may happen — the situation is the same as that for the notion of provability. Nonetheless, in principle at least, *in*consistency is always formally demonstrable by a proof within the given system.)

Should formalized arithmetic be consistent yet formally incapable of proving that fact, then, assuming that the theory is *sound* (that is, that it does not prove any false statements), its consistency must be formally undecidable, that is, neither provable nor disprovable. The converse, however, is not clear: It is conceivable that the consistency of a theory might be internally demonstrable even though other statements within it were undecidable. That, again, is the distinction between the first and second of the two incompleteness theorems.

Such speculations suggest only how undecidable statements might arise, not how they might be detected. Gödel's achievement was to show how to construct a statement whose undecidability could be (informally) demonstrated. He did so by showing in very detailed fashion that a large number of basic

---

[2] In any consistent theory some arithmetical statements — for example those that are *dis*provable — can of course be shown to be unprovable. And in fact, some *nonstandard* statements of the consistency of arithmetic *are* formally provable within the theory. See *Feferman 1960* for a detailed discussion of some examples.

syntactic notions are representable, through their code numbers, by formulas of formalized arithmetic. Specifically, he showed that there is a binary, primitive recursive predicate $B(x, y)$ that formalizes the notion "x is the code number of a proof of the formula whose code number is y", so that if n and m are natural numbers and **n** and **m** are the symbols denoting those numbers within the formal theory, then $B(\mathbf{n}, \mathbf{m})$ is provable in the theory if n really is the code number of a sequence of formulas that constitutes a proof of the formula having code number m, and $B(\mathbf{n}, \mathbf{m})$ is disprovable otherwise. It follows that the monadic predicate $\text{Bew}(y)$, defined as an abbreviation for the formula $(\exists x)B(x, y)$, formalizes the notion "y is provable" (*beweisbar*). The negation of $\text{Bew}(y)$ then obviously formalizes the notion "y is not provable"; and *that* notion, Gödel realized, could be exploited by resort to a diagonal argument reminiscent of Cantor's.

The key idea was to modify the ancient Antinomy of the Liar, in particular the later form of it due to Eubulides ("This sentence is false"). The source of the antinomy is the notion of truth, which, even in number theory, Gödel had recognized to be formally inexpressible. By replacing the notion of falsity by that of unprovability he was able to avoid antinomy while constructing, within formal number theory, an analogous self-referential statement. In essence, he showed that there is a binary formula $Q(x, y)$ such that, given *any* unary formula $F(y)$, if $F(y)$ has code number n, then $Q(x, \mathbf{n})$ formalizes the notion "x is not the code number of a proof of the formula $F(\mathbf{n})$." Thus the formula $(\forall x)Q(x, \mathbf{n})$ asserts the unprovability of $F(\mathbf{n})$. But $(\forall x)Q(x, y)$ is itself a unary formula, so it has some code number, say q. The formula $(\forall x)Q(x, \mathbf{q})$ therefore asserts its own unprovability.

In the introduction to his epochal paper *1931a*, Gödel sketched "the main idea of the [undecidability] proof . . . without any claim to complete precision." His argument can be explained with reference to two tables (below), each infinite in extent [148]. Along the left side of the tables are listed (in some definite order) the unary number-theoretic formulas $F_n(y)$. Along the top are listed the natural numbers m. The entry at position (n, m) in Table 1 is T or F according to whether the formula $F_n(\mathbf{m})$ is true or false when interpreted in the natural numbers (with m taken as the interpretation of $\mathbf{m}$), whereas the corresponding entry in Table 2 is P or N according to whether $F_n(\mathbf{m})$ is or is not provable in formal number theory.[3] Assuming the theory is sound (which entails its consistency), anywhere a P appears in the second table, a T *must*

---

[3] In the hypothetical tables illustrated, the first four formulas in the list may be taken to be $y < 2, y = y$, "y is odd" (expressed say as $(\exists x)(y = 2 * x + 1)$), and $y = 3 \lor y = 4$, respectively.

Table 1: Hypothetical Truth Table for Unary Formulas

Unary Formula	Numerical Argument								
	0	1	2	3	4	5	...	q	...
$F_1(y)$	T	T	F	F	F	F	...	F	...
$F_2(y)$	T	T	T	T	T	T	...	T	...
$F_3(y)$	F	T	F	T	F	T	...	?	...
$F_4(y)$	F	F	F	T	T	F	...	F	...
$\vdots$								$\vdots$	$\vdots$
$F_q(y)$	?	?	?	?	?	?	...	T	...
$\vdots$									

appear in the first. But if a T does appear at position (n, m) in Table 1, must the entry at position (n, m) in Table 2 necessarily be P? If, for some (n, m), it is not, then $F_n(\mathbf{m})$ is true but formally unprovable. On the other hand, since its negation is false, it too (again by the presumed soundness of the theory) is unprovable. Hence $F_n(\mathbf{m})$ must be formally undecidable.

The argument is completed by considering the position (q, q), where $F_q(y)$ is the formula $\forall x Q(x, y)$. If $F_q(\mathbf{q})$ were false, then by the properties of the formal provability predicate noted earlier it would in fact be refutable — that is, formally *dis*provable — and so, by the presumed consistency of the theory, would be formally *un*provable as well. But since that is just what it asserts (that it is unprovable), $F_q(\mathbf{q})$ would then be true, contrary to supposition. Therefore the entry at position (q, q) must be T in Table 1 and N in Table 2.

This informal argument invokes the notion of truth, which many, especially those of the Hilbert school, then considered suspect. Toward the end of his introductory remarks Gödel therefore took pains to stress that "the purpose of carrying out the ... proof with full precision" (as he did in the remainder of the paper) was "among other things, to replace ... the assumption [that every provable formula is true] by a purely formal and much weaker one."[4]

---

[4]Ordinary consistency is all that is needed to prove the *second* theorem and to show that the formula $F_q(\mathbf{q})$ is unprovable, but more is required to ensure that the negation of $F_q(\mathbf{q})$ be unprovable. Since the negation of $F_q(\mathbf{q})$ is logically equivalent to a purely existential formula (one of the form $(\exists x)G(x)$, where G contains no other quantifiers), it suffices to assume that whenever each numerical instance of such a formula is refutable, the formula itself is not provable.

Table 2: Hypothetical Provability Table for Unary Formulas

Unary Formula	Numerical Argument								
	0	1	2	3	4	5	...	q	...
$F_1(y)$	P	P	N	N	N	N	...	N	...
$F_2(y)$	P	P	P	P	P	P	...	P	...
$F_3(y)$	N	P	N	P	N	P	...	?	...
$F_4(y)$	N	N	N	P	P	N	...	N	...
$\vdots$								$\vdots$	$\vdots$
$F_q(y)$	?	?	?	?	?	?	...	N	...
$\vdots$									

Most discussions of Gödel's proof, like that above, focus on its quasi-paradoxical nature. It is illuminating, however, to ignore the proof and ponder the implications of the theorems themselves.

It is particularly enlightening to consider together both the completeness and incompleteness theorems and to clarify the terminology, since the names of the two theorems might wrongly be taken to imply their incompatibility. The confusion arises from the two different senses in which the term "complete" is used within logic. In the semantic sense, "complete" means "capable of proving whatever is valid," whereas in the syntactic sense it means "capable of proving or refuting each sentence of the theory."[5] Gödel's completeness theorem states that every (countable) first-order theory, whatever its nonlogical axioms may be, is complete in the former sense: Its theorems coincide with the statements true in *all* models of its axioms. The incompleteness theorems, on the other hand, show that if formal number theory is consistent, it fails to be complete in the second sense.

The incompleteness theorems hold also for higher-order formalizations of number theory. If only first-order formalizations are considered, then the completeness theorem applies as well, and together they yield not a contradiction, but an interesting conclusion: Any sentence of arithmetic that is undecidable

---

[5] In his original German, Gödel distinguished the two by means of the terms *vollständig* and *entscheidungsdefinit*. The former is properly glossed as "complete," but the latter is better rendered as "decisive" than "incomplete," since "decisive" not only captures the meaning of the German root *entscheid-* but also suggests that the theory is capable of deciding (that is, proving or refuting) each of its sentences. Unfortunately, however, the term "incomplete" is now thoroughly entrenched in the English literature.

must be true in some models of Peano's axioms (lest it be formally refutable) and false in others (lest it be formally provable). In particular, there must *be* models of first-order Peano arithmetic whose elements do not "behave" the same as the natural numbers. Such nonstandard models were unforeseen and unintended, but they cannot be ignored, for their existence implies that *no first-order axiomatization of number theory can be adequate to the task of deriving as theorems* exactly *those statements that are true of the natural numbers.*

◇ ◇ ◇

Even someone as cautious as Gödel could hardly be expected to withhold such a momentous discovery for long, so it seems likely that he obtained the first of his incompleteness theorems only shortly before he confided the result to Carnap during a discussion at the Café Reichsrat on 26 August 1930.

According to Carnap's memorandum of the event [149], the meeting lasted about an hour and a half. Feigl was also present, at least initially, and Waismann came later. The main topic of conversation was the plan for their upcoming journey to the conference in Königsberg, where Carnap and Waismann were to deliver major addresses and where Gödel was to present a summary of his dissertation results. But then, Carnap tersely noted, the discussion turned to "Gödel's discovery: incompleteness of the system of *Principia Mathematica*; difficulty of the consistency proof." How much Gödel told Carnap on that occasion is unrecorded; in any case, it is clear from some of Carnap's later memoranda, as well as his behavior at Königsberg, that he did not completely understand Gödel's ideas. Other sources [150] indicate that Gödel had not yet discovered his second incompleteness theorem and that he did not at first see how to obtain an undecidable proposition within arithmetic. His original example was reportedly an involved combinatorial statement expressed in the richer language of the system of *Principia Mathematica.*

Gödel discussed his results further with Carnap at another meeting in the same café three days later. Then on 3 September the four participants in the earlier meeting traveled together from Vienna's Stettiner Bahnhof to Swinemunde. There they met Hans Hahn and Kurt Grelling, and the six of them boarded a steamer for Königsberg, where they arrived the next day [151].

The Conference on Epistemology of the Exact Sciences ran for three days, from 5 to 7 September. It had been organized by the Gesellschaft für empirische Philosophie, a Berlin society allied with the Vienna Circle, and was held in conjunction with and just before the ninety-first annual meeting of the

Society of German Scientists and Physicians (Gesellschaft deutscher Naturforscher und Ärzte) and the sixth Assembly of German Physicists and Mathematicians (Deutsche Physiker- und Mathematikertagung). The first day featured hour-long addresses on the three competing mathematical philosophies of the time, logicism, intuitionism, and formalism, delivered by Rudolf Carnap, Arend Heyting, and John von Neumann, respectively [152]. Gödel's contributed talk took place on Saturday, 6 September, from 3 until 3:20 in the afternoon, and on Sunday the meeting concluded with a roundtable discussion of the first day's addresses. During the latter event, without warning and almost offhandedly, Gödel quietly announced that "one can even give examples of propositions (and in fact of those of the type of Goldbach or Fermat) that, while contentually true, are unprovable in the formal system of classical mathematics" [153].

Gödel waited until late in the discussion to make his remarks, perhaps unsure how receptive his hearers would be. No doubt he hoped the others would first commit themselves to firm positions. If so, he was not disappointed, though he may have been surprised; for Carnap — despite his prior knowledge of Gödel's results — persisted in advocating consistency as a criterion of adequacy for formal theories! Hahn, too, spoke as though he were quite unaware of Gödel's latest discoveries, even though it seems likely that he had sponsored Gödel's presentation of his thesis results the day before. Could it be that, once again, Gödel had not seen fit to apprise his adviser of the fruits of his work?

The transcript of the session records no follow-up discussion of Gödel's announcement, and the summary of the session prepared after the event by Hans Reichenbach for publication in Die Naturwissenschaften [154] makes no mention at all of Gödel's participation. By the time the session transcript was published, however, Gödel's incompleteness paper had already appeared and its significance was coming to be appreciated. The editors of Erkenntnis therefore belatedly asked Gödel to supply a postscript giving a synopsis of his results.

The evidence thus suggests that few of those present at the discussion grasped the import of what Gödel had said. Indeed, it seems that von Neumann, with his legendary quickness of mind, may have been the only one to do so. Before Gödel spoke von Neumann had already expressed reservations about the adequacy of the consistency criterion (though on very different grounds), and after the session he is said to have drawn Gödel aside to learn further details [155].

Following the conference von Neumann continued to reflect upon Gödel's ideas, and on 20 November he wrote excitedly to him to report that in doing so he had come upon a result that seemed to him "remarkable": that in a consistent system any effective proof of the unprovability of the statement $0 = 1$ could itself be transformed into a contradiction. Von Neumann said that he would be "*very* interested" to hear Gödel's opinion of the matter, and, if Gödel were interested (!), he would be glad to impart details of the proof to him — as soon as he had written them up for publication.

Gödel's letter of reply is no longer extant, but in the meantime, on 23 October, he had submitted to the Vienna Academy of Sciences an abstract (*1930b*) containing statements of both the incompleteness theorems. The manuscript of the full paper (*1931a*) was received by the editors of *Monatshefte für Mathematik und Physik* on 17 November — three days before von Neumann's letter — and on 29 November von Neumann wrote Gödel again to thank him for having sent a preprint of it. Von Neumann was not accustomed to being anticipated, and the tone of his second letter, though gracious, conveys his disappointment: "Since you have established the unprovability of consistency as a natural continuation and deepening of your earlier results, of course I will not publish on that subject." Nevertheless, he could not refrain from explaining how his own proof differed from Gödel's in certain details.[6] If von Neumann was momentarily crestfallen, the affair served in the end only to heighten his respect for Gödel and his abilities. Not long afterward he curtailed his own research efforts in logic, but from then on he was a close friend of Gödel's and a champion of his work.

Indeed, the incompleteness theorems so fascinated von Neumann that on at least two occasions he lectured on Gödel's work rather than his own. Carl Hempel [156] recalled that in Berlin (presumably in the fall of 1930) he took a course from von Neumann that "dealt with Hilbert's attempt to prove the consistency of classical mathematics with finitary means," and that "in the middle of the course von Neumann came in one day and announced that he had just received a paper from a young mathematician in Vienna . . . who showed that the objectives which Hilbert had in mind . . . could not be achieved at all." A year later von Neumann was invited to Princeton to speak to the mathematics colloquium there, and then, too, he chose to talk about Gödel's 1931 paper.

---

[6] Gödel did not, in fact, give a detailed proof of the second theorem in his paper *1931a*, but only a sketch of it. He promised that full details, together with extensions of his results to other systems, would be given "in a sequel to be published soon," but no such paper ever appeared. A full proof of the second theorem was first published eight years later in *Hilbert and Bernays 1939*.

Stephen C. Kleene, who was among the student members of the audience, said that was the first time he ever heard of Gödel [157].

Herman Goldstine tells a story about von Neumann that helps to put his fascination with Gödel's work in perspective [158]. For quite some time von Neumann had attempted to carry out Hilbert's program for proving the consistency of classical mathematics by finitary means. Unlike Gödel, however, he had not recognized the difficulties inherent in formalizing the notion of truth but had persisted in efforts to find a positive solution. Having obtained some partial results he continued, as was his wont when he seemed to be making progress, to work on the problem uninterruptedly. One evening he dreamed that he had surmounted the last remaining obstacle. He stood up, went to his writing table, and carried the proof attempt further — but not quite to the end. The next day he worked intensively on the problem, but again went to bed without having achieved his goal. That night he dreamed once more that he had discovered the key to the solution, but when he went to write it down he was disappointed to find yet another gap in the argument. And so he turned his attention to other matters. Looking back on the affair in light of Gödel's subsequent results he quipped to Goldstine, "How fortunate it is for mathematics that I didn't dream anything the third night!"

As it happened, Hilbert himself was present at Königsberg, though apparently not at the Conference on Epistemology. The day after the roundtable discussion he delivered the opening address before the Society of German Scientists and Physicians — his famous lecture "Naturerkennen und Logik" (Logic and the understanding of nature), at the end of which he declared:

> For the mathematician there is no Ignoramibus, and, in my opinion, not at all for natural science either. ... The true reason why [no one] has succeeded in finding an unsolvable problem is, in my opinion, that there *is no* unsolvable problem. In contrast to the foolish Ignoramibus, our credo avers:
> We must know,
> We shall know [159].

Since Gödel did not leave for Berlin until 9 September, it is very likely that he was in the audience when Hilbert proclaimed that dictum; and if so, one must wonder what his reaction was. For Gödel too believed that no mathematical problems lay beyond the reach of human reason. Yet his results showed that the program that Hilbert had proposed to validate that belief — his proof theory — could not be carried through as Hilbert had envisioned.

Gödel did not attempt to speak to Hilbert then, nor, as he later confirmed [160], did he ever meet Hilbert face-to-face or have any direct correspondence with him. It is possible that von Neumann might have apprised Hilbert of Gödel's results before he left Königsberg, but there is no evidence that he in fact did so. In any case, a few months after the meetings Paul Bernays, in his capacity as Hilbert's assistant, wrote Gödel to thank him for having sent an offprint of his completeness paper. He went on to say, "I heard from Prof. Courant and Prof. Schur that you have recently obtained significant and surprising results in the domain of foundational problems, which you intend soon to publish. Would you be so kind, if possible, as to send me a copy of the galleys" [161]. Gödel did so a week later, and Bernays acknowledged receipt in his letter to Gödel of 18 January 1931. So at that point, if not before, Hilbert must have become aware of what Gödel had done.

There is no need to speculate about Hilbert's reaction, for in a letter of 3 August 1966 to Constance Reid Bernays testified that "some time before" he learned of Gödel's theorems he had himself become "doubtful ... about the completeness of the formal systems" and had "uttered [his doubts] to Hilbert." The very suggestion provoked Hilbert's anger, as did Gödel's results when they were first brought to his attention. But Hilbert soon accepted their correctness, and, as Bernays hastened to point out, in his publications *1931a* and *1931b* Hilbert "dealt with [the consequences of the incompleteness theorems] in a positive way."

The papers to which Bernays referred were those in which Hilbert introduced a form of the so-called ω-rule, according to which, for any quantifier-free unary formula $F(x)$, the generalized formula $(\forall x)F(x)$ may be inferred whenever each of the infinitely many numerical instances $F(1)$, $F(2)$, $F(3)$, ... has been proved. It may be (as Bernays seems to suggest)[7] that Hilbert proposed his rule in direct response to Gödel's results — perhaps after reflecting on the assumption Gödel had invoked in order to circumvent the notion of truth. Certainly Gödel's example of an undecidable statement was a purely universal formula of the sort to which Hilbert's rule referred, so acceptance of the rule

---

[7] Bernays's statement is somewhat ambiguous on this point and is not altogether consonant with statements he made on some other occasions. In addition, neither of Hilbert's papers cites Gödel's work, and his paper was based on a talk he delivered in December 1930 — quite soon after the Königsberg meeting — to the Philosophical Society of Hamburg. For further discussion of the issue see the note by Solomon Feferman, pp. 208–213 in vol. I of *Gödel 1986–*.

would render it provable.[8] But the rule itself was an infinitary proof principle whose practical utility was not at all apparent. Accordingly there were many, including Gödel, who deemed it to be contrary to Hilbert's basic principles [162].

Bernays himself regarded Gödel's incompleteness discovery as "a really important advance in research on foundational problems" [163]. Yet he apparently found neither the proofs nor the implications of the theorems easy to assimilate, for on 18 January, 20 April, and 3 May he wrote Gödel at length to request clarification on points of detail. He was confused in particular by earlier finitary consistency proofs put forward by Ackermann and von Neumann, proofs whose restricted applicability was only belatedly recognized. And at a meeting with Gödel on 7 February 1931, Carnap confessed that he too still found Gödel's work "hard to understand" [164].

In the meantime, on 15 January 1931, Gödel had spoken on his incompleteness results before the Schlick Circle. Rose Rand took minutes of the discussion that followed, and a copy of her transcript is preserved among Carnap's papers at the University of Pittsburgh [165]. It shows that Hahn asked Gödel to go over the *Leitgedanken* of his proof once more; that he saw the analogy with Cantor's diagonal method but apparently failed to recognize that Gödel's argument for establishing the existence of undecidable statements was constructive; and that Felix Kaufmann, another of the Circle's members, failed to grasp the distinction between formal and informal proofs. Gödel's responses are in every case clear and incisive. In particular, in reply to questions from Kaufmann and Schlick, Gödel noted that Brouwer doubted whether any formal system could comprise all of intuitionistic mathematics. The same issue, he pointed out, was also the weak point in von Neumann's conjecture that if there really were a finitary consistency proof, it ought to be formalizable (contrary to the second incompleteness theorem).[9]

---

[8]If, more generally, the ω-rule is taken to apply to *all* unary formulas (not just those that are quantifier-free), then it can be shown that its adjunction to the machinery of Peano arithmetic yields a system whose theorems exactly coincide with the statements that are true in the standard model (the natural numbers themselves).

[9]In a similar vein, Carnap (in a memorandum of 16 October 1930 [166]) recorded Gödel's refutation of a claim by Heinrich Behmann that every existence proof could be made constructive. Gödel's counterexample is much like some that Brouwer had given earlier: Define $F(n)$ to equal 1 if all even numbers $< 2n$ are the sum of two primes, and 2 if not. Then consider $\sum_{n=1}^{\infty} \frac{F(n)}{3^n}$. The series clearly must converge to a *rational* sum, but the value will be $\frac{1}{2}$ if and only if Goldbach's conjecture is true.

A week later, on 22 January, Gödel presented his results to Menger's colloquium [167]. Menger himself was away, spending the academic year 1930–31 as a visiting lecturer in the United States. In his absence he had left Georg Nöbeling in charge of the colloquium, and it was he who, early in February, wrote to Menger to inform him of Gödel's achievement. Then at Rice University in Houston, Menger, like von Neumann, was so excited by the news that he "interrupted [his] lecture on dimension theory and metric geometry and inserted a report on Gödel's epoch-making discovery" [168]. Soon thereafter he sent Gödel a letter of congratulations, in which he also took the opportunity to pose a minor question concerning the propositional calculus [169]. Gödel promptly found the answer, communicated it to Menger, and took it as the subject of his next contribution to the colloquium (on 6 June). Published as *Gödel 1932d,* the result has been described by W. V. O. Quine as "a pioneer venture in the metalogic of indenumerable notation" [170]; in effect, it represented an extension to uncountable sets of propositional formulas of what is known today as "Lindenbaum's lemma."

By 25 March the incompleteness paper (*1931a*) had appeared in print, for on that day 100 offprints of the article were shipped to its author. Gödel wasted no time in sending two of them to Bernays (one for Hilbert), who acknowledged their receipt in his letter of 20 April; but Bernays did not thereupon discard the galleys that Gödel had sent to him earlier. Instead, he passed them on to Jacques Herbrand, who was then visiting Berlin and had already heard of Gödel's work from von Neumann.

Herbrand found the incompleteness results to be of great interest. They stimulated him to reflect further on the nature of intuitionistic proofs and of schemes for the recursive definition of functions. On 7 April he wrote Gödel, enclosing offprints of a few of his own logical works, raising questions about some of the implications of the incompleteness theorems, and suggesting a more general recursion schema. Like von Neumann, Herbrand said that he did "not at all understand how it is possible for there to be intuitionistic[10] proofs that are not formalizable in Russell's system"; but at the same time he expressed his belief that one could never *prove* that all such proofs were so formalizable because he felt it was "impossible to describe all the intuitionistic procedures for constructing functions."

Due, perhaps, to the press of other work, or because Herbrand's ideas led Gödel, in turn, to reflect at length on the issues involved, Gödel did not re-

---

[10] By "intuitionistisch" Herbrand meant something close to "finitary," not "intuitionistic" in the sense of Brouwer.

ply to Herbrand's letter until 25 July — a delay that was most regrettable, for, tragically, Herbrand was killed in a mountaineering accident on the same day that Gödel's letter arrived at his home in Paris. He was thus unable to react to Gödel's differing view that it would be premature to presume that all finitary proofs could be formalized within the system of *Principia Mathematica*. (Gödel had in fact already expressed that opinion near the end of his *1931a*, after saying that he wished "to note expressly" that the second incompleteness theorem did "not contradict Hilbert's formalistic viewpoint." That he reiterated the point both in the discussion at the Schlick Circle and in his letter to Herbrand suggests that Gödel held the view sincerely.[11]) Subsequently the correspondence between the two men was mislaid, to be rediscovered only in 1986. Nevertheless, Herbrand's ideas for generalizing the notion of recursive function were taken up by Gödel, who presented a modification of Herbrand's schema in the lecture course on the incompleteness theorems that he gave at Princeton in 1934.

◇ ◇ ◇

Outside Vienna the first occasion on which Gödel lectured on his incompleteness theorems was at the meeting of the German Mathematical Union (Deutsche Mathematiker-Vereinigung) that took place at Bad Elster in September of 1931.[12] The event was significant because it was there, in the person of Ernst Zermelo, that Gödel encountered his first and most vocal detractor.

Then sixty, Zermelo was a battle-scarred veteran of the conflict over the Axiom of Choice who, some years before, had suffered a nervous breakdown. By the time of the sessions in Bad Elster he had recovered, but Olga Taussky, an eyewitness to the events there, recalls that he remained "a very irascible person" who "felt ill treated ... [and] had no wish to meet Gödel" [171]. He had surely become aware of Gödel's work prior to the conference, but his own outlook on logical matters was so utterly different from Gödel's that he could not properly understand it. Indeed, in his own talk at Bad Elster[172] he launched a reactionary assault against "Skolemism, the doctrine that *every* mathematical theory, even set theory, is realizable in a countable model." He thought of quantifiers as infinitary conjunctions or disjunctions of unrestricted

---

[11] Gödel shared Hilbert's "rationalistic optimism" (to borrow Hao Wang's term) insofar as *informal* proofs were concerned. He addressed the issue explicitly in his Gibbs Lecture of 1951 and in his posthumous note *1972*, where he disputed Turing's contention that "mental procedures cannot go beyond mechanical" ones.

[12] Gödel's talk, "On the existence of undecidable arithmetical propositions in the formal systems of mathematics," was delivered on the afternoon of 15 September.

cardinality and conceived of proofs not as formal deductions from axioms, but as metamathematical determinations of the truth or falsity of a proposition through transfinite induction on the complexity of its construction from primitive constituents having designated truth values. Syntactic considerations thus played no role in his thinking (even though it was he who had first proposed *axioms* for set theory!); he could hardly be expected to appreciate even the completeness theorem, much less the idea of undecidable arithmetical statements.

No wonder then that he resisted when "a small group," hoping to introduce him to Gödel, "suggested lunch at the top of a nearby hill." He offered various excuses — that "he did not like [Gödel's] looks," that "the climb [would be] too much for him," that "there would not be enough food" to share if he joined the group — but in the end, when Gödel showed up, "the two immediately discussed logic and Zermelo never noticed [that] he had made the climb" [173].

Nevertheless, as Taussky goes on to say, "th[at] peaceful meeting was not the start of a scientific friendship between [the] two logicians." Far from it. For just after the conference, on 21 September, Zermelo wrote to Gödel to advise him that he had found "an essential gap" in Gödel's argument. Indeed, he said, by simply omitting the proof predicate (!) from Gödel's construction, one could obtain a sentence asserting its own falsity, "a *contradiction* analogous to Russel[l]'s antinomy" [174]. After a brief delay (occasioned, he explained, by his having been away for a few days), Gödel replied on 12 October [175]. In calm and patient terms he stressed that he had employed the notion of truth only in the informal introduction to his paper and that the alleged "gap" in the proof was in fact filled in the formal treatment that followed. He also noted that the contradiction to which Zermelo alluded arose from the assumption that truth *could* be defined within the formal theory itself — an assumption that, he pointed out, was "yours, not mine." After reviewing the steps in his proof in considerable detail (the letter runs to ten handwritten pages), he expressed the hope that Zermelo would now be convinced of the correctness of the results. He concluded by thanking Zermelo for sending him a copy of one of his papers (*1930*), which, he said, he had read shortly after its appearance and which had stimulated some further thoughts of his own. In turn, he enclosed an offprint of his completeness paper, copies of which he had not brought along to the conference.

Unfortunately, Gödel's hopes remained unfulfilled. For though Zermelo wrote again on 29 October to thank Gödel for his "friendly letter" — from which, he said, he was able to comprehend Gödel's intent better than he had from his lecture or his published paper — he now understood Gödel's proof to apply only to "the provable statements of [the] PM-system." Of those, he said — correctly enough — there would be only countable many; but that, he seemed to think, was due to Gödel's having applied what he called "the finitistic restriction" not to all the formulas of the system (which he mistakenly thought would comprise an uncountable totality), but just to those that were provable.

Zermelo thus continued to be confused about Gödel's results, believing that they, like the Löwenheim-Skolem theorem against which he had campaigned at the conference, somehow depended on artificial cardinality restrictions. Gödel made no attempt to disabuse him of the idea, even after Zermelo went on to express his criticisms in print [176]; he shrank instinctively from public controversy, and even on a personal basis he no doubt thought it pointless to pursue the matter any further. He did, however, show Zermelo's letters to Carnap, who agreed that Zermelo had "completely misunderstood" Gödel's attempt to clarify the issues [177].

It is hardly surprising that the incompleteness theorems should have met with opposition in some quarters, and Zermelo was not alone in misunderstanding them [178]. On the surface, at least, Wittgenstein's posthumously published *Remarks on the Foundations of Mathematics* betrays a particularly glaring lack of comprehension [179]; and Russell, in a letter he wrote in April of 1963, admitted that while he had "realized, of course, that Gödel's work ... [was] fundamental," he was "puzzled by it." His somewhat ambiguous remarks — especially his question "Are we to think that $2 + 2$ is not 4, but 4.001?" — suggest that he thought Gödel had found an *inconsistency* in arithmetic. At any rate, he acknowledged that he had never taken Hilbert's program very seriously because he never imagined that a system could be *proved* to be consistent (nor, presumably, that formal attempts to do so could be proved to be subject to limitations). In the wake of Gödel's results his bewilderment was such that he was "glad [he] was no longer working at mathematical logic" [180].

Like Zermelo, there were also others rash enough to challenge Gödel's results in print. Against them, too, Gödel disdained to respond. Yet, unjustified as he knew their criticisms to be, he must nevertheless have found such articles somewhat stressful. Indeed, Rudolf Gödel reported [181] that "shortly after the publication of his famous work" his brother exhibited signs of depression so serious that his family, fearing that he might become suicidal, caused him to be committed against his will to a sanatorium in Purkersdorf bei Wien for a period of "a few weeks."[13]

---

[13] Allegedly, Adele once helped Kurt escape from such confinement through a sanatorium window; but reports of the incident are indefinite as to date and are not wholly trustworthy.

On the basis of that report it has been claimed that Gödel first entered a sanatorium sometime in 1931 [182]. Such an interpretation is hard to reconcile, though, with his documented record of activities. From his correspondence with von Neumann and Bernays it is evident that after his presentation at Königsberg he worked actively until shortly before the Christmas holidays. In late January, as already noted, he spoke on his incompleteness results on two separate occasions. Between February and June he met several times with Carnap [183] and was involved with Nöbeling in editing the proceedings of Menger's colloquium. That spring he solved the problem Menger had posed to him, on which he spoke at the colloquium session of 24 June; on 2 July he attended a meeting of the Schlick Circle; and on 25 July he replied to Herbrand.

There is a brief gap in the record from then until 3 September, when Gödel wrote to Arend Heyting to accept an invitation to co-author a monograph with him.[14] Neither the Circle nor the colloquium met during August, so it is possible that Gödel could have been hospitalized then without his friends having been aware of it; but, as it happens, in a letter to Heyting of 5 August Otto Neugebauer mentioned that until 15 August letters to Gödel should be sent to the Hotel Knappenhof in Edlach, a village in the Raxalpen. So it seems that he simply went on vacation.

In October Gödel attended the meeting in Bad Elster; the next month he resumed his participation in Menger's colloquium and Hahn's seminar; and on 28 November he made his presentation to the Vienna Mathematical Society.

His almost continuous record of activity from then until his well-documented admission to Purkersdorf in 1934 is discussed in detail in chapter V. Suffice it to say that even granting that he might have been able to carry on some correspondence while convalescing, the chronology of his face-to-face encounters with other individuals leaves little room for him to have been secretly confined for several weeks during the years 1930–33.

The absence of any confirmatory evidence for Rudolf Gödel's claim is particularly telling in view of the limelight that was focused on his brother during just that period. Very likely Rudolf's recall of events that had taken place more than fifty years before was affected by psychological time compression. In particular, his imprecise dating of the episode ("shortly after the publication of his famous work"), together with his remark that he and his mother became aware of Kurt's fame only when Menger informed them of it much later, suggests that the "publication" he remembered was probably that of the

---

[14] An account of that ultimately abortive venture is given in chapter V.

mimeographed notes for Gödel's 1934 lectures at Princeton — an event that *did* shortly precede his hospitalization during the fall of that year.

Rudolf Gödel also stated that "due to his weak nerves" his brother had "twice" been in sanatoria — at "Purkersdorf and Rekawinkel" [184]. That number accords with other evidence of Gödel's admissions to those institutions *after* his return from America in 1934. So there is no reason to believe that he was confined there before then.

V
Dozent in absentia
(1932–1937)

ONE MIGHT EXPECT Gödel to have submitted the incompleteness paper as
his *Habilitationsschrift* almost immediately after the resumption of the academic
year in October 1931. In fact, however, he did not do so until 25 June 1932.
The reason for the delay is not entirely clear — perhaps there were residency
requirements to be satisfied or other conditions to be met — but during the
intervening eight months Gödel contributed prolifically both to Hahn's seminar on mathematical logic (which ran from 26 October through 4 July) and
to Menger's colloquium. Mimeographed notes of the former show that Gödel
spoke at length at eight of the twenty-two sessions, on such varied topics as
Heyting's formalization of intuitionistic logic (18 January); the interpretation
of Heyting's calculus within that of *Principia Mathematica* (25 January); Herbrand's thesis (13 June) and his consistency proof for a subsystem of arithmetic
(27 June); and his own completeness and incompleteness theorems (20 June
and 4 July, respectively). In the vita he submitted with his application for the
Habilitation [185], Gödel noted that he had also been responsible for choos-
ing the topics to be discussed in the seminar and for helping to prepare the
speakers.

Aside from brief remarks on the presentations of others, Gödel's published
contributions to the colloquium included answers to two questions of Hahn
(2 December and 25 February), on classical and intuitionistic propositional

logic, respectively; three minor results in geometry (two on 18 February and the third on 25 May); an important paper (*1933a*) on the consistency of classical arithmetic relative to its intuitionistic counterpart (28 June); and an undated note (*1933b*, published as part of the *Gesammelte Mitteilungen* for 1931/32) that, in hindsight, may be considered the first contribution to the modal logic of provability.

All of these results were overshadowed by the incompleteness theorems, but their lasting philosophical significance deserves further comment. In *1933a* Gödel introduced a translation of the classical connectives and quantifiers into corresponding notions of Heyting's intuitionistic logic and showed that under that interpretation "*the classical propositional calculus is ... a subsystem of the intuitionistic one.*" He then went on to show that "something similar holds *for all of arithmetic and number theory ... so that all propositions provable from the classical axioms hold for intuitionism as well*";[1] consequently, if classical arithmetic is inconsistent, so is intuitionistic arithmetic.[2] Gödel traced the cause of the subsumption of classical theories within intuitionistic ones to

> the fact that the intuitionistic prohibition against restating negated universal propositions as purely existential [ones] ceases to have any effect, because the predicate of absurdity can be applied to universal propositions, and th[at] leads to propositions that formally are exactly the same as those asserted in classical mathematics.

In his article on David Hilbert in the *Encyclopedia of Philosophy*, Paul Bernays noted that one repercussion of Gödel's paper *1933a* was that it showed, "contrary to the prevailing views at the time," that "intuitionistic reasoning is not identical with finitist reasoning" and hence that "proof theory could be fruitfully developed without fully keeping to [Hilbert's] original program." The paper also had an important effect on Gödel's own career, for it was delivered, at Menger's request, to a session of the colloquium at which Oswald Veblen was present as an invited guest. Veblen was just then deeply involved with plans for the organization of the Institute for Advanced Study in Princeton, and he was so impressed by Gödel's talk that a few months later he

---

[1] A variant translation that can be used to establish the same result had been given earlier by A. N. Kolmogorov (*1925*), without full details. Gödel was apparently unaware of Kolmogorov's work, which was published only in Russian and attracted little attention at the time. Independently of Gödel and almost simultaneously Gerhard Gentzen obtained very similar results, but on learning of Gödel's work he withdrew his paper at the galley stage [186].

[2] Whence — to a *classical* mathematician — intuitionism may appear "much ado about nothing." For intuitionists, however, the issue is not so much consistency as it is constructive content and methodology.

extended a tentative invitation to him to work at the institute during its first year of operation (1933–34) [187].

The note *1933b* received much less attention at the time. In contrast to what he had done in *1933a*, Gödel there showed how, if classical propositional logic were extended by a monadic predicate symbol B representing the *in*formal notion "provable by some correct means," intuitionistic propositional logic could be interpreted within the resulting system. Gödel characterized B by the three axioms

1. $Bp \rightarrow p$

2. $Bp \rightarrow .B(p \rightarrow q) \rightarrow Bq$

3. $Bp \rightarrow BBp$,

which, he noted, were nearly identical with axioms that C. I. Lewis had proposed to characterize a quite different notion (that of "strict implication"). He further noted that no B satisfying axiom 1 could be taken to represent provability within a given formal system, since 1 expresses a soundness principle that implies the provability of consistency (in conflict with the second incompleteness theorem). Today Lewis's system of propositional modal logic is known as S4, and Gödel's provability interpretation has become the basis for a considerable body of investigations (by Martin Löb, Saul Kripke, Robert M. Solovay, and others) of the syntax and semantics of modal statements involving the *formal* notion of provability (so-called "provability logic"). In particular, for those statements valid under all interpretations a complete axiomatization has been found [188].

In addition to his contributions to Hahn's seminar and Menger's colloquium, between 1931 and 1936 Gödel also served actively as a reviewer for *Zentralblatt für Mathematik und ihre Grenzgebiete* (founded in 1931) and, to a lesser extent, for *Monatshefte für Mathematik und Physik*; altogether he contributed thirty-three reviews, twelve of them during 1932. And as mentioned already in chapter IV, he had agreed some months earlier to collaborate with Arend Heyting in writing a survey of research on the foundations of mathematics.

The latter project had been proposed by the managing editor of *Zentralblatt für Mathematik und ihre Grenzgebiete*, Otto Neugebauer. In a letter of 25 June 1931 to Heyting, Neugebauer mentioned that the publisher, Springer, was planning to initiate a series of monographs surveying recent developments in specific areas of mathematics. Each report should run to "perhaps 5–7 sheets" and should include important references to the literature. Would Heyting be willing to write such a report on foundational research in mathematics?

Heyting replied that he would, but because he did not feel at home with logicism or with collateral work by such figures as Wittgenstein or Frank Ramsey he suggested that a co-author be found to treat those topics.

On 5 August Neugebauer advised Heyting that Gödel had agreed to serve as collaborator and proposed April 1932 as a target date for completion of the manuscript. Heyting then wrote Gödel to suggest a tentative sectional outline:

1. Short historical introduction; Poincaré's criticism

2. The paradoxes, and attempts to clarify them outside the three principal philosophical traditions

3. The calculus of logic and its further development; logicism

4. Intuitionism

5. Formalism

6. Other standpoints

7. Relations between the different directions

8. Mathematics and natural science

How all that was to fit within the compass of ten to fourteen pages was left unsaid. In any case, Heyting asked Gödel whether he would be willing to write the first three parts.

On 3 September, Gödel wrote to express his essential agreement with the plan, with three exceptions: He thought the paradoxes and Poincaré's criticism should be treated in the section on logicism, "in whose development they played a decisive role"; he suggested treating the "outside" approaches to the paradoxes in section 6; and he felt that in addition to the short historical survey something should be said at the outset about the goals and problems of foundational research. He further remarked that he was unsure where to place some of the recent research on "logical calculi for their own sake," mentioning as examples the work of Post and Bernays on the propositional calculus and of Tarski, Leśniewski, Łukasiewicz, Skolem, and Herbrand on the functional calculus. Since his own work fell in that area, he offered to treat such contributions in a separate section to be placed after that on logicism.

Correspondence between the two collaborators then apparently lapsed until June of 1932, when Heyting wrote to suggest a few further modifications of the plan. In the meantime, the deadline had been extended until 1 September.

By 20 July Heyting's draft of the section on formalism was nearly finished. Gödel, however, confessed that because he had been so busy during the two previous semesters his own progress had been very slow. Realizing that he could not meet the new deadline, he urged Heyting to request a further three-month extension.

Gödel finally reported progress on 15 November. By that time Heyting's work was almost completed, and Gödel claimed that he was half done with the section on logicism and had written a few paragraphs for section 4. Not until 16 May 1933, however, did he write again, and then only to report that because of the press of other work (especially that involved in earning his *Dozentur*) and having been sick for a time[3] he had had to interrupt his writing after having drafted about three-quarters of his assigned portion. And now, he announced, he needed to finish the work before departing for America in October.

That he was unable to do so is evident from Negebauer's letter to Heyting of 26 September, in which Neugebauer said that he had secured Gödel's promise to deliver his portion of the manuscript by the end of the year. But Neugebauer was beginning to lose patience, and so he broached to Heyting the possibility of publishing his part separately. Both continued to wait until, on 3 January 1934, Gödel wrote from Princeton to say that he would hand over the manuscript in *July*. At that, Neugebauer moved swiftly and with un-concealed exasperation to remove him from the project. Shortly thereafter Heyting's contribution was published as the monograph *Mathematische Grund-lagenforschung. Intuitionismus. Beweistheorie* [189].

Of Gödel's efforts on behalf of that ill-fated venture, all that remains is a small, very messy shorthand notebook [190], filled with overwritten and crossed-out passages. The draft is so fragmentary as to defy reconstruction, and as such — if it represents all that he had in fact done— makes manifest just how little progress he had actually made.

---

[3] After the discussion in chapter IV, one might wonder whether greater significance should be attached to this remark. In view of Gödel's hypochondriacal tendencies and his other activities during that period it seems likely that he suffered nothing more than a routine bout of illness. But on the other hand, among his library requests for 1932 were a text on psychology, a book on psychiatric disturbances caused by war injuries, and— perhaps most noteworthy— Emil Kraepelin's classic text on the effects of drug treatment on psychic functions (*Über die Beeinflussung einfacher psychischen Vorgänge durch einige Arzneimittel*). So it may be that he had begun to question his own mental stability.

In the long run the episode was but a minor embarrassment for him. It would hardly be worth noting were it not a foretoken of his behavior in several later instances when he accepted invitations to contribute essays to collective works. Worth noting, too, is that he published only one joint article after the fiasco with Heyting (a brief note on coordinate-free differential geometry, co-authored with Menger and Abraham Wald).

Gödel's correspondence with Menger shows that even while he struggled to produce the promised chapters for the monograph with Heyting, Gödel had agreed to review several draft chapters of a book that Menger was then writing on geometry. At almost the same time he was looking over the manuscript for the second part of Carnap's book *Metalogik* (an early draft of his *Logical Syntax*), in the course of which he discovered a serious error in the definition Carnap had given of what it meant for a formula to be "analytic";[4] and with the resumption of classes at the university in October of 1932 Gödel also served as a grader for one of Hahn's courses on elementary algebra and geometry, for which he presumably received a small stipend. (A few pages of unclaimed homework survive among Gödel's papers [192], replete with his admonitions "vereinfachen!" [simplify!] and "algebraisch ausdrücken!" [express algebraically!]. Students, it seems, have changed little over the years.)

It may be that Gödel felt an obligation to his mentors and so was reluctant to say no when asked to undertake such additional responsibilities. But he may also have begun to feel a bit of a financial pinch[5] — a circumstance that could also explain his decision to apply for the *Dozentur* at just that time. To be sure, the fees a *Privatdozent* could expect to receive from students were not enough to live on, but they would provide a modest supplemental income; and as Hermann Weyl remarked in describing the situation in Germany, "Sometimes, and much more frequently in the days of the Republic when fortunes had been wiped out by inflation, the *Privatdozentur* was combined with an assistantship or a salaried *Lehrauftrag* (teaching assignment) for some special field" [193].

In any case, the *Dozentur* was an essential prerequisite for entering into an academic career. Its significance in that regard deserves some further explanation, for the institution of the *Privatdozent* was a cornerstone of academic freedom within the German university system. That was so because whereas

---

[4] Carnap's original (recursive) definition was so unrestrictive as to permit an infinite regress. Gödel informed him of the problem in his letter of 11 September, to which Carnap replied with thanks on 25 September. He wrote Gödel again two days later, proposing a solution [191].

[5] In particular, there is evidence that during 1932–33 he endeavored unsuccessfully to cash in some long-term Czech mortgage bonds prior to their maturity dates.

"a state examination was the entrance to all [other] academic professions," the *Privatdozent* was granted the *Venia legendi* (the right to lecture) "on the ground of an examination by the faculty." Thus "the keys to the career of a university teacher were in the possession of the university itself, subject to no supervision by the state or administrative officers outside the university's teaching staff." It was for that very reason that *Privatdozenten* received no salary: They were not appointees of the state, as the professors were [194].

As Weyl's description suggests, there was more to earning the *Dozentur* than simply writing a second dissertation. A vita had to be submitted, as well as a list of topics on which the candidate was prepared to give a probationary lecture (*Probevortrag*); one or more professors were expected to sponsor the application; the *Habilitationsschrift* had to be defended before a faculty colloquium; and each step in the approval process required the vote of the entire philosophical faculty.

The commission set up to consider Gödel's application met on 25 November 1932. Chaired by the dean, Dr. Heinrich Srbik, it had as its other members Professors Furtwängler, Hahn, Himmelbauer, Menger, Prey, Schlick, Tauber, Thirring, and Wirtinger, though Furtwängler and Tauber were absent from the meeting because of illness [195]. Hahn served as Gödel's sponsor and as the commission's secretary. In his offical report of the meeting [196] he declared that Gödel had already obtained results of great scientific worth in his dissertation, and he deemed the *Habilitationsschrift* to be "an achievement of the first rank" that had "attracted the greatest attention in scientific circles" and made its mark on the history of mathematics. He also mentioned the note "Zum intuitionistischen Aussagenkalkül" (*1932a*), in which Gödel had established that Heyting's system of intuitionistic propositional logic could not be realized as a many-valued logic with only finitely many truth values. On the basis of all those works Hahn judged Gödel's accomplishments to have exceeded by far what was usually expected of one seeking the *Habilitation.* The commission accepted his recommendation and certified unanimously that the candidate had met the established criteria of worthiness, both on personal and scientific grounds.

At the faculty meeting of 3 December, after presentation of the commission's report, it was moved that Dr. Gödel be approved to receive the *Habilitation* (that is, the degree "Dr. habil.") and to undertake the further steps toward the *Dozentur.* Separate votes were taken on his personal and scientific suitability for such advancement. The results, as recorded by the dean in his report of 17 February 1933 to the ministry of instruction, were, respectively, fifty-one "yes," one "no," no abstentions, and forty-nine "yes," one "no," no abstentions (two faculty apparently having left in the meantime).

The "no" vote on Gödel's technical merit is surprising, to say the least. In a private communication [197], Dr. Werner Schimanovich has reported that the naysayer was Professor Wirtinger, who thought that the incompleteness paper overlapped too much with the dissertation! The claim seems doubtful since Wirtinger was a member of the commission that had recommended Gödel unanimously. But on the other hand, Wirtinger's field was far removed from logic (he had been an authority on abelian functions), and in his later years he apparently became somewhat estranged from his colleagues. According to Olga Taussky–Todd [198], "when Furtwängler ... proved the group-theoretic part ... of the *Hauptidealsatz* [a central result of algebraic number theory] and received a big prize for [it], Wirtinger moved into the background. He moved towards retirement, he cancelled his lectures frequently, his hearing was almost nil and ... he became a disappointed and even angry person." It is conceivable, then, that he might have changed his vote.

Gödel defended his results at a colloquium on 13 January 1933. The commission again unanimously approved his performance, and at the faculty meeting held eight days later a simple majority (all that was officially required) approved his taking the final step toward the *Dozentur*: His *Probevortrag* was scheduled for 3 February.

Gödel had listed eight topics on which he was prepared to speak: symbolic logic; logical foundations of arithmetic and analysis; foundations of geometry; the axiomatization of set theory; the problem of the consistency and completeness of formal theories; the three principal directions of research on mathematical foundations (logicism, formalism, and intuitionism); the calculus of classes of Boole and Schröder; and recent results in measure theory. The commission, however, chose none of those, but elected instead to hear Gödel lecture on the theme "Über den intuitionistischen Aussagenkalkül" (On the intuitionistic propositional calculus). The title of the lecture is very similar to that of Gödel's paper *1932a*, which Hahn had mentioned in his commission report. However, the content could also have been that of his note *1933b* (discussed earlier), which would not yet have appeared in print.[6]

The *Probevortrag* was officially approved, again by a simple majority, at the faculty meeting of 11 February, after which the final vote, for awarding the

---

[6] In a letter to von Neumann of 14 March 1933 (GN 013032) Gödel mentioned his provability interpretation but said that he had so far been unable to demonstrate its "complete equivalence with Heyting's system."

Venia legendi, was taken. This time there were no negative votes: Forty-two of the faculty voted to grant the right to lecture, with a single abstention. The appointment as *Privatdozent* was confirmed by the dean exactly one month later.

◇   ◇   ◇

On that same day Gödel's priority in obtaining the incompleteness results was challenged by Paul Finsler, a senior lecturer in mathematics at Zürich. Writing privately to Gödel, Finsler declared that he wanted to become better acquainted with his works, as he had looked at them only "fleetingly"; but, he said, so far as he could see, what Gödel had done was "in principle somewhat similar" to ideas he had presented in an earlier paper of his own (*1926*), on which he had lectured in Düsseldorf. "However," he went on,

> you have taken as starting point a narrower and therefore sharper formalism, whereas I, in order to shorten the proof and be able to stress just the essentials, adopted a general formalism. It is of course of value actually to carry out the ideas in a special formalism, yet I had shied away from the effort of doing so since the result seemed to me already established, and therefore I couldn't muster sufficient interest in the formalisms themselves [199].

In fact, as Gödel was quick to point out, Finsler had little clear idea at all of what constituted a formal system. In his reply of 25 March, Gödel stated bluntly that

> The system ... with which you operate is not really [well] defined, because in its definition you employ the notion of "logically unobjectionable proof," which, without being made more precise, allows arbitrariness of the widest scope. ... [In particular,] the antidiagonal sequence you define on p. 681, and therefore also the undecidable sentence, is *never* representable in the same formal system one starts out with [200].

Much later, in reply to a graduate student's query, Gödel admitted that at the time he wrote his paper he was unaware of that of Finsler, while "other mathematicians or logicians probably disregarded it because it contain[ed such] obvious nonsense" [201].

The justness of Gödel's uncharacteristically harsh judgment is readily confirmed by a study of Finsler's other papers [202]; but, like Zermelo, Finsler himself never understood Gödel's point. After a delay of three months, during which he studied the offprints Gödel had sent him, he retorted angrily that he could just as well object to Gödel's proof on the grounds that he had employed axioms (Peano's) that could not be proved consistent (!); and while noting that the truth of Gödel's undecidable sentence could only be established metamathematically, he still failed to see the importance of making such distinctions. "To make statements about a system," he contended, it was "not at all necessary" that the system be "sharply defined"; it sufficed that one "take it as given" and recognize "a few of its properties, from which the desired consequences may be drawn" [203].

Gödel made no further reply. Finsler was not a figure of Zermelo's stature, and his protestations apparently caused few ripples within the mathematical community. In the midst of the far more ominous events that were then taking place in Germany and Austria, the dispute was but a minor distraction.

◇ ◇ ◇

Politically, Austria was in a state of increasing tension and chaos. Engelbert Dollfuss had become head of the government the previous May, but he had been unable to forge a workable coalition to address the country's growing economic crisis or to lessen strife among the various political parties and the several paramilitary organizations controlled by them, especially the fascist Heimwehr and Front Fighters and the socialist Schutzbund. Though anti-Nazi and determined to resist *Anschluß* by Germany, Dollfuss nevertheless embraced fascist ideology. In the fourteen months between his accession to power and his assassination by Nazis in July of 1934 he established a clerical-fascist police state and effectively "pawned his country to Mussolini" [204].

Austria's political disintegration accelerated after Hitler was named chancellor of Germany. Factional struggles within the Austrian parliament culminated in the impetuous resignation of its president and two vice-presidents on 4 March 1933 [205]. The following day the Nazis swept to victory in the German *Reichstag* elections, and on 7 March Dollfuss announced that he would henceforth govern Austria without a parliament. At the same time he instituted press censorship and banned public marches and assemblies.

Of Gödel's reaction to those developments Menger recalled [206]:

> Gödel kept himself well informed; he was very interested in events and spoke with me a good deal about politics. Yet his political views were always noncommittal and usually ended with the words "Don't you suppose?" I never could discern any strong emotional engagement on his part . . . .

Whether such detachment was genuine or another example of Gödel's innate caution or naiveté is hard to say. In any case, it alienated those at both ends of the political spectrum: On the one hand, Gödel's "apolitical" stance was to prove vexing to the Nazi bureaucrats who, a few years later, vetted his request for a leave of absence abroad. On the other, his legalistic attitude toward violations of his own academic rights, in the face of the far more serious injustices suffered by those around him, eventually cooled Menger's friendship with him [207].

Gödel was not anti-Semitic — indeed, Olga Taussky–Todd recalled that he "had a friendly attitude toward people of the Jewish faith" (*1987*, p. 33). Nevertheless, at times he displayed monstrous purblindness to the plight of European Jews. Gustav Bergmann, for example, recalled that shortly after arriving in America in October 1938 he was invited to lunch with Gödel, who inquired, "And what brings you to America, Herr Bergmann?" [208].

Undoubtedly Gödel was preoccupied not only with his research and his quest to become a *Dozent* but with preparations for his upcoming trip to America — which would of course allow him to escape for a time the turmoil in Vienna. As noted earlier, it was Oswald Veblen who first broached the idea of Gödel's spending the academic year 1933–34 at the Institute for Advanced Study (IAS). Because at the time (November 1932) he had "no authority to make any definite offer," he did not write directly to Gödel, but rather to Menger, saying that Gödel "might" be offered a one-year position with a stipend of "something like $2500" and that it "might" also be possible to provide "a small additional amount to cover his traveling expenses" [209]. Menger passed the invitation on to Gödel and acknowleged receipt of the letter in a postcard to Veblen, but Gödel waited to reply until he was sure that a sojourn in America would not delay his *Habilitation.*

Unfortunately, Veblen never received Menger's card. On 7 January, fearing that his letter had been lost, he sent Menger an urgent telegram with prepaid reply. By then it was clear to Gödel that no further obstacles lay in his path, so he promptly cabled his interest in accepting such an offer, explaining in a follow-up letter the reason for his delayed response. An official invitation, incorporating just the terms Veblen had proposed, was tendered almost immediately thereafter by Abraham Flexner, the institute's director and guiding force. Simultaneously Veblen also wrote to Gödel to describe the nearly nonexistent duties of the position. It was intended, he said, "primarily as an opportunity for you to continue your scientific work in cooperation with ... colleagues here." He mentioned that he and von Neumann intended to conduct a joint seminar, probably on the subject of quantum theory, and that Gödel would be free to offer a seminar of his own, should he desire to do so. But, he went on, "Perhaps the most useful contribution ... you could make to the local mathematical situation would be [to] cooperat[e] with Church,"[7] who "I expect ... will be giving a course in mathematical logic at the University. ... I think it would be very interesting to bring your criticism to bear directly upon Church's formalism" [210].

Before Gödel found time to reply to Veblen he heard from von Neumann, who had just returned to Europe from Princeton. Von Neumann expressed his pleasure that Gödel was to be one of his colleagues during the coming year, and he offered to provide further information about the institute, at Gödel's request. Perhaps, he suggested, they might try to get together sometime before his return to America.

Gödel answered exactly one month later (14 March), saying that he would of course welcome further information and that he expected to be in Vienna at least until the end of June, since during the summer semester he intended to exercise his newly won *Venia legendi* by offering his first course. With regard to his stay at the IAS he added that he would gain a great deal from the opportunity to hear von Neumann lecture or direct a seminar on quantum mechanics, a subject in which he had a "lively interest" but to which he had not theretofore found time to devote serious attention [211].

To Veblen, too, Gödel expressed his interest in the seminar, and also his thanks that he was to be granted so much freedom in his work. Writing in English, he said he felt it "would be laborious" to deliver a course of lectures during the first months of his visit; it would be better "first [to] improve my knowledge of English." As to "Dr. Church," Gödel said he was "of course ... greatly interested in [his] formalism."[8] His one serious concern, voiced also in his letter to von Neumann, was financial: In the event that the dollar were devalued, would his stipend be increased accordingly?

To that Veblen had to say no. He was perhaps a bit taken aback by the question, but it was nevertheless an astute one, since the United States had declared a three-day bank holiday just the week before. The collapse of the Austrian Credit-Anstalt two years earlier was no doubt still fresh in Gödel's

---

[7] Alonzo Church (see appendix C) had received his doctorate under Veblen and was on his way to becoming one of America's most eminent logicians. That he and Gödel would see much of each other during Gödel's visit was virtually assured, since until 1939 the IAS had no building of its own. For the first two years of its existence it was housed with Princeton's mathematics department in the university's (old) Fine Hall (now Jones Hall).

[8] About which he had already corresponded with Church the previous summer.

mind, and he was surely aware as well that the United States was about to go off the gold standard (as it did on 14 April). As it turned out, his concern proved to be justified, for the dollar *was* in fact revalued the following year.

Gödel's summer course, "Foundations of Arithmetic," met two hours per week, beginning on 4 May. From the fragmentary traces that survive in one of his notebooks it is difficult to tell much about its content, but it appears to have been a popular offering, with at least fifteen students enrolled seven of them women. The conditions under which it was taught, however, were highly disruptive: The university was temporarily closed on 27 May on account of Nazi activities, and during the week of 12–19 June the Nazis carried out a series of bombings in and around Vienna [212].

Events in Germany were even more alarming: On 7 April the Law for the Restoration of the Professional Civil Service sanctioned the dismissal from its ranks of all officials of "non-Aryan descent," as well as all those "who because of their previous political activity do not offer security that they will act at all times and without reservation in the interests of the national state" [213]. Jewish university professors were among the first to be purged. Overall, during the first year after enactment of the law, "close to 15 percent of the faculty was removed .... The universities of Berlin and Frankfurt ... lost 32 percent of their faculty [while] the ousters ran to between 18 and 25 percent at Göttingen, Freiburg, Breslau, and Heidelberg" [214]. At Freiburg also, on the very day that the University of Vienna was forced to close, Martin Heidegger was installed as rector. In his inaugural address, delivered seventeen days after the Nazis had held massive public book burnings, Heidegger "celebrate[d] the new turn in Germany's destiny" [215].

What notice Gödel may have taken of such happenings is unrecorded, though it is of interest that Menger recalls "post-Kantian German metaphysics" as one of the topics to which Gödel devoted much study "in those years" [216]. (There is no mention of his opinion of Heidegger.) In due course imposition of the "New Order" in Austria would affect Gödel's own *Dozentur*, and even before that (in 1935) he would be forced to join Dollfuss's "Fatherland Front," founded during the waning days of that same eventful May [217].

At some point before his departure for America, presumably shortly after the end of his summer course, Gödel vacationed for a time with his mother in Bled, a Yugoslavian spa just across the Austrian border [218]. No doubt it offered a welcome respite from his labors, which, in addition to all that has already been mentioned, had included the publication of yet another major paper (*1933c*).

Like the brief note *1932b*, the later article *1933c* was concerned with a special case of the decision problem for satisfiability; both papers dealt with *prefix classes* of prenex formulas of the restricted functional calculus. In the earlier paper, Gödel showed that there is an effective way to determine the satisfiability of any prenex formula whose quantifier prefix has the form $\exists \ldots \exists \forall\forall \exists \ldots \exists$, whereas in the later paper he showed that the satisfiability of any formula of the restricted functional calculus can be reduced to that of a prenex formula whose quantifier prefix has the form $\forall\forall\exists \ldots \exists$ (and, moreover, that every such prenex formula that is satisfiable has a finite model). The former theorem strengthened an earler result of Ackermann, the latter one of Skolem. Taken together, they established "a reasonably sharp boundary between decidable and undecidable [prefix classes] ... : two contiguous universal quantifiers [do not suffice] to yield undecidability, whereas three [do]" [219].

Notably, the later paper was published in *Monatshefte für Mathematik und Physik*, the journal edited at the University of Vienna in which both the completeness and incompleteness papers had appeared. Its results were never presented to Menger's colloquium, nor, apparently, given any other oral presentation, perhaps because the details of the proof are rather intricate and somewhat subtle — so subtle, in fact, that Gödel himself was led to make an erroneous claim that went unchallenged for three decades and was not definitely refuted until after his death.

The mistake was in the final sentence of the paper, where Gödel claimed that the result on finite satisfiability, though established therein only for formulas of the *pure* functional calculus (without an equality symbol), could "be proved, by the same method, for formulas that contain the equality sign." In the mid-1960s it was realized that, at the very least, the extension "by the same method[s]" was problematic. Subsequently Gödel suggested how the proof might be carried out, but that argument, too, was soon seen to be flawed. Finally, in 1983, Warren Goldfarb proved that there is in fact *no* effective decision procedure for the "Gödel class" *with* identity [220].

◊ ◊ ◊

Although the Institute for Advanced Study did not begin formal operations until 1933, its certificate of incorporation had been signed on 20 May 1930, as an outgrowth of discussions between the renowned educational reformer Abraham Flexner and the department-store magnates Louis Bamberger and his sister, Mrs. Felix Fuld. In 1929, just weeks before the stock market crash, Bamberger and Fuld had sold their business to R.H. Macy and Co. in order to devote the funds realized from the sale to the establishment of a philanthropic educational foundation. With that aim in mind, in 1925 they had sought advice from Flexner, who had become famous in 1908 for an influential (and damning) critique of American medical education and whose penetrating comparative study *Universities: American, English, German* was just then in proof.

Flexner's brother Simon was director of the Rockefeller Institute for Medical Research, an institution that likely served as the model for Flexner's vision of an institute for advanced study [221]: a haven to which younger scholars from throughout the world could come to study with some of the world's greatest intellects, unfettered by academic responsibilities or contractual research grants. The realization of Flexner's dream — "whose timing could not have been more propitious" — was made possible by Bamberger and Fuld's wealth. But the character of the institute they created was shaped in large part by "the accidental confluence of . . . the American philanthropic tradition [they] exemplified . . . and the rise of Nazism and Fascism which sent to America's shores a flood of refugee scholars" [222].

Plans for the institute took shape as the result of a long series of negotiations among the founders. Flexner began by dissuading Bamberger and Fuld from establishing a new medical school. He proposed instead that they endow an academic institution whose "staff and students or scholars should be few" and whose "administration should be inconspicuous, inexpensive, [and] subordinate" to the faculty [223]. Bamberger and Fuld initially proposed that the institute be located in the Washington, D.C., area, but Flexner ultimately persuaded them that Princeton, with its established academic traditions and peaceful surroundings, would be a more suitable site.

In selecting the faculty, the primary question concerned judgments of academic worth. It was agreed that not all disciplines could be represented, so the initial appointments should be made in a field in which merit could be judged unequivocally. Consequently, during its first two years of operation the institute consisted solely of a School of Mathematics (broadly interpreted to include mathematical physics as well). Flexner was named director in 1930, and in October 1932 Albert Einstein and Oswald Veblen were appointed as the first professors.[9] They were joined the following year by John

---

[9]The preceding March George D. Birkhoff of Harvard had accepted an offer to become the first IAS professor, but, apparently having second thoughts, "asked to be released [from the commitment] eight days later" [224].

FIGURE 10. The original mathematics faculty at the Institute for Advanced Study, 1933. Left to Right: James Alexander, Marston Morse, Albert Einstein, IAS director Aydelotte, Hermann Weyl, Oswald Veblen. Missing from the group is John von Neumann.

von Neumann, James Alexander, and Hermann Weyl[10] (Figure 10).

The institute began life in rented quarters, as a paying guest within the building that housed the Princeton mathematics department. Administratively, though, the two institutions were entirely distinct, and negotiations between them were at first a matter of delicate diplomacy, since it was feared that the institute might lure away too many of the Princeton mathematics faculty. It had, after all, already hired Veblen, von Neumann, and Alexander, who, though they remained with their former colleagues, thereafter earned greater

---

[10] Incongruously, one social scientist was also appointed as a professor within the school: David Mitrany, a specialist in international relations. In addition, Walther Mayer, who had been a *Privatdozent* at the University of Vienna from 1923 to 1931, was given a long-term appointment, initially as Einstein's assistant; but he was not named a professor.

salaries and were not required to teach. The arrangement inevitably created envy and resentment among some of those not appointed, and relations between Veblen and Solomon Lefschetz, the chair of the Princeton mathematics department, became noticeably strained [225]. For the most part, though, conflicts remained submerged, and such strains as there were probably had little effect on the twenty-four individuals, including Gödel, who were invited to serve as visiting scholars that first year [226]. They bore no official title; in the institute's first annual report they are referred to simply as "workers." (The designation "temporary member" was adopted later, at the faculty meeting of 8 October 1935 [227].)

The IAS term began on 1 October, and Gödel intended to be present for the opening. In a radiogram he sent to Veblen on 18 September [228] he stated that he would arrive in New York on the 29th aboard the Cunard liner *Berengaria*, which was scheduled to depart from Southampton on the 23rd. Among those who came to see him off at Vienna's Westbahnhof was Olga Taussky, who recalled that "he boarded a Wagon Lits compartment on the Orient Express. A fine-looking gentleman, presumably his Doctor brother, stood apart from us and moved away as soon as the train started, while we others waved a little longer."

In fact, it proved to be a false start; for as Taussky later learned, Gödel was "taken ill before reaching his boat, took his temperature and decided to return home" [229]. After a few days "his family persuaded him to try again," and so on the 25th he radioed Veblen that his departure had been "postpone[d] ... on account of illness" and that he would be travelling instead on the *Aquitania*, which would depart Southampton on 30 September and arrive in New York on 6 October [230].

The second attempt was successful, and Gödel was met at the dock in New York by Edgar Bamberger, who transported him to Princeton. There is no record of his reaction to the crossing, which was presumably uneventful; but in view of his chronic digestive troubles (real, imagined, and self-inflicted) in later years, it would not be surprising if he suffered somewhat from seasickness. During Gödel's stay Veblen's wife reportedly became concerned about his diet[11] — even preparing some meals for him — because she found that, unaccustomed to the early closing times of Princeton's restaurants, he would

---

[11] Rudolf Gödel has attested that during the years he lived with his brother they often went to restaurants together and that Kurt usually had a good appetite. He recalled hearing nothing about a diet during Gödel's guest visits to Princeton and believed that his brother's overly strict dieting began only about 1940 [231].

often become so involved with his work that he missed dinner [232]. Apropos of Gödel's general state of health at that time it is also worth noting a remark that Flexner made two years later, after Gödel abruptly resigned in the midst of his second visit: "I was under the impression that your health had greatly improved [in the interim], for you certainly looked much better than when you came to Princeton two years ago" [233].

Shortly after his arrival Gödel found lodgings in a house at 32 Vandeventer Avenue, where he remained throughout his stay. Unfortunately, though, from October to February there are few records of his activities; his notebooks cast little light on that period, and almost nothing in the way of library request slips has been preserved from the years 1933–34. Events during the second semester have been well chronicled in several memoirs by S. C. Kleene, who was then a graduate student completing his dissertation under Church [234]. But Kleene was absent from Princeton in the fall of 1933 and so did not witness what took place then. Presumably Gödel did just what he had told Veblen he intended to do: participate in the seminar on quantum mechanics, improve his English, and consult with Church.

It would be interesting indeed to know more about the extent of contact between those two logicians. What is known is that Gödel had written to Church in June of 1932 to pose two questions about the system of logical postulates that Church had proposed in his papers *1932* and *1933*: (1) How was it possible to prove absolute existence statements, such as the Axiom of Infinity, in that system? (2) Assuming the system to be consistent, why couldn't its primitive notions be interpreted in type theory or set theory? Indeed, could its consistency be made plausible other than by such an interpretation [235]?

Church responded a few weeks later, outlining a proof for the Axiom of Infinity and remarking that as to the consistency of his system,

> I cannot see that a proof of freedom from contradiction which began by assuming the freedom from contradiction of ... *Principia Mathematica* would be of much value, because the freedom from contradiction of *Principia Mathematica* is itself doubtful, even improbable. In fact, the only evidence for ... [its consistency] is the empirical evidence arising from the fact that ... [it] has been in use for some time, [that] many of its consequences have been drawn, and [that] no one has [yet] found a contradiction. If my system be really free from contradiction, then an equal amount of work in deriving its consequences should provide an equal weight of empirical evidence for its [consistency] [236].

At the time Church thought that his system — which he regarded as a "radically different formulation of logic" — might escape the strictures of Gödel's incompleteness results, for he "thought that Gödel['s] ... theorem[s] might be found to depend on peculiarities of type theory," a theory whose "unfortunate restrictiveness" he sought to overcome [237].

By the fall of 1933 the consistency of Church's full formal system had begun to be suspect, and in the spring of 1934 Kleene and J. B. Rosser, another of Church's graduate students, showed that it was, in fact, inconsistent — a discovery that forced Kleene to have to rewrite his dissertation. Nevertheless, the subsystem known as the λ-calculus came through unscathed, and with it all the results that Kleene had obtained on the λ-definability of a great variety of effectively calculable functions [238].

Reportedly, those results were "circulating among the logicians at Princeton" when Gödel arrived there [239], and it was on the basis of them that Church was emboldened to propose his "Thesis" that *all* effectively calculable functions are λ-definable (and hence that the *informal* notion of effective calculability is equivalent to the *formal* one of λ-definability).[12] Gödel, however, was skeptical, as Church himself noted in a contemporary account of the evolution of the theory of λ-definability [240]: "My proposal ... he regarded as thoroughly unsatisfactory. I replied that if he would propose any definition of effective calculability which seemed even partially satisfactory I would undertake to prove that it was included in lambda-definability."

In fact, that was just what happened. For though Gödel's "only idea at the time was that it might be possible ... to state a set of axioms which would embody the generally accepted properties" of effective calculability, "it occurred to him later that Herbrand's definition of recursiveness ... could be modified in the direction of effective calculability" [241]. He presented the idea at the end of the course of lectures he gave at the IAS the following spring.

◊   ◊   ◊

Following the conclusion of the fall term, Gödel traveled to Cambridge, Massachusetts, to deliver an invited address at the annual meeting of the Mathematical Association of America (held jointly with that of the American Mathematical Society). Titled "The Present Situation in the Foundations

---

[12]Church did not publicly announce his Thesis until 19 April 1935, in a contributed talk to the American Mathematical Society, and (contrary to what might be inferred from the remarks quoted in the next sentence) he apparently did not mention it to Gödel until the early months of 1934. A detailed consideration of the underlying chronology of events is given in *Davis 1982.*

of Mathematics," the lecture was given on 30 December as part of a session that also included a talk by Church on Richard's paradox. Its text, published as item *1933o in volume III of Gödel's *Collected Works*, is of particular interest for the evidence it provides of changes in his philosophical outlook.

After discussing the system of axioms and rules set up for the simple theory of types and for axiomatizations of set theory, Gödel noted some of the difficulties that had to be confronted in seeking to justify such principles — in particular, the problems of nonconstructive existence proofs and of impredicative definitions. Then, most remarkably in view of his subsequent advocacy of mathematical Platonism (and contrary to his statements on several occasions [242] that he had held such views at least since 1925), he stated that the axioms, "if interpreted as meaningful statements, necessarily presuppose a kind of Platonism, which cannot satisfy any critical mind and which does not even produce the conviction that they are consistent."[13] Following an outline of his incompleteness results he further declared — in contrast to his cautious remark at the end of his paper *1931a* and the view he had set forth in his letter to Herbrand — that

> all the intuitionistic [i.e., finitary] proofs ... which have ever been constructed can easily be expressed in the system of classical analysis and even in the system of classical arithmetic, and there are reasons for believing that this will hold for any proof which one will ever be able to construct .... So it seems that not even classical arithmetic can be proved to be non-contradictory by ... [such] methods ... [243].

◇ ◇ ◇

When Gödel's course of lectures began Kleene had just returned to Princeton after a seven-month absence, and he and Rosser were recruited by Veblen to serve as note-takers for the course. The notes themselves were distributed in mimeographed installments to those who subscribed for them, as well as to a few libraries [244]. Subsequently they circulated widely, but they were not published until 1965, in Martin Davis's anthology *The Undecidable*.

Although the subject matter of the 1934 lectures was essentially that of Gödel's paper *1931a*, a comparison of the titles of the two works suggests the broader scope of the later treatment: "On Formally Undecidable Propositions of *Principia Mathematica* and Related Systems, I" (*1931a*) versus "On

---

[13] By making the following clause unrestrictive, the comma before "which" is critical to the interpretation of this passage. It is a matter for speculation whether it reflects Gödel's intention or was a subtle point of English grammar to which he was insensitive at the time.

Undecidable Propositions of Formal Mathematical Systems" (*1934*). Since even so distinguished a logician as Church was among those who thought that the incompleteness theorems were somehow dependent on the particularities of formalization, it was surely one of Gödel's principal aims in the 1934 lectures to dispel doubts about the generality of the incompleteness phenomena. Accordingly, although he did discuss one particular formal system as an example of those to which his considerations would apply, in the very first paragraph of the 1934 paper he defined a formal mathematical system quite generally, as "a system of symbols together with rules for employing them," where the rules of deduction and the definitions of what were to constitute meaningful formulas and axioms were to be specified constructively (that is, in terms of finitary decision procedures).

As he had in the earlier paper, Gödel proceeded to define the class of functions now called primitive recursive but that he simply called recursive — functions that, he stressed, "have the important property that, for each given set of values of the arguments, the value of the function can be computed by a finite procedure." He was aware, however, that not all functions with that property were primitive recursive, since during the period 1927–28 Wilhelm Ackermann (and, independently and almost simultaneously, Gabriel Sudan) had given examples of effectively computable functions, defined by means of multiple or nested recursions, that grew too fast to be primitive recursive [245]. Gödel mentioned Ackermann's example in the final section of his 1934 paper, as a way of motivating the concept of "general recursive function" that he defined there; but earlier, in footnote 3, he had already conjectured (as "a heuristic principle") that all finitarily computable functions could be obtained through recursions of such more general sorts.

The conjecture has since elicited much comment. In particular, when Martin Davis undertook to publish Gödel's 1934 lectures he took it to be a variant of Church's Thesis; but in a letter to Davis [246] Gödel stated emphatically that that was "not true" because at the time of those lectures he was "not at all convinced" that his concept of recursion comprised "all possible recursions." Rather, he said, "The conjecture stated there only refers to the equivalence of 'finite (computation) procedure' and 'recursive procedure.'" To clarify the issue Gödel added a postscript to the lectures [247], in which he indicated that what had finally convinced him that the intuitively computable functions coincided with those that were general recursive was Alan Turing's work (*1937*).[14]

---

[14]Turing's conception of a universal finite-state machine and his reformulation of the first incompleteness theorem as the so-called halting problem are by now well known.

Gödel's reluctance to regard either general recursiveness or λ-definability as adequate characterizations of the informal notion of effective computability has been examined in detail by several authors [248]. There is a consensus that, in fact, neither Gödel's nor Church's formalisms were so perspicuous or intrinsically persuasive as Turing's analysis, and Wilfried Sieg has argued that the evidence in favor of Church's Thesis provided by the "confluence of different notions" (the fact that the systems proposed by Church, Gödel, Post, and Turing all turned out to have the same extension) is less compelling than has generally been supposed. Hence, quite apart from Gödel's innate caution, there were good reasons for his skepticism. But what, then, *was* he attempting to achieve through his notion of general recursiveness?

Gödel's conspicuous absence from the further evolution of recursion theory, in sharp contrast to the roles played by Church, Kleene, Rosser, and Post, among others, may provide one "Holmesian" clue to his intentions. In part, such lack of involvement was characteristic of much of Gödel's work: He was essentially a pathbreaker, one who tackled big problems, made incisive breakthroughs, and left the detailed development of his ideas to others. Beyond that, however, his hindsightful remarks at the Princeton Bicentennial Conference (*Gödel 1946*) suggest that he simply did not expect it to be *possible* to give "an absolute definition of an interesting epistemological notion." That general recursiveness turned out to do just that— in particular, that "the diagonal procedure [did] not lead outside" the class of (partial) recursive functions — seemed to him "a kind of miracle." Thus, in defining the class of general recursive functions it was not his aim to capture all possible recursions, for to attempt to do so would, in his view, have been to pursue a phantom.

Rather, Gödel obtained his definition through modification of Herbrand's ideas [249]; and Wilfried Sieg has argued that his real purpose in the final section of the 1934 paper was "*to disassociate recursive functions from* [Herbrand's] *epistemologically restricted notion of proof*" by specifying "*mechanical* rules for deriving equations."   What was more *general* about Gödel's notion of "general" recursiveness was, Sieg suggests, that Herbrand had intended only to characterize those functions that could be *proved* to be recursive by *finitary* means [250].

Gödel's lectures extended from February until May. During April, however, he made excursions to New York and Washington to present two popular lectures on the incompleteness theorems. The first, titled "The Existence of Undecidable Propositions in Any Formal System Containing Arithmetic," was delivered on the evening of Wednesday, 18 April, to the Philosophical Society of New York University. Two days later he addressed the Washington Academy of Sciences on the topic "Can Mathematics Be Proved Consistent?."[15]

On 24 April Gödel returned to New York, where he remained for several days. While there he may have conferred with U.S. and Austrian officials concerning his visa status; for on 7 March Flexner had informed him that the sum of $2,000 would be put at his disposal by the institute "as a grant-in-aid for a half year or more during the academic year 1934–1935" [251].

◊   ◊   ◊

Gödel departed for Europe on 26 May, aboard the Italian liner *S.S. Rex*. The ship reached Gibraltar five days later, docked at Naples on 2 June, and went on the next day to Villefranche and then to Genoa, where Gödel debarked. After spending the night in Milan he took time for a three-day sojourn in Venice before traveling home to Vienna on 7 June. He left no account of the journey, but its leisurely character— especially the excursion to Venice— suggests that, at least to some extent, it served as a short vacation.

No doubt he was in need of a holiday just then; but whatever relief his travels provided was short-lived, for the university, city, and country to which he returned were in profound crisis. On 12 February the first of two episodes of civil war had taken place in Austria. By June matters were rapidly approaching a violent climax, and in July a series of disturbing events, partly unrelated, occurred in close succession: On the 8th the minister of education, Kurt Schuschnigg, decreed that elected rectors and deans of the universities would be recognized by the government "only if they joined the Fatherland Front" [252]. At about the same time Gödel's mentor, Hans Hahn, was forced to cut short a planned vacation in the countryside when he was unexpectedly discovered to require urgent cancer surgery. He underwent the operation, but died on the 24th at the age of fifty-five— one day before a group of Nazis stormed the chancellery in Vienna and assassinated Dollfuss.

The attempted putsch failed, after which "the situation soon became stabilized — to the outward eye— under the immense pressure of the new ... regime" (with Schuschnigg as its head). But as a result of the "propaganda and pressure, fear and suspicion, irresolution and incipient emigration," Vienna's intellectual life began rapidly to wither, and within the University "the radical nationalistic majority ruled in the faculty as well as in the student body" [253].

---

[15] A manuscript for the former, but not the latter, survives. Still unpublished, and likely to remain so, its text is of interest only insofar as it exemplifies Gödel's ability to explain his ideas to a lay audience.

The situation was exacerbated by a law of 7 August, one of whose provisions "empowered the minister [of education] to pension temporarily professors and associate professors in *all* publicly supported colleges and universities" whenever it was deemed "necessary in connection with reductions in expenditures ... or with reorganization measures" [254]. Unlike the German statute discussed earlier, the law did not single out Jews for such action; but in fact, "some of those pensioned were Jews who had never been politically active" — among them Professor Gomperz, from whom Gödel had taken his courses in the history of philosophy [255].

In the midst of these ominous developments Gödel's physical and mental health began to deteriorate. According to his own later account to Veblen [256] the trouble started with "an inflammation of the jawbone from a bad tooth," an incident that happened, he later recalled, while he and his mother were vacationing together in a small village. He blamed the whole affair on the dentist who had filled the tooth — "or rather, infected it" — the same individual who had performed a clumsy jaw operation on his friend Marcel Natkin [257]. Afterward, he told Veblen, he felt "wretched for a long time" and lost a good deal of weight, partly due to his "usual indigestion"; by September he had "almost recovered," but in October he again ran a fever and suffered from "insomnia, etc." [258].

What he did not say was that he had spent the week of 13–20 October, and perhaps more, as a patient at the Sanatorium Westend in Purkersdorf bei Wien,[16] an institution founded by the industrialist Viktor Zuckerkandl for "healing waters and physical therapy."

Built in 1904–5 on the site of a former spa of the same name, the sanatorium building (Figures 11 and 12) — designed by Josef Hoffmann,[17] with interior furnishings produced by the Wiener Werkstätte on designs of Hoffmann and Koloman Moser — is a landmark of fin-de-siècle architecture that remains standing today [259]. In accord with the institution's purpose the ground floor was occupied by two large bathing rooms, for men and women, respectively; a

---

[16] The dates are based on financial receipts in Gödel's *Nachlaß*. It is noteworthy too that in his letter to Veblen Gödel did not mention Hahn's death (though he *had* noted it in his draft), nor any of the other turmoil then going on around him. On the contrary, he reported that "Economic conditions in Austria seem to be improving" and that "At the university everything is quiet and almost nothing has altered through the change in government"! He noted only that Nöbeling — who until then had served with him as co-editor of the proceedings of Menger's colloquium — had "become a Nazi and got a position in Germany."

[17] One of the leaders of the Secession movement, of which Zuckerkandl's sister-in-law Bertha was a devoted patron.

FIGURE 11. The Purkersdorf Sanatorium, Purkersdorf bei Wien, Austria, ca. 1906. View from the east.

room for exercise therapy, filled with precursors of today's Nautilus equipment; and the doctors' "surgery" (examining room). The first floor contained the dining hall, music room, game rooms for table tennis, billiards, and board games, rooms for reading and writing, and a wide veranda. Patients were accommodated in about fifteen private rooms on the top floor.

The Purkersdorf sanatorium was an establishment for the well-to-do, part spa, part clinic, part rest home, with facilities for diet and rehabilitation therapy. It could serve those frail in mind or body and those who thought themselves to be. It was not a mental institution in the modern sense, but it could provide a restful atmosphere and a balanced regimen suited to recovery from physical or mental stress. As such it was an appropriate refuge for someone like Gödel, strained from overwork, who was of unstable temperament, hypochondriacal, and able to afford the sort of treatment it offered.

Among those called in for consultation on Gödel's case was Julius Wagner-Jauregg, a world-famous psychiatrist whose treatment for general paresis had won him the Nobel Prize in 1927. In his judgment Gödel had suffered a

FIGURE 12. The Purkersdorf Sanatorium, Purkersdorf bei Wien, Austria, ca. 1906.
Interior hall.

"nervous breakdown" brought on by overwork, from which he could soon
be expected to recover [260]; and in fact, he did recover quickly enough to
attend the meeting of Menger's colloquium on 6 November (the first of the new
academic year), to which he contributed a remark on mathematical economics
(*1936a*).   Nevertheless, in his letter to Veblen of 1 January he wrote that
though he "felt much better" he was "still somewhat sensitive and fear[ed]
that a stormy crossing or ... sudden change in conditions might bring about
a relapse."   Accordingly, he requested that his return visit to Princeton be
postponed to the autumn [261].

(Veblen replied with concern, declaring it "a pity" that Gödel could not
come that year, since Church was on leave from his teaching duties and would
have been free to discuss logical matters with him in depth. Nonetheless he

spoke to Flexner about Gödel's request, and they agreed to carry Gödel's stipend over to the next year.)

Documentation of Gödel's activities during the next few months is sparse, but he apparently remained quite active in Menger's colloquium, whose proceedings he continued to edit (assisted now by Abraham Wald). Menger himself has recalled:

> Gödel was more withdrawn after his return from America than before; but he still spoke with visitors to the Colloquium, ... especially with Wald and Tarski, the latter working the first half of the year 1935 at the Colloquium with a fellowship. To all members of the Colloquium Gödel was generous with opinions and advice in mathematical and logical questions. He consistently perceived problematic points quickly and thoroughly and made replies with greatest precision in a minimum of words, often opening up novel aspects for the inquirer. He expressed all this as if it were completely a matter of course, but often with a certain shyness whose charm awoke warm personal feelings for him in many a listener [262].

Menger also noted that "Gödel now spoke more about politics," but that "even in the face of the horrible, seemingly inescapable dilemma" then confronting Europe and the world "he remained impassive" [263].

Gödel's library request slips and notebooks from that time show that he was then much occupied with physics. Among the many books he requested were works of Eddington, Planck, Mach, Born, Nernst, Schrödinger, Lorentz, Sommerfeld and Dirac, as well as Hermann Weyl's *Mathematische Analysis des Raumproblems*. Three physics notebooks, all dated 1935, contain material on such topics as statistical mechanics and optics, while a fourth, undated one is devoted to quantum mechanics. Since several loose pages of examination excerpts were found along with the notebooks, it may be that Gödel sat in on some physics classes that spring, perhaps as a follow-up to von Neumann's lectures at the IAS the previous year [264].

Equally remarkable is the near absence of library requests for readings in philosophy. In particular, though Menger has attested that Gödel had become deeply involved in the study of Leibniz some years before, he apparently owned only a single work by Leibniz (the *Kleinere philosophische Schriften*), and none of Leibniz's works show up among his request slips from 1935. The one item among them that is of interest is Edmund Husserl's *Vorlesungen zur Phänomenologie des Bewusstseins* (Lectures on the Phenomenology of Consciousness). Perhaps Gödel's interest in that work was aroused by Husserl's appearance in Vienna that May, when he delivered a pair of lectures on "Die

Philosophie in der Krisis der europäischen Menschheit" (Philosophy in the crisis of European humanity) [265].

That same month Gödel resumed lecturing himself. His second offering as a *Dozent*, a two-hour course entitled "Selected Chapters in Mathematical Logic," met on Thursdays and Fridays, commencing on 9 May. Again, only fragmentary notes for the course remain, and only nine enrollment slips for it have been found, though more students than that may actually have attended.

It was presumably while teaching that course that Gödel obtained another important breakthrough: a proof for the relative consistency of the Axiom of Choice with the other axioms of set theory. When he first turned his attention to set theory[18] is uncertain, but had he obtained the result earlier he would surely have presented it to Menger's colloquium; as it was, the first person to whom he is known to have communicated the result was von Neumann, when the two met again the following autumn.

Quite apart from that discovery, on 19 June Gödel did make a short presentation to the colloquium — his last published contribution thereto [266]. Titled simply "On the Length of Proofs," its main result was stated without proof: that by passing to a logic of higher type not only would certain previously undecidable propositions become provable, but, if the length of a formal proof were taken to be the number of formulas (not symbols) in the proof sequence, then it would become *"possible to shorten, by an extraordinary amount, infinitely many of the proofs already available."* With the development of computational complexity theory some thirty years later Gödel's result was recognized, despite some ambiguities in its statement, as an early example of what are known today as "speed-up" theorems — a major topic in theoretical computer science [267].

On 1 August Gödel wrote to Flexner that he had been in his "normal state of health for some months" and so was "looking forward with pleasure to being in Princeton for the next term." He stated that he expected to arrive in New York at the end of September aboard either the *Georgic* or the *President Roosevelt* and requested an advance of $150 to cover his travel expenses [268].

Flexner replied on 22 August, noting his "delight" in Gödel's recovery and enclosing a draft for the $150 [269]. He was unaware that at nearly that moment Gödel was once again staying at a sanatorium, this time in Breitenstein am Semmering. It seems, however, that Sanatorium Breitenstein was more an alpine health resort than a mental treatment facility, and as already noted,

---

[18] At which time he allegedly exclaimed "Jetzt, Mengenlehre" (And now, [on to] set theory).

when Gödel arrived in Princeton some five weeks later Flexner thought he looked much better than he had two years before.

For his second voyage to America, Gödel made it to the dock without incident; and this time he did not lack intellectual companionship during the crossing, for among his fellow passengers aboard the *Georgic* when it left LeHavre on 20 September were Wolfgang Pauli and Paul Bernays, both also headed for the IAS. Pauli had not met Gödel before, and on learning from Bernays that Gödel was on board he sent a note saying that he would like very much to make Gödel's acquaintance [270]. What the two of them may have discussed together is unknown, but between Gödel and Bernays there was an exchange of considerable importance: For as Bernays later told Kreisel [271], it was during that voyage and the few weeks following their arrival in Princeton that Gödel explained to him the details of the proof of the second incompleteness theorem, as they were subsequently presented in Hilbert and Bernays's text.

Of Gödel's work in Princeton in the autumn of 1935, two notebooks survive [272]. The first contains some differential geometry, followed by a systematic development of set theory similar to that in his later monograph (*1940*). The second, largely in shorthand and rather messy, includes a section on the Axiom of Choice labeled *rein* (that is, fair copy), as well as material on the Continuum Hypothesis bearing the annotation *halbfertig* (half-finished). Those distinctions, coupled with the evidence from Gödel's correspondence with von Neumann (with whom he discussed his work on the former topic but, apparently, not the latter), suggest that at that time Gödel was preparing to publish the *Reinschrift* material separately.

On arrival Gödel found lodging in a house at 23 Madison Street, where he presumably intended to stay at least until the end of the fall term. Abruptly, however, on 17 November he notified Flexner that he felt compelled to resign for reasons of health (apparently suffering from a recurrence of depression). The two met on the 20th to discuss the situation, and within a few days of that conversation Veblen accompanied Gödel to an appointment with a physician, who agreed that a return to Vienna would be "the wisest course." Flexner accepted the verdict with sympathy and regret, but he had to advise Gödel that, in view of the timing of his resignation, he could not authorize payment of more than half his stipend (which, it will be recalled, was for "a half year or more.") He expressed the hope that Gödel would make "a complete and prompt recovery" and that he would soon "be able to resume" his "important scientific activities." Veblen concurred, and reminded Gödel that "the professors in the Institute ha[d] already agreed" to invite him back as soon as he had recovered and wanted to return [273].

Gödel left New York on 30 November aboard the *Champlain*. Veblen accompanied him to the dock and promised not to alarm Gödel's family by informing them of his condition. But he soon had second thoughts. On 3 December he wrote Gödel that on further reflection he felt he "dared not risk" failing to alert his brother Rudolf, in case "some accident [might] befall you on the way without any of your friends on either side of the Atlantic ... [learning of] it for several days or ... weeks." Accordingly, he had wired Rudolf that Kurt was returning home because of health and would arrive at Le Havre on 7 December [274].

It would appear that Veblen was seriously worried about Gödel's state of health and mind. Yet, on the other hand, he apparently expected a prompt recovery, for just a week after sending his letter to Gödel he wrote to Professor Paul Heegard in Oslo to suggest that Gödel be invited to give one of the principal addresses at the upcoming International Congress of Mathematicians. Veblen made no mention of Gödel's breakdown, nor did he say anything specific about his latest discovery, but only remarked, vaguely and somewhat tantalizingly, that Gödel's "recent, as yet unpublished work" was, "like his published work, ... interesting and important" [275].

From Le Havre Gödel made his way to Paris, where, in a state of mental distress, he telephoned his brother and asked him to escort him back to Vienna. But Rudolf was unable to do so on short notice, so Kurt was forced to wait. He stayed three days at the Palace Hotel and then, on the 11th, found that his condition had stabilized enough that he was able to take the train to Vienna by himself [276].

By that time conditions in Austria had deteriorated so much that Menger had made up his mind to seek a position in America. In an undated letter to Veblen written from Geneva not long before Gödel's return, Menger described the situation: "Whereas," he said, "I ... don't believe that Austria has more than 45% Nazis, the percentage at the universities is certainly 75% and among the mathematicians I have to do with ... [apart from] some pupils of mine, not far from 100%." Their interference was mainly in administrative affairs that Menger himself was "only [too] glad to stay away [from]"; the real problem was the "general political atmosphere, whose tensions and impending dangers" exerted an ever-increasing strain on the nerves that had by then become "unsupportable. ... You simply *cannot* ... find the concentration necessary for research if you read twice a day [about] things ... which touch the basis of the civilization of your country as well as your personal existence" [277].

Veblen must already have begun to make inquiries on Menger's behalf by the time of Gödel's departure, for Menger wrote to thank him on 17 December; and in that letter Menger noted that he had already seen Gödel twice since his return. "It is too bad," he said, "that he overworked himself to a degree that he needed soporific things, but perhaps still worse that he actually started easing [up on] them. But I think that after a few weeks of complete rest the depression will be gone" [278].

In fact, recovery was to take much longer. In a passage that he crossed out from the draft of a later letter to Veblen, Gödel mentioned having stayed "in a sanatorium for nervous diseases [at Rekawinkel, just west of Vienna] for several months in 1936," [279] and among his papers there are also receipts dated June 1936 from an institution in Golling bei Salzburg. Its nature is unclear, but Gödel may well have sought solace away from Vienna just then, for on 22 June Moritz Schlick was assassinated.[19]

In addition, on three occasions later in the year (17–29 August, 2–24 October, and 31 October–21 November) Gödel stayed at hotels in Aflenz, a spa that in earlier years had been a favorite retreat of the Gödel family. There is no record of anyone having accompanied him on the first or third of those visits, but on the second his companion was Adele — registered as "Frau Dr. Gödel" [280].

Whether the two had lived together in Vienna prior to that is unclear,[20] but in any case, Adele contributed more than romance to the sojourn in Aflenz. By her own later testimony [281], she served as a taster for Kurt's food — a task that foreshadowed her role as protectress in years to come, when Gödel's dietary concerns and obsessive fear of being poisoned, especially by gases that might be escaping from his refrigerator, became ever more manifest, symptoms of an underlying mental disturbance that was increasingly to threaten his physical well-being.

---

[19] Schlick was shot ascending the steps to the main lecture hall at the university, on his way to the final lecture of the term. The assassin was a disturbed former student, Dr. Hans Nelböck. The Nazis attempted to portray the crime as having resulted from Schlick's "subversive" activities, but it seems rather to have been a personal psychopathic act. The crime is discussed in detail in *Siegert 1981* (extracted from the book *Attentate, die Österreich erschütterten*), where it is suggested that it was perhaps fomented by a friend of Nelböck's (Sylvia Borowicka) who had also undergone psychiatric treatment and had developed a hatred of Schlick.

[20] In guarded responses during an interview with Werner Schimanovich and Peter Weibel, Rudolf Gödel reluctantly admitted that there were "a few dwellings" where his brother had stayed in Vienna that he had never become acquainted with; but he gave no dates and said he "didn't know" whether Kurt had lived in any of them with his "Freundin." Among Gödel's papers there is, however, a contractor's receipt dated November 1937, made out to "Herrn und Frau Dr. Gödel."

Further evidence of Gödel's mental state is provided by the list of books he sought in succeeding months. In December 1936, for example, while staying for five nights at the Hotel Drei Raben in Graz, he tried to buy a handbook on toxicology and a treatise on intoxication; and during the next two months his library request slips show that he read widely on such subjects as pharmacology, normal and pathological physiology, psychiatry, neurology, and laws pertaining to mental illness. (Among the titles requested were *Neue Behandlungsmethode der Schizophrenie, Differentialdiagnostik in der Psychiatrie, Die Irrengesetzgebung in Deutschland, Handbuch der österreichischen Sanitätsgesetze und Verordnungen,* and *Geschichte der Geisteskrankheiten* — readings that may well have made Gödel all the more difficult a patient to treat.) The title of one technical volume, requested several times, is especially suggestive: *Die Kohlenoxydgasvergiftung* (Carbon Monoxide Poisoning). In view of Rudolf Gödel's recollections discussed at the end of chapter IV, his brother's interest in such a topic might be taken as evidence of suicidal intent. But once again, Gödel's latent *fear* of poisoning seems an equally plausible explanation; and in fact, since the apartments in which he lived in Vienna were heated with coal and coke, such fears may not have been wholly unreasonable.

◊　◊　◊

Looking back forty years later Gödel declared that 1936 had been one of the three worst years of his life [282]. For the history of logic, though, it was a momentous time: it was then that *The Journal of Symbolic Logic* was founded; then that Alonzo Church published his papers "An Unsolvable Problem of Elementary Number Theory" (in which he used his Thesis to establish the undecidability of the notion of arithmetic truth) and "A Note on the Entscheidungsproblem" (wherein he showed the decision problem for first-order logic to be unsolvable); then that Kleene's investigations on "General Recursive Functions of Natural Numbers" and on "λ-definability and Recursiveness" appeared, containing his recursion and normal form theorems; and then that Gerhard Gentzen gave the first proof-theoretic demonstration of the consistency of formalized arithmetic. It was in 1936, too, that Alan Turing went to Princeton to obtain his doctorate under Church. He expected to meet Gödel there as well, but, by the fateful quirk of Gödel's incapacitation, was denied the opportunity.[21] In the end, the two never did meet.

By the beginning of 1937 Gödel had recovered enough to announce plans once again to offer a course on axiomatic set theory. Lectures were scheduled to begin in February, but they were postponed at least twice. They were finally offered that summer, heralding Gödel's return both to (relatively) good health and to preeminent work in logic.

---

[21] He might have missed him even so, for it is not clear that Gödel ever intended to stay at the IAS during the second term of the 1935–36 academic year. Indeed, records from the University of Vienna show that Gödel had at one time announced that he would offer a course on set theory there during that winter.

VI

"Jetzt, Mengenlehre"

(1937–1939)

THE EARLY HISTORY of set theory, from its founding by Cantor as an outgrowth of his studies of trigonometric series to its axiomatization by Zermelo in 1908, has been briefly sketched in chapter III. It will be recalled that it was Cantor who first conjectured (and later attempted unsuccessfully to prove) both that every uncountable set of real numbers can be put in one-to-one correspondence with the set of all real numbers and that every set can be well-ordered; Hilbert who first called those two problems to the attention of the mathematical community, when he listed them together as the first of the twenty-three open problems that he posed to mathematicians of the coming century; and Zermelo who succeeded in deriving the well-ordering property from the Axiom of Choice, a principle whose fundamental character he was the first to recognize.

A definitive history of the controversy that ensued thereafter, together with a detailed examination of the complexly interwoven ramifications of the Axiom of Choice and its numerous variants, has been given by Gregory H. Moore [283]. His account is the basis for the discussion given in the next few pages, which is devoted to a summary of the developments in set theory during the years 1908–30 that are essential to a proper understanding of Gödel's later contributions to the subject.

Zermelo's first axiomatization for set theory appeared in his paper *1908b*. He was not the first, however, to suggest the need for such an axiomatization: The Italian mathematician Cesare Burali-Forti had already done so in 1896, well before the set-theoretic paradoxes came to the fore. Indeed, Burali-Forti had proposed two possible candidates for axioms; but he had not given an

axiom *system.* Similarly, "certain propositions which now seem similar to ...
Zermelo's Axioms of Union and Separation" had also appeared in a letter that
Cantor sent to Dedekind in 1899; in that case, however, "there is no evidence
that Cantor regarded [them] as *postulates*" [284]. Zermelo was thus the first to
propose an explicit axiomatization for set theory as such.

Zermelo's axiomatization preceded by twenty years the precise delineation
of first-order logic (the framework in which his axioms are now usually recast),
so his system was not a formal one in the modern sense. Rather, his seven
axioms were augmented by a number of "fundamental definitions," among
them that "Set theory is concerned with a *domain* ... of individuals ... call[ed]
objects ... among which are the *sets*"[1] and that between some of those objects
"certain *fundamental relations* of the form $a \in b$ " hold (in which case b was
understood to be a set and a an element of it). In addition to the usual notion
of subset Zermelo introduced the notion of *definite property*: one for which, "by
means of the axioms and the universally valid laws of logic," a determination
"whether it holds or not" could be made from "the fundamental relations
of the domain" [285] (in effect, a property decidable on the basis of those
relations). The axioms themselves (lightly paraphrased in modern notation)
were the following:

> I. (Axiom of Extensionality) If M and N are sets for which both $M \subset N$
> and $N \subset M$ (that is, every element of M is an element of N, and vice-
> versa), then $N = M$. Sets are thus completely determined by their
> elements.

> II. (Axiom of Elementary Sets) There exists a set (the *null set*) that con-
> tains no elements at all; if a is any object of the domain, there exists a
> set $\{a\}$ containing a as its only element; if a and b are any two objects of
> the domain, there exists a set $\{a, b\}$ containing just a and b as elements.

> III. (Axiom of Separation) Whenever the propositional function $F(x)$
> is *definite* for all elements x of a set M, M possesses a subset whose
> elements are those x in M for which $F(x)$ is true.

> IV. (Axiom of the Power Set) To every set M there corresponds another
> set (its *power set*) whose elements are (all) the subsets of M.

---

[1] In particular, Zermelo allowed the domain to contain objects that were not themselves sets.
Such *urelements* could occur as elements of sets, but beyond that the axioms said nothing about
them. Their ontological status was left entirely unspecified.

V. (Axiom of the Union) To every set M there corresponds a set (its *union*) whose elements are (all) the elements of the elements of M.

VI. (Axiom of Choice) If M is a set whose elements are all sets, and if those elements are all mutually disjoint and not null, then the union of M includes at least one subset having one and only one element in common with each element of M.

VII. (Axiom of Infinity) There is at least one set in the domain that contains the null set as an element and that contains $\{a\}$ as an element whenever it contains a itself as an element.

According to Moore, "Zermelo ... [gave] no explicit rationale for his [particular] choice of axioms except that they yielded the main theorems of Cantorian set theory" [286]. He recognized that Cantor's set theory — in particular, Cantor's conception of a set as "a collection, gathered into a whole, of certain well-distinguished objects of our perception or ... thought" — had been undermined by "certain contradictions," but he thought that by restricting Cantor's principles he could exclude the contradictions while retaining everything of value. Zermelo admitted that he had "not yet ... been able to prove rigorously" that his axioms were consistent, but he endeavored nonetheless to show how "the antinomies discovered so far vanish one and all if [my] principles ... are adopted"; in particular, he noted that his Axiom of Separation did not allow sets to be "*independently defined*" but only to be "*separated* as subsets from sets already given," and hence that "contradictory notions such as 'the set of all sets' or 'the set of all ordinal numbers'" would be excluded. Likewise, he pointed out that the requirement that the separating property be definite prevented antinomies such as that of Richard from arising. With regard to the "deeper problems" of consistency and independence he expressed the hope that he had "at least ... done some useful spadework ... for subsequent investigations."

In fact, it was to be many years before such questions were taken up. Meanwhile, Zermelo's attempt to secure the acceptance of his well-ordering theorem by embedding the Axiom of Choice within a larger axiomatic framework did little to silence his critics. Instead, his Axiom of Separation — and especially the notion of "definite property" that occurred within it — served only to generate further controversy. Bertrand Russell and Philip Jourdain were among those who criticized Zermelo's axiomatization privately, while Poincaré and Weyl did so publicly (the latter constructively, by making the concept of "definite property" more precise). Outside Germany Zermelo's axiomatization found no acceptance whatever.

Among Zermelo's compatriots, one of the few to embrace it was Abraham Fraenkel. In 1919 Fraenkel published an introductory textbook on set theory in which he adopted Zermelo's system without qualification — apparently unconcerned (or unaware) that although the Axiom of Separation prevented the construction of Russell's paradoxical set $\{x : x \notin x\}$, it did not rule out sets that were self-members (nor, more generally, those that stood at the top of an infinite descending $\in$-sequence: $\ldots \in M_n \in \ldots \in M_3 \in M_2 \in M_1$).

The Russian Dimitry Mirimanoff had pointed out that possibility two years earlier. He had called such sets "extraordinary," but he apparently regarded their existence more as a curiosity than a defect. In 1922, however, Fraenkel and, independently, Skolem, discovered a more serious flaw in Zermelo's system: Its axioms did not permit the construction of certain sets of mathematical interest (in particular, the set $\{\mathbb{N}, \mathfrak{P}(\mathbb{N}), \mathfrak{P}(\mathfrak{P}(\mathbb{N})), \ldots, \}$ where $\mathfrak{P}$ denotes the power-set operation).

To remedy that oversight Fraenkel, in his paper *1922*, proposed a new axiom — that of Replacement, according to which, whenever each element of a given set is replaced by an object of the domain, the entity so obtained is also a set. At almost the same time Skolem, in his *1923b*, formulated the same principle more precisely, within the framework of what is now called first-order logic. Specifically, by a "definite" property Skolem understood one defined by a first-order formula of the language of set theory (having just $\in$ and $=$ as relation symbols). Given such a formula $A(x, y)$, with the free variables x and y, Skolem's axiom stated that if A were *functional* on a given set M (so that for each $x \in M$ there was at most one y for which $A(x, y)$ was true), then the image of M under A (that is, $\{y : \exists x (x \in M \wedge A(x, y))\}$) was also a set.[2]

Two other papers of Fraenkel dealt with issues involving urelements. In *1921* he proposed an Axiom of Restriction that would rule out such nonsets, while in *1922b* he made essential use of urelements in demonstrating the independence of the Axiom of Choice from the other axioms. Eventually the advantages of doing without urelements were acknowledged, and Fraenkel's Axiom of Restriction became implicit in the underlying formalization: Anything not mandated by the other axioms was presumed not to exist. As a result, however,

---

[2] Actually, in his 1917 paper Mirimanoff had anticipated both Fraenkel and Skolem, by taking as axiomatic that any collection of ordinary sets that was in one-to-one correspondence with a given set was itself an ordinary set; and though he had stated it as a fact rather than an axiom, Cantor had made essentially the same assertion in another of his letters to Dedekind.

the independence of the Axiom of Choice re-emerged as an open and much more difficult problem — one on which Gödel expended much effort in the early 1940s, with only limited success.

Further modifications were introduced by von Neumann in his paper *1925*. Harking back to Cantor's distinction between "consistent" multitudes (sets) and "inconsistent" (or "absolutely infinite") ones, von Neumann noted that the latter did not give rise to antinomies so long as they were not permitted to be elements; he characterized them as those multitudes that could be put in one-to-one correspondence with the class of all sets — a criterion that subsumed the schemas of Separation and Replacement and could be expressed as a single axiom. He also initiated the study of so-called inner models of set theory — those contained and definable *within* a given model — and used them to show that if the axioms of his set theory were consistent they would remain so when augmented by an additional axiom that prohibited infinite descending ∈-sequences of sets. Gödel would later employ the same method to demonstrate the relative consistency of the Axiom of Choice and the Generalized Continuum Hypothesis.

Eventually, in his paper *1930*, Zermelo published a revised axiomatization that he called ZF', acknowledging Fraenkel's contribution. Like his original formulation, ZF' permitted urelements. Following von Neumann, however, Zermelo added a new axiom — that of Foundation — to rule out descending ∈-sequences. He retained his earlier formulations of the axioms of Extensionality, Power Set, and Union, but replaced the Axiom of Elementary Sets with an Axiom of Pairing, due to Fraenkel, that asserted the existence of the set $\{a, b\}$ for any two objects a, b of the underlying domain. The axioms of Infinity and Choice were dropped, while those of Separation and Replacement were altered along the lines suggested by Skolem and Fraenkel, rendering the notion "definite property" more precise. However, since he found the Skolem Paradox "thoroughly repugnant . . . [and] believed that the Löwenheim-Skolem Theorem revealed the limitations of first-order logic rather than the inadequacy of axiomatic set theory," [287] Zermelo expressed the axioms of Separation and Replacement in terms of *second-order* propositional functions.

Such was the background to his polemical address at Bad Elster in September of 1931. In the end his views on first-order logic did not prevail, but the system ZF' — reformulated as a first-order theory with the Axiom of Infinity but without urelements — soon became the standard axiomatization for set theory, known thenceforth simply as ZF (or, with the Axiom of Choice adjoined, as ZFC).

Coincidentally, Zermelo first proposed his axioms for set theory in the same year that Hausdorff enunciated the Generalized Continuum Hypothesis.

But the ensuing axiomatic development did little to resolve the continuum problem. Indeed, prior to 1926 the only significant result concerning the power of the continuum, aside from Cantor's own theorems, was obtained by Julius König in 1904. In an address that year to the Third International Congress of Mathematicians he showed, in effect, that $2^{\aleph_0} \neq \aleph_{\alpha+\omega}$. (More generally, his arguments could be extended to show that $2^{\aleph_0} \neq \aleph_\beta$ for any β cofinal with ω.) König, however, mistakenly thought that he had *refuted* both the Continuum Hypothesis and the Axiom of Choice.

By contrast, in his famous address "On the Infinite" (*1926*), Hilbert proclaimed just the opposite: As a consequence of "the demonstration that every mathematical problem can be solved," he claimed to have *proved* the CH. He offered what he called "a brief intuitive presentation ... of the fundamental idea of the proof," but his argument was vague and unconvincing. (Nevertheless, as Gödel himself was later to point out [288], Hilbert's approach to the problem bore a "remote analogy" to Gödel's own later conception of "constructible" sets.)

Shortly thereafter Adolf Lindenbaum and Alfred Tarski (*1926*) investigated various statements of cardinal arithmetic that were equivalent to the Axiom of Choice, the GCH, or both. Among other results they noted the remarkable fact that the Axiom of Choice follows from the assertion that for an infinite set A, no set B both properly contains A and is properly contained within the power set of A.

◇ ◇ ◇

Gödel's own contributions to set theory were made within the context of these developments. The evidence [289] suggests that he began to work on Hilbert's first problem at about the same time that he tackled the second: In 1928 he had requested Skolem's Helsingfors Congress paper (*1923b*) from the University of Vienna Library, and during 1930 he requested it again, along with Fraenkel's *Einleitung in die Mengenlehre*, the volumes of *Mathematische Annalen* and *Mathematische Zeitschrift* containing von Neumann's papers *1928a* and *1928b*, and the volume of the *Göttinger Nachrichten* in which Hilbert's 1900 address had been published. Following the Königsberg conference Gödel took time to visit the Preußische Stadtbibliothek in Berlin in quest of various Scandinavian and Polish journals, and in 1931, with the incompleteness paper behind him, he requested an extended loan of Fraenkel's paper *1922a*.

That Gödel should have undertaken a combined assault on Hilbert's first two problems is not at all surprising given that he had started out to give a relative model-theoretic consistency proof for analysis. The arithmetical comprehension axiom that arose in that context is quite similar to Skolem's formulation of the Axiom of Separation (which Gödel once described as "really the only essential axiom of set theory" [290]), and there is a striking parallel between Gödel's attempt to interpret second-order number-theoretic variables in terms of *definable* subsets of the integers and his construction of a model of set theory through iteration of a *definable* power-set operation. Moreover, the failure of the former effort (which the second incompleteness theorem had shown to be unavoidable) had made it clear that any proof of the consistency of ZFC must be a relative one.

Beyond such conceptual affinities, there is also an overall tactical similarity between the arguments in Gödel's incompleteness paper (*1931a*) and those in his set-theoretic monograph (*1940*): In each case a fundamental property of predicates is introduced — in the former instance, that of being primitive recursive; in the latter, that of being "absolute for the constructible submodel" — and the proofs proceed via a long sequence of lemmas in which a large stock of predicates are shown to possess the given property. In each case, however, one crucial notion — that of being provable in formal number theory, or of being a cardinal number within a model of set theory — conspicuously *fails* to have the specified property, and the rest of the proof hinges on that failure. In particular, central to both proofs is a distinction between internal and external points of view: in the incompleteness paper, between mathematical and meta-mathematical notions; in the set-theoretic consistency proofs, between those functions that exist within a given submodel and those that exist outside it.

A concept essential to Gödel's set-theoretic work is that of the *cumulative hierarchy* of sets, a notion that Zermelo introduced in his paper *1930*.[3] Since both extraordinary sets and urelements are disallowed in ZF, the cumulative hierarchy may be defined as the transfinite sequence of levels $R(\alpha)$ (where α ranges over all the ordinal numbers) given inductively by the stipulations:

$$R(0) = \emptyset \text{ (the empty set)},$$

$$R(\alpha + 1) = \mathfrak{P}(R(\alpha)),$$

and (for limit ordinals λ)

$$R(\lambda) = \bigcup_{\alpha < \lambda} R(\alpha)$$

(Here again, $\mathfrak{P}$ denotes the power-set operation).

---

[3] The underlying idea had, however, been anticipated long before by Mirimanoff, who, in his paper *1917*, had shown how to assign an ordinal *rank* to each ordinary set.

The levels of Gödel's *constructible hierarchy*, now usually denoted by the symbols $L(\alpha)$, are defined in the same way, with L in place of R, except that $\mathfrak{P}$ is replaced by the operation that assigns to a set s not the set of *all* its subsets, but just those that are *definable* by means of first-order formulas of the language of set theory (augmented by constant symbols for elements of s) when the variables are restricted to range only over the elements of s. Clearly $L(\alpha) \subset R(\alpha)$ for each α, and Gödel showed that the class L, comprising the union of all the levels $L(\alpha)$, is a submodel of the universe V of all sets (the union of all the levels $R(\alpha)$). That is, L satisfies the axioms of ZF whenever V itself does. But since, at each level α, the new sets that appear may be *definably* enumerated according to the least code numbers of the formulas that define them and the enumerations already established for the sets of lower level that appear as parameters within the defining formulas, the entire class L possesses a definable well-ordering. Thus L satisfies the Axiom of Choice as well, so ZFC is consistent if ZF is.

That much, presumably, Gödel had already obtained by the time of his visit to Princeton in 1935. Had it not been for the extended bout of depression he then suffered he would likely have published an announcement of his results shortly thereafter. As it was, however, it was not until 1937 that he first presented them in public.

Gödel's summer course "Axiomatik der Mengenlehre," which began in May of that year, met weekly for a total of twelve sessions. A relatively coherent picture of its content is provided by Gödel's shorthand notes for his lectures and by the testimony of Professor Andrzej Mostowski, one of the students in the course [291]. Both sources confirm that Gödel made no mention then of the Axiom of Constructibility (the statement $\forall x \exists \alpha(x \in L(\alpha))$, which played a prominent role in his later treatments of the subject), nor did he mention the Continuum Hypothesis. However, in a letter to Menger of 3 July Gödel stated that he had reported a partial result about the latter to a meeting of (what remained of) Menger's colloquium.[4]

What Gödel did *not* tell Menger was that just three weeks before — thus, during the time he was presenting his lectures — he had in fact made the final breakthrough in his relative consistency proof for the CH. An annotation on the inside cover of the first of his sixteen *Arbeitshefte* (mathematical workbooks) indicates that the event occurred during the night of 14–15 June 1937 [293].

---

[4]Menger himself had left Austria in January of 1937 to take up a position at Notre Dame. After his departure Franz Alt and Abraham Wald arranged a few more meetings, but no records of their content were kept [292].

The first person to whom Gödel confided the news of his success was von Neumann, who had written from Budapest on 13 July to express his pleasure that Gödel's health had recovered enough for him to resume lecturing. Von Neumann suggested that Gödel come to the IAS again in 1938–39, and he invited him to publish his paper on the Axiom of Choice in the *Annals of Mathematics*, the journal edited jointly by mathematicians at Princeton University and the IAS. In closing, von Neumann mentioned that he expected to visit Vienna within the next few weeks, and it was on that occasion that Gödel sketched the details of his new proof to him [294].

Von Neumann was obviously excited by the result, for he wrote again on 14 September to report that he had conferred with Lefschetz about the matter and could now "officially promise" that Gödel's paper on the GCH would appear in the *Annals* in the first issue to be printed "fourteen or more days after the arrival of the manuscript" [295]. He was unable, though, to say anything definite about the possibility of Gödel's returning to the IAS since Flexner and Veblen had gone away for the summer and neither had yet returned.

In the meantime, through Menger's efforts, Gödel had been invited to visit Notre Dame during the upcoming winter and spring. In a letter of 22 May Menger told him that Notre Dame was seeking to upgrade its mathematics and physics departments. He said that he had mentioned Gödel's name to Father John Cardinal O'Hara, Notre Dame's president, and that O'Hara had immediately expressed interest. Consequently, he was writing to inquire whether Gödel would consider coming to Notre Dame if an offer were to be made [296].

Gödel replied on 3 July, saying that the idea appealed to him but that he could not come there before the summer term of 1938 at the earliest. Because of the health problems he had experienced on his previous visits to America he preferred not to commit himself for more than a single semester, and he had already received a tentative invitation from Philipp Frank to spend the semester just ahead in Prague.

Gödel's reply took weeks to reach Menger, and in the interim President O'Hara extended an official invitation to him to stay at Notre Dame from February to June of 1938. He offered him a stipend of $2,000, plus living expenses and $400 for travel [297].

Menger had still heard nothing from Gödel when Veblen stopped off at Notre Dame en route back east at the end of the summer. Menger told Veblen of O'Hara's offer, but neither of them was yet aware of von Neumann's contacts with Gödel.

Finally, on 12 September, Menger sent Gödel a radiogram. In it he expressed his hopes of seeing Gödel in February, and, as further inducements for him to come, mentioned that the algebraist Emil Artin would be visiting Notre Dame that year, that the university attracted some good students, that the work atmosphere was peaceful, and that the cost of living in South Bend was lower than on the East Coast [298].

In the end there was no conflict among the various invitations: Frank's offer fell through, and von Neumann's plan from the beginning had been to invite Gödel for the *following* year, since IAS stipends for the academic year 1937–38 had already been awarded. Veblen did write to Gödel on 1 November to say that an unexpected resignation had freed up funds to pay for a series of five to ten lectures, should Gödel care to visit Princeton "incidental to a more extensive stay in Notre Dame" [299]. By then, however, Gödel had decided to remain in Vienna. He wrote President O'Hara that he could not come to Notre Dame that academic year, but added that he might be able to do so the following September if the invitation could be deferred.

His uncertainty was partly, of course, a reflection of the steadily worsening political and academic climate in Austria. With the deaths of Hahn and Schlick and the emigration of Carnap and Menger, Gödel was left as almost the only one in Vienna still working in logic. That he was able to work at all in such circumstances indicates how focused he was on his research and how detached he remained from the larger crisis taking place around him. In his letter to Menger of 3 July he mentioned that he was still undecided as to whether, during the coming year, he "should lecture on something introductory, on something advanced, or ... not at all," noting that "in the second case there is the danger that I won't have any students" [300].

With Hahn and Menger both gone there was also little in the way of seminars or colloquia for Gödel to take part in. There was, however, a small discussion group, originally under the leadership of Heinrich Gomperz, that had met for some time at the home of the philosopher Viktor Kraft. Gomperz had been forced to retire from the University of Vienna in 1934, and in 1938 he left for America, where he took up a position at the University of Southern California. In his absence his student Edgar Zilsel undertook to keep the circle going. An organizational meeting was held on 2 October 1937, and Gödel was among those invited to it.

An informal shorthand record [301] of the proceedings of the meeting reveals that those present included Edith Weisskopf (sister of the well-known physicist Victor), who had just earned her Ph.D. in psychology and, with her teacher Else Frenkel, was co-author of the book *Wunsch und Pflicht in Aufbau des menschlichen Lebens* (Desire and Obligation in the Structure of Human Life); Walter Hollitscher, a medical student, Austrian Communist, and follower of Schlick, who went on to edit Schlick's writings after the war; Heinrich Neider, who had joined the Vienna Circle while still a student; Franz Kröner, a student of Gomperz who was acquainted with several members of the Circle; and Rose Rand (who, according to Gödel's account, slept through most of the meeting!), a refugee from Lemberg who had also been a student member of the Circle. It was agreed that the group would meet every other Saturday, and Zilsel suggested to Gödel that he report at an upcoming meeting on the status of consistency questions in logic, a topic in which Gentzen's consistency proof had sparked renewed interest.

Gödel at first demurred, saying that he would need time to prepare such a talk and might not be able to attend the sessions after the first few weeks. Eventually, however, he was persuaded. So far as is known his lecture to the Zilsel circle on 29 January 1938 was his last presentation to a Viennese audience. It remained unpublished during his lifetime, but after his death a shorthand draft of the text was found among his papers. Though incoherent in places, the content proved to be of such historical and philosophical interest as to warrant editorial reconstruction in volume III of Gödel's *Collected Works*, where it appears as item *1938a*.

Gödel began his lecture by asserting that a consistency proof for a theory is meaningful only in the sense of a reduction either to the consistency of a proper part of it or to that of "something which, while not truly a part of it, is somewhat more evident, reliable, etc., so that one's conviction is thereby strengthened" (as, for example, the reduction of nonconstructive arguments to constructive ones). After pointing out the difficulties inherent in characterizing the notions "finitary" and "constructive," he concluded that a truly finitary proof of consistency was impossible for arithmetic and so devoted the remainder of the lecture to a consideration of how finitary methods might be extended so as to obtain meaningful proofs of consistency. The extensions he considered were of three kinds:

- 1. Those involving the introduction of higher types of functions (an idea that Gödel would develop further in the years ahead, but would only finally publish in his *Dialectica* paper of 1958)

- 2. Those involving modal-logical considerations

- 3. Those, like Gentzen's, involving transfinite induction up to "certain concretely defined ordinal numbers of the second number class."

The paper is typically Gödelian: incisive, thoroughgoing, and forward looking. One can only wonder what impression it made on the mathematically unsophisticated audience to which it was delivered.

◇ ◇ ◇

In mid-November 1937 Gödel moved out of the building on Josefstädterstrasse and took up residence in a third-floor apartment at Himmelstrasse 43/5 in the Viennese suburb of Grinzing. The move seems to have been prompted by financial concerns, considerations that also led Gödel's mother to return to the villa in Brno (an act of considerable courage since from the first she had been a vehement opponent of the Nazis and was frequently incautious in her remarks about them).[5]

Despite the move, his precarious economic and academic situation, and the ominous events taking place all around him, Gödel managed over the next few months to fill three notebooks with material on the Continuum Hypothesis [303]. Contrary to what one might expect, they do not contain drafts for a detailed presentation of the results he had already obtained, but rather attempts to extend them still further. In a letter to Menger dated 15 December he confided that "Right now I am trying also to prove the *independence* of the Continuum Hypothesis, but don't know yet whether I will succeed with it" (emphasis added). Until then Gödel had told no one but von Neumann even of his *consistency* result for the GCH, and he requested Menger "for the time being to please tell no one else about it" [304].

Shortly thereafter von Neumann returned briefly to Europe, and while there he took the opportunity to set up an appointment with Gödel. The two met at the Hotel Sacher on Sunday morning, 23 January 1938 [305]. Von Neumann wanted to discuss further his plan that Gödel visit the IAS the next year, and he no doubt inquired as well about Gödel's progress in writing up his consistency results. He learned that in the interim, among other things, Gödel had obtained a theorem concerning one-to-one continuous images of co-analytic sets.[6]

---

[5] Marianne shared the house with Aunt Anna until Anna's death (about the time the war began), and with her sister Pauline until the latter's death in 1942. She returned to Vienna only in 1944, when her son Rudolf persuaded her that it was too dangerous to remain in Brno any longer [302].

[6] A subset of $\mathbb{R}^n$ is *analytic* if it is the projection (with respect to the last coordinate) of a Borel subset of $\mathbb{R}^{n+1}$, and *co-analytic* if its complement is analytic.

Unfortunately, the "theorem" later turned out to be incorrect. In a letter written on the eve of his return to America the following autumn, Gödel told von Neumann that it had been contradicted by a recently published theorem of Stefan Mazurkiewicz and that since then he himself had obtained a few results "tending in the opposite direction"; specifically, he said, one could consistently assume that there exist nonmeasurable sets that are one-to-one continuous images of co-analytic sets, and also that there exist uncountable co-analytic sets that contain no perfect subset [306]. In the same letter Gödel announced that about two months before he had "essentially finished" writing up his consistency results. But he still wanted to make "a few changes," and he felt that it would save sending the proof sheets back and forth if he waited to submit his paper until after his arrival in America.

Presumably, Gödel devoted the winter and spring of 1938 to the preparation of his manuscript and to making arrangements for his upcoming year abroad. In December Veblen had written again to propose that he spend the fall term of 1939 at Notre Dame and the spring term in Princeton; but in March Gödel replied that he was unsure he could obtain a leave of absence for the entire year, and even if he could, he thought it might be better if he were to visit Princeton before Christmas and Notre Dame afterward.

In due course that arrangement was agreed on, to the satisfaction of all concerned: Gödel would lecture on his set-theoretic work at the IAS during November and December and would then be in residence at Notre Dame from February until June. Menger suggested that during his stay at Notre Dame Gödel should offer "a very elementary introduction to mathematical logic ..., something of the Hilbert-Ackermann type," in addition to "an advanced discussion either of the continuum problem, or of the *Entscheidungsproblem*, for two hours once a week" [307]. Gödel, however, felt that the set-theoretic material would be hard to fit into a two-hour format, and he also expressed reservations about teaching an introductory course because of his "insufficient English, lack of experience in introductory lecturing and [limited] time for preparation" [308]. After further negotiation, it was agreed that he would lecture three hours per week on the advanced material and that the two of them together would conduct an introductory seminar on logic.

Of Gödel's letter to Veblen only a burnt fragment has survived; it is dated 26 March, just thirteen days after Hitler's *Anschluß* [309]. It would be interesting to know what, if anything, Gödel had to say about that event, or what immediate effect it had on his life or work, but, incredibly, there is no mention of the Nazi takeover in any of Gödel's correspondence. His papers do, however, contain several documents from Nazi functionaries at the University of Vienna, which reveal that on 23 April his *Lehrbefugnis* (authorization to teach) officially lapsed.

There is no evidence that Gödel made any attempt at the time to protest the change in his teaching status, nor did he mention the fact in his correspondence with colleagues at Notre Dame or the IAS. Neither, however, did he request a leave of absence for the year ahead. On the contrary, only on 31 October — two weeks after arriving in Princeton — did he inform the dean of Vienna's philosophical faculty that he had accepted an invitation to spend the winter semester of 1938–39 in Princeton. (He made no mention at all of Notre Dame.) It may be that the withdrawal of his right to teach was what finally convinced him to return to America. In any case, he informed Menger and Veblen of his decision only in late June.

The state of Gödel's finances at that time may be inferred from his request to Veblen for an advance of $300 to cover his travel expenses. Because of restrictions on currency exchange that were then in effect, Veblen was unable to oblige. He offered instead to purchase Gödel's ticket for him, but in the end Gödel was able to come up with the necessary funds on his own. He was not permitted to take any money out of Germany, however, and so had to ask that $25 be sent to the Hapag bureau in New York, since he had been told he would need to have that amount in his possession in order to be allowed ashore [310].

Remarkably enough, he seems to have had no trouble obtaining the necessary exit and entrance visas, but he told Flexner that the boats were "so overcrowded" that he had found it "very difficult to get a ticket" [311]. He finally booked cabin class passage aboard the German liner *Hamburg*, expecting to arrive in New York on 7 October. Later, though, something went awry, for on 29 September he sent an urgent radiogram from Hamburg. Could the IAS, he asked, obtain a ticket for him from Holland or England? Otherwise, he declared, he could not come [312].

Flexner replied the next day that he could "do nothing at the moment" [313], and the *Hamburg* did indeed sail without Gödel. Eventually he managed to obtain passage on the *New York*, which left Cuxhaven on 6 October and arrived in New York on the 15th.

What had happened? If, as his letters of 3 September seem to imply, he had already secured a ticket at that time, why was he unable to embark on the *Hamburg*? Had the shipping line overbooked? Or was there some other explanation?

One can only speculate, but there is good reason to suspect that what Gödel failed to obtain was a *second* ticket; for on 20 September he and Adele had

Figure 13. Wedding portrait of Kurt and Adele Gödel, Vienna, 1938.

been quietly married in Vienna (Figure 13). Whether she subsequently accompanied him to Hamburg is unknown, but it is of interest that on the 29th, the very day he wired Flexner, he also signed two restricted powers of attorney, authorizing Adele to publish an announcement of their marriage and to obtain an official marriage certificate [314]. Surely he would have wanted his bride to accompany him to America (especially in view of the "frustrations of his bachelor life in Princeton" that he recalled to Kreisel more than twenty years later [315]), and that is perhaps why he stated in the radiogram that he "could not come" if the IAS were unable to help him get another ticket. If so, he (or she) must have changed his mind during the week prior to his departure, for she remained behind at their apartment in Grinzing.

The wedding itself was a private, civil ceremony attended by the couple's immediate families and a few other acquaintances. Two witnesses signed the marriage certificate: Karl Gödel, a cousin of Kurt's father, identified as an "akademische Maler" (Academy painter) living at Amalienstrasse 55; and Hermann Lortzing, a "Buchsachverständiger" (book expert) residing at Domanigasse 3. Although Kurt's brother was among those present, when interviewed later he recalled little about the event. He did say, though, that Kurt had not introduced him to Adele until they decided to marry, even though the two had been betrothed for several years [316].

Neither Menger nor Veblen had any inkling of Gödel's impending marriage, though Menger had heard, just a few months before, of his engagement. As he wrote to Veblen, "I got the same announcement [you did]. . . . I never met [Gödel's] bride, and only know that 3 years ago when he was ill somebody with the first name Adele visited him. I, too, think that marriage may be quite good for him" [317]. One can only marvel that Gödel succeeded for so long in concealing his relationship with Adele from even his closest acquaintances.[7]

◇   ◇   ◇

Although Gödel had not planned to begin his IAS lectures until around 1 November, his delayed crossing undoubtedly made life more hectic than he wished during the first few weeks after his arrival. Evidently he was in a hurry to settle in quickly, for he did not rent an apartment as he had on his previous visits to Princeton, but lodged at the Peacock Inn near the university campus throughout his stay. He must have resumed work almost immediately on arrival, for by 9 November he had already sent an announcement (*1938*) of his consistency results to the *Proceedings of the National Academy of Sciences* [318].

In the meantime, in late October, he had also traveled to New York City to attend a regional meeting of the American Mathematical Society (AMS). His primary purpose in doing so seems to have been to speak further with Menger about the courses to be taught at Notre Dame. But the occasion also provided an incidental opportunity for him to make the acquaintance of Emil Post.

As noted in chapter III, Post had demonstrated the syntactic completeness of the propositional calculus in his doctoral dissertation of 1920. At that time he had also come close to obtaining the incompleteness theorems — a fact recognized only much later, when his "Account of an Anticipation" was published posthumously in *Davis 1965*. So encountering Gödel face-to-face was a very emotional experience for him. "For fifteen years," he wrote later that same day, "I had carried around the thought of astounding the mathematical

---

[7]To say nothing of biographers: Not a single letter between the two has turned up.

world with my unorthodox ideas, and meeting the man chiefly responsible for the vanishing of that dream rather carried me away" [319]. His attitude toward Gödel was, however, one of admiration rather than resentment. "As for any claims I might make," he went on, "perhaps the best I can say is that I would have *proved* Gödel's Theorem in 1921 — had I been Gödel."

His conversation with Gödel so haunted Post that he wrote again the next day, attempting to explain further just what he himself had done. "My dear Dr. Gödel," he began,

> In our conversation of yesterday each time a comparison of your Theorem and absolutely unsolvable problems arose you kept emphasizing that in the former a particular proposition appeared as undecidable. I therefore want to point out that that is exactly what arose in my procedure of getting your Theorem as a corollary of the existence of absolutely unsolvable problems. . . .
>
> Of course your particular undecidable proposition for a logic has intrinsic interest in its interpretation as a statement of the consistency of the formal system. I'm afraid its very interest has led both to misinterpretations of the meaning of your Theorem and its relation to possible proof[s] of consistency. . . .

Post's poignant outpouring of his ideas ran to four handwritten pages. In the end he acknowledged that "nothing that I achieved could have replaced the splendid actuality of your proof." He had failed to attain that actuality

> chiefly [because] I thought I saw a way of so analyzing "all finite processes of the human mind" . . . that I could establish the above conclusions [concerning the existence of undecidable propositions] in general and not just for *Principia Mathematica*. . . . As for not publishing my work as outlined above, the whole force of my argument depended on identifying my "normal systems" with any symbolic logic, and my sole basis for that were the reductions I mentioned.

In view of the difficulty he had experienced in getting others of his papers published, Post could see

> no hope of getting those reductions into print. And then came [manic-depressive] illness, and a sort of preparatory regime for work which would gradually make me less excited by those general ideas of mine. And that regime took and is taking so much more time than I had planned for it that finally you and Church and even Turing sped by me.

Post apologized for his "egotistical outbursts," assuring Gödel that "any resentment I may have is at the Fates if not myself ... and after all, it is not ideas but the execution of ideas that constitutes a mark of greatness" [320].

After their meeting Gödel sent Post a "sheaf of reprints," but he did not tell him of his new results in set theory. Post learned of them only from the abstract (*1939a*) for the address that Gödel gave to the annual meeting of the AMS in Williamsburg on 28 December, following the end of his lectures in Princeton. The news prompted Post to write Gödel a letter of congratulation, in which he opined that the reason for Gödel's success was his "genius for producing a simplified logical system which has the earmarks of generality ... yet is manageable" [321].

For his 1938 lectures Gödel prepared detailed manuscript notes [322]. Yet, as he had in 1934, he requested that a graduate student be assigned to take notes for publication. The individual chosen for the task was George W. Brown, a student of Samuel S. Wilks in mathematical statistics. Brown had been exposed to logic at Harvard through the lectures of E. V. Huntington and W. V. Quine, and according to his own recollections it was "probably" Alonzo Church who recruited him to become the note-taker. In any case, Brown "was still interested in logic and was flattered by the opportunity to assist Gödel" [323].

Brown found Gödel to be "a strict taskmaster," but he nevertheless testified that he "enjoyed the experience thoroughly" and found "the lectures ... [to be] models of precision and of organization." Remarkably, he did "not recall ever seeing Gödel's own notes." His own duties were to "take notes, put in the connective language, and ... supervise preparation of the manuscript." The final typescript was prepared using a machine called a "Varityper," which had interchangeable type wheels similar to those of a modern daisy-wheel typewriter [324].

In contrast to the treatment that he gave on several other occasions, in his lectures at Princeton Gödel did not work directly in ZF set theory. Rather, he adapted a class formalism for set theory that had been developed by von Neumann and Bernays. Their formulation had two important technical advantages: It provided a *finite* axiomatization, and it allowed the notion of first-order definability over a structure to be expressed syntactically, in terms of closure under eight fundamental set-theoretic operations. It also allowed the Axiom of Constructibility to be expressed by the simple class equation $V = L$. As Gödel himself later acknowledged, however [325], while the eight operations simplified some of the considerations involved in the proofs, they tended to obscure the underlying intuitions. Consequently, most modern expositions have followed the approach Gödel outlined in his abstracts *1939a* and *1939b*.

Either way, the key steps in the proof are the same. After defining the constructible hierarchy of sets, three facts must be established: first, that the universe of such sets satisfies all the axioms of ZF set theory; second, that it also satisfies the Axiom of Constructibility; and third, that the latter implies both the Axiom of Choice (as indicated earlier) and the Generalized Continuum Hypothesis.

The proof of the first fact is relatively straightforward. The second, however, involves the subtlety that *some* definable notions, such as that of cardinality, depend crucially on the underlying universe: To know that a one-to-one correspondence between two sets exists within V (the "ground model" of ZF within which the construction is carried out) does not guarantee that such a correspondence also exists within the submodel L. Notions that do have the same meaning in L as in V are said to be *absolute* for the constructible submodel, so the proof of the second fact amounts to showing that the notion of constructible set is *itself* absolute (that is, that every set constructible within V is also constructible within L).

So expressed, the concept of absoluteness appears to be a semantic notion and the consistency proof for the Axiom of Choice and the GCH to rest not only on the *consistency* of ZF but on the existence of a *model* for the axioms — which, by the second incompleteness theorem, is not provable within ZF even if ZF *is* consistent. To circumvent that problem Gödel gave a *syntactic* characterization of absoluteness, via the notion of the "relativization" of a formula of ZF to the definable class L.[8] In terms of that characterization the arguments can then be carried out entirely within the framework of ZF.

The third step in the proof — that of showing that the Axiom of Constructibility implies the GCH — is the most difficult. It rests on the notion of the *order* of a constructible set x, defined as the least ordinal α such that $x \in L(\alpha + 1)$. By the very definition of the constructible hierarchy, it is clear that $L(\alpha)$ is always a proper subset of $L(\alpha + 1)$ (in particular, $\alpha \in L(\alpha + 1) - L(\alpha)$), so for any ordinal α there will always *be* constructible sets of order α; by applying Cantor's theorem on the cardinality of power sets one can then obtain the stronger conclusion that for each ordinal β between ω_α

---

[8]Without entering into details, the relativization Φ^L of a formula Φ of ZF is obtained inductively by restricting all quantifiers in Φ to refer to elements of L. Thus if Φ has the form $\exists x \Psi$, Φ^L has the form $\exists x (L(x) \wedge \Psi^L)$, and if Φ has the form $\forall x \Psi$, Φ^L has the form $\forall x (L(x) \rightarrow \Psi^L)$.

and $\omega_{\alpha+1}$ (the least ordinals of cardinalities $\aleph_\alpha$ and $\aleph_{\alpha+1}$, respectively) there are constructible subsets of $L(\omega_\alpha)$ of order β or greater. But Gödel showed that there are *no* constructible subsets of $L(\omega_\alpha)$ of order $\omega_{\alpha+1}$, whence, within L, the power set of ω_α — a subset of the power set of $L(\omega_\alpha)$ — must have cardinality $\aleph_{\alpha+1}$, that is, $2^{\aleph_\alpha} \leq \aleph_{\alpha+1}$. It was presumably the proof of that fundamental fact that Gödel discovered during the night of 14–15 June 1937.

$$\Diamond \quad \Diamond \quad \Diamond$$

At Princeton Gödel could count upon an audience well versed in mathematical logic. Kleene and Rosser were gone by then, but Church was still there, and Haskell Curry, the American logician who pioneered the study of combinatory logic, was spending the year as a visiting member at the IAS. At Notre Dame, however — and indeed, throughout "the Catholic universities in America in the mid-thirties" — logic "was completely dominated by the writings of [Jacques] Maritain and the philosophical school of Laval University." [9] Both were "quite opposed" to mathematical logic, which was regarded as "merely formal" and was identified by many with Bertrand Russell, toward whom "there was [much] ... antipathy [in] religious circles" [326].

Spring term at Notre Dame began in February. Gödel arrived there the month before, following his lecture at Williamsburg, but accounts differ as to where he stayed and what his state of health was. In a memoir written some forty years later Menger recalled that Gödel "lived on campus ... for a large part of the semester" [327], a report affirmed as well by a Notre Dame graduate student, who wrote that "Gödel spent [the] semester brooding in the Lyons Hall [dormitory] annex" [328]. But other evidence shows that he rented a room at the Morningside Hotel from 27 January until 31 May, just prior to his departure. As to Gödel's health, Menger stated in his memoir that "during his stay at Notre Dame Gödel appeared healthy but not particularly happy," while in a letter to Veblen written at the time he reported that "Gödel became ill with a bad flu, which kept him in bed for more than a week, in his room for another, and very weak for a third" [329].

Memories fade, of course; but if Gödel did not, in fact, live on campus, it is hard to know what to make of Menger's further claim that Gödel "had quarrels with the prefect of his building [described as "an old priest, very set in his

---

[9] Maritain himself was a frequent visitor to Notre Dame, though it seems unlikely that he attended Gödel's lectures. In later years, by an odd coincidence, he lived in Princeton on the same street as Gödel, only about two blocks away. There is no record, however, of any contact between the two.

ways"] for all sorts of trivial reasons" [330]. The story accords perfectly with Gödel's legalistic concern for his "rights" and with the fears and hypochondria that caused him to change his domicile on several later occasions. Given his generally dour aspect and the fact that the German invasion of Czechoslovakia occurred in the very midst of his stay, there is also little reason to doubt that he may have spent a good deal of his time "brooding." Yet Menger reports that he avoided discussing political matters with Gödel *not* for fear of adding to his worries but because, despite everything, Gödel "remained his equanimous self," seemingly oblivious to what was happening in Europe except for his protestations about the violation of his own rights in the matter of his *Lehrbefugnis.*

Probably, as in Vienna, Gödel took refuge in his work. His teaching responsibilities at Notre Dame were significantly greater than they had been in Vienna, and they included participation in the symposium that Menger had established as a continuation of his Vienna colloquium. Indeed, Menger seems to have expected Gödel to lead its weekly sessions. But as one observer has recalled, the fact was that "Menger was a dynamo" whereas "Gödel was ... shy and reserved." Consequently, "if Menger was there, as he usually was, Gödel would scarcely broach the matter at hand before Karl would dart in and take over" [331].

The bulk of Gödel's effort was devoted to his two lecture courses. Titled "The Theory of Sets" and "Introduction to Logic," they were offered as electives for graduate students only. The former course was a reprise of the lectures Gödel had just given at Princeton, so it should have required relatively little preparation. Nevertheless, he prepared a separate set of notes for it that fill five spiral-bound notebooks.

For the introductory course, taught jointly with Menger, there were initially "about 20 students," half of whom were "young instructors and doctoral candidates with good preparation in mathematics" who "attended [the] lectures to their conclusion." The other half consisted of "older philosophers and logicians and occasionally one or another member of the physics department," most of whom displayed "no particular interest" in the lectures. Professor Yves Simon, a student of Maritain, was one of the few who "made a special effort to take advantage of Gödel's presence" [332].

At the outset mimeographed notes were distributed to the students. That practice does not seem to have continued for long, however, so the content of the course must instead be ascertained from Gödel's own notes (preserved in seven more spiral-bound notebooks).

The opening lecture was given by Menger, who, no doubt to the dismay of the second group of listeners, pointed out some of the inadequacies of traditional logic. In particular, he noted that Aristotelian logic did not suffice to yield all logically true propositions, nor, from among the various inferential laws it treated, did it attempt to isolate an independent set from which all the others could be derived. Gödel then announced in the following lecture that the aim of the first part of the course would be "to give, as far as possible, a complete theory of logical inferences and logically true propositions and to show how they can be reduced to a certain number of primitive laws." He went on to develop propositional logic in a systematic and leisurely fashion: After introducing the usual connectives via numerous examples in natural language, he discussed truth tables and their realization, gave an axiomatization for propositional logic (essentially derived from Russell) based on four axioms and three rules of inference, and showed that the axioms were independent and sufficed for the deduction of all tautologies.

He then paused to note that although Russell's axiomatization had become the standard for textbooks, it was nonetheless open to criticism on aesthetic grounds. In particular Gödel pointed out that statements taken as axioms "should be as simple and evident as possible ... [certainly] simpler than the theorems to be proved, whereas in [Russell's] system ... the very simple law of identity $p \supset p$" was a theorem rather than an axiom. An alternative formulation that he regarded as more satisfactory had recently been developed by Gentzen, to whose system of sequents (schematic proof figures) he then paid brief attention.

The latter part of the course was devoted to a development of the functional calculus of predicates and quantifiers, as well as to brief discussions of the calculus of classes, the logical paradoxes, and Russell's theory of types. Gödel mentioned both the completeness of predicate logic and the unsolvability of its decision problem, but he did not enter into the proofs of either.

Overall, the course provided a remarkably thorough introduction to modern logic. Gödel's notes are eminently readable, both in terms of clarity and style, and could be used today with little alteration. They are written in English, with one amusing exception: For security, he rendered the examination questions in Gabelsberger shorthand.

◇   ◇   ◇

Apart from his teaching, Gödel also found time for reading and writing. Early in the semester he composed a brief note (*1939b*) on his consistency proof for the GCH, which he communicated to Veblen for transmission to the National Academy of Sciences; by mid-April he had completed the manuscript for his monograph *1940*, in which full details of his set-theoretic results were presented; and in his spare time, according to Menger's testimony, he indulged his preoccupation with Leibniz, some of whose "important writings ... [he believed] had not only failed to be published, but ... [had been] destroyed in manuscript" [333].

Before leaving Notre Dame Gödel accepted invitations to return to the IAS the following fall and to deliver a major address at the next International Congress of Mathematicians, scheduled to be held in Cambridge, Massachusetts, in September of 1940 [334]; but as it turned out, events in Austria delayed his return to Princeton, and the 1940 congress was canceled after the outbreak of World War II.

Gödel surely did not foresee all that was about to happen to him. But he did advise Flexner that he was "not yet absolutely sure" he would "be able to stay [at the IAS] for both terms" of the upcoming academic year, and he expressed regret that he had been unable to speak with Flexner before going on to Notre Dame since he "should have liked very much" to obtain his advice about "some personal matters" [335].

Near the end of his stay in South Bend Gödel wrote Flexner again [336]. Gödel planned to stop off in Princeton prior to his return to Europe, in order to confer with Veblen and von Neumann about arrangements for the publication of his monograph (of which von Neumann had written, "You have settled this enormous problem with truly masterful simplicity and have reduced the unavoidable technical details of the proof to a minimum .... Reading your treatise was really an aesthetic delight of the first class" [337]). Might there be an opportunity then, Gödel inquired, for the consultation he had been unable to arrange earlier?

Evidently there was, for on 7 June Flexner advised him, "I have made inquiry as I promised ... [but] find to my great regret that there is nothing I can do on this side of the Atlantic either through the Department of State or the Department of Labor." In a postscript he added, "I return herewith your passport" — the first hint that difficulties had arisen regarding Gödel's visa [338].

VII
Homecoming and Hegira

(1939–1940)

GÖDEL'S DIFFICULTIES WITH American immigration authorities were the result, apparently, of his having been admitted to the United States in 1933 or 1935 as one seeking to establish permanent residence in the country. Unfortunately, his reentry permit and his Austrian passport both lapsed during the interim before his next trip, and when he returned in 1938 he came with a German passport and a visitor's visa [339]. The problem he faced in the spring of 1939 was thus that of having his visa extended.

Despite the failure of Flexner's efforts, there is no indication that Gödel foresaw any serious obstacles to his returning to the United States that fall. He was no doubt accustomed to bureaucratic red tape, and when he boarded the *Bremen* on 14 June for the voyage back to Europe the thought of being reunited with his wife was probably foremost in his mind.

To judge from his correspondence with colleagues, he was also more concerned about the publication of his Princeton lectures than he was about visa matters. Indeed, on the very day he arrived in Bremen Gödel dispatched a letter to Bernays — the first, apparently, since their parting in November of 1935 — in which he noted that just as his lectures were about to be mimeographed he had discovered that the axiom system he had employed differed in one respect from that which Bernays had published in the *Journal of Symbolic Logic*. He asked Bernays to clarify the roles that he and von Neumann had played in the formulation of the class formalism for set theory, and, despite the long lapse in their correspondence, requested an immediate reply to avoid further delaying the duplication of his lectures [340].

A month later von Neumann wrote Gödel to ask whether he had any objection to having his lectures published as a monograph, using a "technically more complete" method of reproduction. Gödel answered on 17 August, following a week's vacation with Adele in Baden: He agreed to the idea, and in closing remarked casually, "There is not much new with me; recently I had to attend to a lot of things with the authorities. I hope to be in Princeton again by the end of September" [341].

He repeated the same hope in a letter to Menger of 30 August — a letter that Menger thought established "a record for non-involvement on the threshold of ... historic events":[1] "Since the end of June," Gödel wrote, "I have again been here in Vienna, and in the last few weeks I've had a lot of running around to do, so that up to now it has unfortunately been impossible to write up anything for the colloquium. How did the exams over my logic lectures turn out?" [342].

One must wonder whether Gödel was really as blithely unconcerned about his return to the United States as his letters suggest. Was he naively optimistic? Oblivious to what was going on? Or just trying to mask his fears? And what was the nature of the affairs that occasioned so much "running around" dealing with "the authorities"?

First of all, there was trouble of some kind concerning the transfer of funds from his bank account in Princeton to that in Vienna. A total of $1,084 remained uncredited as of the end of July, when he wrote the Devisenstelle Wien in an attempt to straighten things out. The matter dragged on for several months before it was finally resolved. It is worthy of note because, in contrast to his practice in earlier years, and despite his professed expectation of returning to Princeton the very next month, Gödel closed out his Princeton account on 20 September. (Even during 1936 that account had been left largely untouched. Indeed, in June of 1936 it was credited with a $400 deposit.) In addition, the salutation on his letter to the Devisenstelle is the only known instance in which he prefaced his signature with the words "Heil Hitler" [343].

There was also the matter of his absence from the University of Vienna during the previous academic year. For Gödel had informed the dean of his sojourn in Princeton only *after* his arrival in the United States, an action that stood in violation of a decree that had been handed down by the Austrian Minister of Education on 19 August 1938, about six weeks prior to Gödel's departure. According to that edict, copies of which were circulated to the

---

[1] *Menger 1981*, p. 21. Two days later, Hitler would invade Poland, an action for which, in signing his infamous non-aggression pact with Russia, he had prepared the way a week before. Gödel may well have been unaware of the latter event, but within a few months it would prove crucial to his future.

rectors of all Austrian universities, it was "the duty of all academical teachers" — including those who had been retired or dismissed — "not to negotiate with a foreign university without the approval of the ministry." Should the rectors be informed of a teacher's attempt "to obtain an appointment abroad," they were "obliged to advise the Minister at once" [344].

Consequently, on receipt of Gödel's letter of 31 October the dean had forwarded it to the Ministry of Instruction. It arrived there on 23 November and was then sent on to the Ministry for Internal and Cultural Affairs. But no further action was taken until 4 July of the following year. On that date an official of the latter ministry wrote to the rector to inquire "whether Dr. Gödel is staying on in the U.S.A. for personal reasons, or on account of a promised teaching position" [345]. The rector, in turn, passed the inquiry on to the dean, who replied on 12 July that he could shed no light on the matter because Gödel had supplied no further details about the nature of his activities in Princeton. Clearly unaware that Gödel was then back in Vienna, the dean proposed that Gödel's *Lehrbefugnis* (his official authorization to teach) be rescinded since Gödel had not requested a leave of absence for the summer semester [346].

Again the correspondence was passed upward to the Ministry of Internal and Cultural Affairs, where another official advised the rector on 12 August that the action recommended by the dean seemed unnecessary. Gödel's *Lehrbefugnis*, he noted, was already in abeyance, and it would officially expire on 1 October unless Gödel submitted an application in the meantime to be named *Dozent neuer Ordnung* (*Dozent* of the new order). In the latter case the application might be denied; it would be up to the rector, in forwarding such an application, to see that the dean's recommendation was carried out.

It is unclear whether Gödel was aware of all that activity behind the scenes. Very likely, though, another matter was the focus of his attention, for shortly after his return to Austria he had received notice to report for a physical examination to determine his fitness for Nazi military service.

While still in America Gödel had apparently anticipated receiving such a summons; it was probably one of the things he discussed with Flexner just prior to his return. But, as he later explained to Veblen, "When I came back to Germany I expected that I would be sent back for good by the military authorities ... [as he later was by an American draft board, which unhesitatingly declared him 4-F]. Instead, my examination ... was put off from month to month, so that it was impossible to leave before the war broke out. Finally I was ... found to be fit for garrison duty ... [a circumstance that made it] much more difficult to obtain leave of absence" [347].

It was a rude surprise. Yet, if the German authorities had refused to believe that Gödel's heart had been damaged by his childhood bout with rheumatic fever, it was most fortunate that they had *also* overlooked his episodes of mental instability. Had they taken notice of them he might indeed have been "sent back for good" — perhaps to a concentration camp or a clinic for "mental defectives." (As Gödel was well aware, it was also important that such problems be concealed from American immigration authorities: In a passage that he crossed out from the draft of his letter to Veblen he expressed his "fear that an immigrant's visa may not be granted, because I was in a sanatorium for nervous diseases for several months in 1936.")

By mid-September Gödel was thus in a distressing predicament: Subject to imminent induction, he was unlikely to be allowed to leave the country; yet, with the lapsing of his *Lehrbefugnis* he was also left without employment.

Faced with that dilemma, he pursued three courses of action: On 25 September he applied to be named *Dozent neuer Ordnung*, an appointment that, in contrast to the earlier *Dozentur*, carried with it a small salary; at the same time he inquired of military and university authorities about the possibility of obtaining leaves of absence; and he continued to negotiate with the American consulate regarding his eligibility for a visa.

The first action subjected him once more to the scrutiny of the Nazi academic bureaucracy. His application was first vetted by the *Dozentenbundsführer*, Dr. A. Marchet, who reported to the dean on 30 September that Dr. Gödel was "well recommended scientifically" but that his *Habilitation* had been carried out under the direction of "the Jewish Professor Hahn." It "redound[ed] to his discredit" that he had "always travelled in liberal-Jewish circles," but, on the other hand, "mathematics was at that time strongly 'Jewified' [*verjudet*]." Marchet was unaware, he wrote, of any "direct utterances or action" on Gödel's part "against National Socialism," nor were colleagues in his discipline able to supply any more detailed information about him. Consequently he found it impossible either expressly to approve Gödel's application or to suggest substantive grounds for its rejection [348].

Gödel described the results of his other inquiries in a wire to the IAS. As von Neumann interpreted it [349], Gödel's "essential difficulty" at that moment lay in obtaining a U.S. visa; were one to be granted to him, "the German military authorities w[ould] release him." The situation was therefore "considerably better than ... [had been] expected," for "while it would have been difficult ... to intercede with the German authorities," von Neumann thought that "there should be a way to persuade the American consul in Vienna to grant a visa."

But what kind of visa? In his letter to Veblen of 27 November Gödel explained that "the American consul refuses to give me a visitor's visa under the prevailing conditions. On the other hand a (nonquota) immigrant's visa is likely to cause trouble with the German inspector of passports, since I can at the best obtain a leave of absence [only] for a limited time." It was a dilemma familiar to all too many refugees. Laura Fermi, for example, in describing Italian passport practices at that time, has written that "Italian authorities would not issue a passport to the United States until the applicant had obtained an American visa. And American consular officials would not grant, or promise, a visa to a person who had not yet secured an Italian passport valid for travel in the United States. Only occasionally was the impasse broken" [350].

In Gödel's case there was another problem as well. U.S. immigration statutes of 1921 and 1924 had instituted a quota system for immigrants not from the Western Hemisphere. The total number of immigrants who could enter the United States was determined by presidential proclamation; by the time Hitler came to power the number was set at 153,849, of which 25,957 spaces were reserved for Germans [351]. Surprisingly, during the 1930s that quota was seldom filled. Yet even so, for those that qualified it was much quicker to obtain a *nonquota* visa. Such visas were made available under section 4(d) of the 1924 statute [352], which exempted from the quota restrictions, among others,

> an immigrant who continuously for at least two years immediately pre- ceding the time of his application for . . . admission to the United States has been, and seeks to enter the United States solely for the purpose of, carrying on the vocation of . . . professor of a college, academy seminary or university; and his wife . . . if accompanying or following to join him.

In his article "American Refugee Policy in Historical Perspective" [353], Roger Daniels has commented that while "such a provision . . . [might seem to have been] almost providential for . . . scholars" like Gödel, in fact "the State Depart- ment, and especially Avra M. Warren, head of its Visa Division, continually raised — one is tempted to say invented — difficulties." Daniels mentions the example of a teacher at a Berlin *Hochschule* who was denied a nonquota visa (and ultimately perished in Bergen-Belsen) because the school where he had taught was subsequently downgraded by the Nazis to a *Lehranstalt* (academy).

Similar objections were raised against granting Gödel a nonquota visa. In

response to a plea by Flexner on Gödel's behalf [354], Warren enclosed a copy
of a State Department memorandum interpreting section 4(d). It advised that

> Ordinarily a person applying for a nonquota visa as a "professor" will be
> required to show that he has actually been engaged in giving instruction
> to students as a member of the faculty in a recognized college, academy,
> seminary, or university and that this has constituted his principal occu-
> pation. . . . In cases involving conditions varying from those . . . , con-
> sideration may be given to all of the facts pertaining to the nature of the
> applicant's educational activities and the type of institution with which he
> has been associated in the past . . . . The applicant must establish that he
> has followed the vocation of professor continuously for at least two cal-
> endar years immediately prior to applying for admission into the United
> States, except that consideration may be given to the attendant circum-
> stances in cases where the applicant's vocation may have been interrupted
> for reasons beyond his control.

At the same time, Warren requested further information about the position
the IAS had offered Gödel. Would he be a member of the faculty? To what
extent would he be involved in teaching? And was the appointment likely to
be renewed?

The issues raised in the memorandum had already been anticipated by von
Neumann in his letter to Flexner of 27 September, and after studying both the
memorandum and Warren's letter von Neumann wrote Flexner again to suggest some possible responses [355]. He stressed Gödel's unique stature within
the mathematical community and focused on the final "exception clause" in
the passage quoted, pointing out that Gödel's suspension from teaching had
been at the behest of the German government. But he refrained from noting
the awkward fact that even before his *Venia legendi* had been withdrawn Gödel
had only infrequently exercised his right to lecture.

Von Neumann noted that Gödel's stipends had steadily increased,[2] but he
did not otherwise address the questions regarding Gödel's status and duties at
the IAS — questions that on the surface seemed innocent enough but that,
given the unique character of the IAS, were especially problematic, as Warren
no doubt realized. Of course, teaching in the usual sense of the word did
not take place at the institute;[3] and while Gödel had proved theorems of

---

[2] They had been $3,000 for the two terms of 1933–34, $2,000 for the fall of 1935, and $2,500
for the fall of 1938. The invitation for the two terms of 1939–40 carried an offer of $4,000.

[3] Early on, the IAS was accredited to award graduate degrees, a right it retains to this day; but
it has never done so.

the utmost significance — a fact that carried little weight under the terms of section 4(d) — he had never been offered a faculty position at the IAS, nor even a long-term appointment.

In the end the task of drafting a formal reply to Warren fell not on Flexner but on his successor, Frank Aydelotte. It had not been Flexner's intention to step down as IAS director, especially since the institute had just moved into the new quarters (Fuld Hall) that he had helped to bring into being. But "as the Institute grew, differences of opinion [had arisen] between the director and some trustees on [the] one hand, and the faculty on the other." Among the former "there was a rather widespread feeling that ... it was wrong for ... [professors] to get involved in administrative matters," while the professors themselves "understandably ... wanted to have at least a strong consultative voice in important academic matters." Thus "when Flexner appointed two professors in economics without any faculty consultation" there was "such an uproar that ... [he] had to resign" [356].

In response to Warren's questions Aydelotte stated forthrightly that the institute's trustees had not yet acted on the matter of renewing Gödel's appointment. But he believed there was "every hope" of their doing so, as the renewal had his own approval and was "the desire of all the members of the School of Mathematics."

With regard to Gödel's duties, however, he was somewhat disingenuous. "As a member of the Institute," he declared, "Professor Gödel's responsibilities [do] involve ... teaching. The instruction here is on a very advanced level and is consequently less formal than in the ordinary university or graduate school" [357]. In fact, of course, Gödel had never held the rank of professor and was under no obligation to lecture at all.

On 24 November Warren advised Aydelotte of "a telegraphic report ... from the American consular officer at Vienna" stating that Gödel's case had "been approved upon preliminary examination as that of an applicant apparently entitled to nonquota status." But, the report continued, though Gödel had been "invited to file ... formal application for a visa" as far back as 1 September, he had up to that time "not made application for the final examination of his case." The reason, according to the consul, was that Gödel believed he would be unable to obtain permission to leave Germany, "although he apparently has not endeavored to obtain such permission" [358].

The consul's report is in agreement with Gödel's own account of the matter in the letter that he wrote to Veblen three days later (quoted above). By that time he had apparently become resigned to remaining in Vienna, for in another crossed-out passage he stated, "I am beginning now to look for a position here (perhaps in the industry) since my savings are not sufficient to live on for a longer period of time." Nevertheless he did not withdraw his request for a leave of absence from the university. Indeed, to judge from a letter the dean addressed to the rector that very same day, Gödel invoked his impecunious situation as an argument in support of his application.

In his letter the dean weighed Gödel's scientific reputation against his political views. He noted that "as to character, Gödel makes a good impression": He was well mannered and, if permitted to go abroad, would "certainly commit no social blunders that would reflect ill upon his homeland." But since he also appeared to be "a thoroughly apolitical person" who possessed "hardly any inner commitment to National Socialism," he could be expected to have trouble coping with those "more difficult situations that will surely arise in the U.S.A. for a representative of the new Germany." On the other hand, if he were denied permission to leave on political grounds, the question would arise as to how he would support himself. He presently had "no income at all" and desired to accept the invitation to the United States "only in order to be able to earn a living. The whole issue of a trip abroad would vanish were it possible to offer [him] a corresponding paid position within the Reich" [359].

How precarious Gödel's financial situation really was may be gauged from a statement of assets that he drew up in mid-December of 1939 [360]. He listed two bonds, one for 1,250 Reichsmarks and the other for 1,500 Swiss francs; two bank accounts, amounting to 1,854 Reichsmarks and 1,684 Czech crowns, respectively; 4.5 shares of stock in a Czech ironworks; 1 share in an Austrian *Waggonfabrik*; and his part ownership of the house in Brno, which he valued at 10,000 Reichsmarks. Altogether, apart from the few shares of stock, his net worth was about $5,600 (the equivalent of around $66,000 today [361]), of which nearly two-thirds was tied up in the house in which his mother and great aunt were then living. His liquid assets thus amounted to about half the offer then outstanding to him from the IAS. In one respect, though, his economic plight was not quite so dire; for in early November he and Adele had moved from their rented apartment in Grinzing to one they had purchased back in the city.[4]

To have made such an expenditure at a time of such great uncertainty seems quite remarkable, and must be counted as strong evidence that, despite all his

---

[4] At Hegelgasse 5, not far from the Staatsoper. The two retained ownership of the property until several years after the end of the war, renting it out to tenants in their absence. The resulting income probably helped to supplement the support payments they made to their families.

inquiries, Gödel still did not seriously expect to emigrate. Undoubtedly the city of Vienna continued to hold its attractions for him, and he and Adele were surely reluctant to forsake their family ties there as well. Yet barely two months later they were to leave Austria for good. What then prompted their precipitate departure?

Very likely the answer is to be found in an incident that occurred around that time. While walking with Adele one day in the vicinity of the university, Gödel was assaulted by a band of young Nazi rowdies. For whatever reason — whether he was mistaken for a Jew, recognized as one who had fraternized with Jewish colleagues, or was simply targeted as an intellectual[5] — the youths seized him, struck him, and knocked off his glasses before Adele managed to drive them off with blows from her umbrella.

Gödel was not injured, but the incident must have made it clear to him just how perilous his situation really was, for by early December he had resolved to travel to Berlin to seek a way out. He continued, however, to display remarkable otherworldliness. In his letter to Veblen of 27 November he inquired, rather offhandedly, "Do you think it is dangerous at present to cross the Atlantic on an Italian boat?" And on 5 December he wrote to Helmut Hasse, then head of the mathematics institute at Göttingen, to say that he had some things to do in Berlin the following week and that he intended to stop off in Göttingen on his way back. Perhaps, he suggested, he could use that opportunity to lecture on his proof of the consistency of the Continuum Hypothesis, should the mathematicians there be interested in the matter [363]. He sent particular regards to Gerhard Gentzen.

Hasse replied two days later to advise Gödel that he had scheduled his lecture for 8:30 P.M. on Friday, 15 December, and had written to Gentzen, who was then away soldiering in Braunschweig, to suggest that he obtain leave to attend the presentation. (It is not known whether Gentzen was in fact in the audience.) He himself looked forward to the talk, but he regretted that he could not offer Gödel accommodation in the university's guest rooms since they were already full.

Gödel lodged instead at the Hotel zur Krone, where he arrived on the 14th.

---

[5] All three explanations seem equally plausible. In particular, it must be borne in mind that such gangs were quite indiscriminate in their attacks. According to one account of the situation in Vienna at that time, "Mobs of teenage boys and girls roamed the streets molesting anyone who looked the least bit Jewish in their eyes" [362]; so Gödel's habit of dressing in black hat and long overcoat might have made him a target. But he was also mistaken for a Jew by some who should have known better. Bertrand Russell, for example, did so in the second volume of his *Autobiography* (p. 326).

In the interim he had rearranged his itinerary so as to visit Berlin after, rather than before, his lecture — perhaps in order to allow himself more time both to prepare his talk and to negotiate with the immigration authorities. The lecture was, in fact, the only time he spoke on his GCH results to a European audience, and he took care in drafting the shorthand text.

Typically, his presentation was a model of clarity. After briefly surveying the historical background to the continuum problem, Gödel proceeded to sketch the basic steps in his proof. He avoided entering into too many technical details, but he nevertheless gave a well-motivated account of the underlying ideas; and, as befit the occasion, he paid homage to Hilbert by stressing certain analogies between the structure of his own argument and that which Hilbert had proposed but been unable to carry out in his 1925 lecture "On the Infinite" (published as *Hilbert 1926*). Specifically, Gödel reminded his listeners that Hilbert had singled out "a certain class of functions of integers, namely those that are defined recursively" and had attempted to prove, first, that they could be put in one-to-one correspondence with the ordinals of Cantor's second number class (those $< \omega_1$) and, second, that "the other definitions that occur in mathematics, namely those involving quantification, ... can [be] consistently assume[d] ... not [to] lead outside the domain of the recursively definable functions." Likewise, Gödel said, he himself had singled out a certain class of sets (those that are constructible) and had proved two analogous facts about them: to wit, that the set of constructible subsets of the natural numbers has cardinality $\leq \aleph_1$ and that the class of constructible sets is closed under the various methods of definition that are applied in mathematics (even those that are impredicative). Moreover, he noted, the proof of the first fact rested on the same idea that Hilbert had tried to use to establish *his* first lemma, namely, that in the definitions in question variables of too high a type can be avoided [364].

Gödel left Göttingen on the morning of 17 December and arrived a few hours later in Berlin; and there, on the 19th, he at last obtained German exit visas for himself and Adele, valid for a single departure (and presumed return) until 30 April 1940. That he was ultimately granted leave was, in his own view, probably the result of a plea that Aydelotte had written on his behalf to the chargé d'affaires at the German Embassy in Washington. In that letter, dated 1 December 1939, Aydelotte had emphasized that Gödel was an Aryan who was one of the greatest mathematicians in the world. "His case could hardly create a precedent," he argued, "because there are so few men in the world of his scientific eminence." He hoped, therefore, "that the German government w[ould] think it more important for [Gödel] to continue his scientific work" than to enter into the military service for which, Aydelotte "supposed," he would be liable [365].

Gödel expressed his gratitude to Aydelotte in a letter he wrote from Vienna on 5 January. Though he had yet to obtain the American immigration visas, he hoped they would be issued without further delay (as indeed they were, just three days later) so that he would be able to leave Vienna right away. "The only complication which remains," he advised, "is that I shall have to take the route through Russia and Japan [the trans-Siberian railway]. The German certificate of leave makes explicitly this requirement and in addition I am told in all steamship bureaux that the danger for German citizens to be arrested by the English is very great on the Atlantic" [366].

Before embarking on the long train journey there were still transit visas to be obtained. The Russian visas were issued in Berlin on 12 January and were valid for fourteen days — three of which passed before Gödel was able to obtain permits for himself and Adele to cross Lithuania and Latvia. Finally, late in the afternoon of the 15th, he wired the IAS from Berlin to advise that they expected to leave Moscow on the 18th and arrive in Yokohama on the 30th. He requested that an outside cabin for two be booked on the *S.S. Taft*, scheduled to depart Yokohama on 1 February, and that confirmation be sent to the Hotel Metropole in Moscow.

The Gödels crossed the border into Lithuania on the 16th. They entered Latvia the next day, and on the 18th their passports were stamped at Bigosovo, a rail junction near modern Druya on the Russian side of the Latvian border. From there — already behind schedule — they traveled to Moscow and thence over the trans-Siberian route via Novosibirsk, Chita, and Otpor (modern Zabaykalsk, on the border of what is now Manchuria but was then the Japanese puppet state of Manchukuo) to the terminus at Vladivostok.

Gödel left no account of that grueling journey, either in his correspondence or his private papers. Adele did briefly recall the event in conversations she had years later with her friend Elizabeth Glinka; but her only enduring memories were that they had traveled a great deal at night and were in constant fear of being stopped and turned back.[6]

◇ ◇ ◇

---

[6]Because of the short time that the German-Soviet non-aggression pact endured, relatively few émigrés escaped via the trans-Siberian route. One other mathematician who did so was the geometer, topologist, and group theorist Max Dehn.

By the time the Gödels arrived in Yokohama on 2 February the *Taft* had already sailed, a circumstance that forced them to remain there until the arrival of the *President Cleveland* on the 20th. The change in plans no doubt added to their anxiety, but the delay also provided a welcome respite from traveling. Gödel found the American hotel in which they stayed "very beautiful," and Adele took advantage of the opportunity to do some shopping. When they finally departed she reportedly took with her a "boxful" of purchases [367]. (There is no record of what belongings the two had brought along from Vienna, but they could not have taken much with them. After the war Gödel's brother shipped him a trunkload of his books and papers, including an assortment of bills, receipts, and memoranda of little value except to biographers.)

Yokohama was the second port of call for the *President Cleveland.* It began its voyage in Manila on 10 February and stopped first at Shanghai, where it took on most of its trans-Pacific passengers. It reached Honolulu eight days after leaving Yokohama, and docked at last in San Francisco on 4 March.

In accord with U.S. immigration regulations, the ship's master was obliged to submit a manifest listing the names, citizenship, and destination within the United States of all arriving passengers. The list also described the passengers' physical features, recorded the captain's judgment of their physical and mental condition, and gave their responses to a series of questions regarding their immigration status.

The name "Gödel" appears on the roster of second-class passengers, where it stands out amid a long list of Chinese names [368]. Kurt is described as being 5'7" tall, of light complexion and medium build, with brown hair and blue eyes, while Adele, whose maiden name is badly garbled, is listed as 5'2" in height, also of light complexion (apart from her birthmark), with light hair and gray eyes. The captain deemed both of them to be in good physical and mental health, and they both answered "no" in response to the questions "Have you ever been a patient in an institution for the care and treatment of the insane?" and "Whether coming by reasons of any offer ... to labor in the United States" — responses that, from an immigration standpoint, were truthful.[7] Both stated that they did not intend to return to the country from which they came, but expected instead to become permanent residents of the

---

[7]The intent of the latter question was to exclude "aliens under contract" and those "likely to become public charges." But "artists, musicians, teachers, and members of learned professions" were excepted from the category of "laborers," so in escaping the Catch-22 provisions of the immigration laws intellectuals "were doubly privileged" [369].

United States; they also indicated, however, that they did not intend to become U.S. citizens.

After landing Gödel still had to purchase the tickets for their transcontinental rail trip; but that evidently presented no difficulties, for on 5 March he wired the IAS that they expected to arrive in New York the following week.

En route during that final leg of the journey he and Adele could at last relax a bit more and enjoy the passing scenery. They had only a fleeting glimpse of the grandeur of the American West, but it left a lasting impression: In reply to a picture postcard that his mother sent him from Semmering a decade later, Gödel remarked that such beautiful landscapes existed in the United States only in its far western regions [370].

VIII
Years of Transition
(1940–1946)

AMONG THE FIRST to see Gödel after his arrival in Princeton was Oskar Morgenstern, who had made Gödel's acquaintance years before at meetings of the Schlick Circle. In the interim the two had had little contact, in part because of Gödel's breakdown and his trips to America. Morgenstern himself had come to the United States in January of 1938 to undertake a series of lectures as a Carnegie visiting professor. The *Anschluß* took place just two months later, and almost immediately thereafter Morgenstern was dismissed as director of the Austrian Institute for Economic Research. He wisely chose to remain in America and was fortunate to be offered a position at Princeton.

Morgenstern may have encountered Gödel briefly during the latter's visit to the IAS in the fall of 1938. At that time there was little opportunity for them to become reacquainted, but when Gödel returned in 1940 Morgenstern was eager to speak with him to learn more about conditions inside Austria. Morgenstern was taken aback, though, by the response he received: "Gödel," he reported, "has come from Vienna. ... In his mix of profundity and otherworldliness he is very droll. ... When questioned about Vienna, he replied 'The coffee is wretched'" [371].

Further surprises were in store for Morgenstern as he became better acquainted with his countryman over the next few months. He was astonished, for example, to learn that Gödel took an interest in ghosts, and he was very much dismayed by Gödel's choice of wife, whom he described as "a Viennese washerwoman type: garrulous, uncultured, [and] strong-willed." He foresaw how difficult it would be for Adele to gain acceptance in Princeton society, and he found it almost impossible to talk to Kurt when she was present. The attraction between the two remained "a riddle" to him, but he could not deny that Gödel seemed in better spirits in Adele's company than at any other time. She had "probably saved his life," he surmised; and then, too, he confided, Gödel himself was "a bit crazy" [372].

Immediately after their arrival the Gödels moved into an apartment at 245 Nassau Street. Evidently it was sparsely furnished, for in a letter to Veblen written the following July Gödel mentioned that his wife had "bought a surprisingly pretty rug for 10$ [sic], an excellent sewing machine for 25$ and a vacuum cleaner (new) for 30$" and said that "Next month she wants to buy a table and some chairs." He added that Adele had recently dismissed a Negro maid they had hired — a bad idea, in his opinion, because he thought "she should not overwork herself in the hot season and should use more of her time for other things, e.g. learning English" [373].

Such comments suggest that finances were not a serious problem for the couple, even though Gödel ended up receiving only half the stipend the IAS had originally offered him for the 1939–40 academic year. (At a meeting of the School of Mathematics in December 1939 it had been recommended that "in view of [Dr. Gödel's] enforced absence ... a sum of $1500 [be diverted] from the funds reserved for [him], ... to be used in assisting Drs. [Paul] Erdös and [Paul] Halmos.") The full stipend of $4,000 was paid to him during 1940–41 and was renewed for the following year [374].

The source of some of those funds was external. On 2 August 1940 Aydelotte wrote the Rosenwald Foundation, seeking support for a number of émigrés, including Gödel. He was referred by them to the Emergency Committee in Aid of Displaced Foreign Scholars, an organization founded in May 1933 to assist colleges and universities in providing temporary positions for refugee scholars of high professional standing. The response was prompt — perhaps in part because Oswald Veblen had served on the committee since its inception: On 24 December the committee's treasurer remitted a grant of $11,000 to the IAS to provide partial support for Gödel and nine others (Valentin Bargmann, Alfred Brauer, Paul Frankl, Felix Gilbert, Anton Raubitschek, Herbert Rosinski, Carl Siegel, and Kurt Weizmann). Gödel's share came to $1,222.22 [375].

Of more immediate concern to Gödel was his immigration status. On 25 April Aydelotte wrote both the German consul general in New York and the chargé d'affaires at the German embassy in Washington to ask their help in obtaining an extension of Gödel's leave of absence, which was due to expire on 31 July [376]. But when Gödel wrote to Veblen a week before the deadline he had still heard nothing about his application.[1] Nor, he said, had he "heard anything yet about [his] declaration of intention" to become a U.S. citizen [377].

From those remarks it appears that Gödel was shrewdly following two contrary courses of action. On the one hand, as noted earlier, he had stated on his arrival in San Francisco that he did not intend to apply for citizenship, no doubt realizing that to do so might compromise his request for an extension of his leave. On the other, once he had finished applying for that extension he wasted no time taking steps toward becoming an American; he and Adele received their first naturalization papers on 12 December 1940. Of course, his prospects at that time were uncertain in both countries. His position at the IAS was not yet a permanent one, and it was only on 28 June 1940 that his application to become *Dozent neuer Ordnung* at Vienna was finally approved. Thereafter, until the end of the war, he was listed in the *Personal- und Vorlesungs-Verzeichnis* of the University of Vienna as a *Dozent* who "would not lecture."

For all his acumen, however, Gödel made one mistake that was to be a source of considerable inconvenience over the next few years: Because their departure from Austria had occurred after the *Anschluß*, he and Adele had been issued German, rather than Austrian, passports, and that had led them to believe, erroneously, that the United States no longer recognized Austria as a separate country. When confronted by the Alien Registration Act of 1940 they had therefore registered as German citizens, and were consequently regarded as *enemy* aliens who had to carry certificates of identification and obtain permission if they wished to travel outside Princeton. The misunderstanding was not rectified before the spring of 1942.

Another matter that claimed some of Gödel's attention during his first months back in Princeton was the publication of his monograph on the consistency results. On 24 July Gödel wrote to Veblen (then on vacation in Maine) that George Brown, the note-taker for the lectures, was in Boston and that the notes themselves were "apparently still in the state of proof-reading," though Professor Tucker at Princeton had promised him that "the multigraphing would be finished ... most certainly by July 15." It is unclear when work on the notes was finally completed, but the volume did appear before the end of the year. Published by Princeton University Press as the third of the

---

[1]The wheels of the Austrian educational bureaucracy ground so slowly that an extension of Gödel's leave (to 31 July 1941) was not finally approved until 7 March 1941. It was then extended once more (on 21 June) to 1 December 1941. The relevant correspondence is preserved in Gödel's personal file in the *Dekanats-Bestand* of the philosophical faculty at the University of Vienna.

prestigious [378] *Annals of Mathematics Studies*, it has remained in print ever since.

While the proofreading was still in progress, Gödel once again lectured on his consistency results to an audience at the IAS. Later, on 15 November, he also presented them in an address to the mathematics colloquium at Brown University. On both occasions, as was his habit, he wrote out his remarks anew, and at the conclusion of the Brown lecture he expressed his belief that the negation of the Continuum Hypothesis, too, would turn out to be consistent with the axioms of set theory. Two reasons for believing so, he said, were that the inconsistency of that negation "would imply [the] inconsistency of the notion of a random sequence ... and would yield a proof for the axiom of choice," neither of which he deemed likely.

◇   ◇   ◇

Gödel had begun to seek a proof for the independence of the Continuum Hypothesis as early as the fall of 1937, and it had continued to engage his attention ever since. Nevertheless, according to the description of his activities given in IAS *Bulletin* no. 10, published at the end of the 1939–40 academic year, it was not the only item on his research agenda: More broadly, his plans for 1940–41 were to lecture on certain "problems related to the foundations of mathematics," and in the spring of 1941 he did just that. The problems in question were an outgrowth of the ideas he had first broached in his lecture at the Zilsel circle.

Once again, Gödel presented the results of his investigations to two different audiences: in a course of lectures at the IAS, and in a single lecture at Yale on 15 April (published posthumously as item *1941* in volume III of Gödel's *Collected Works*). The ideas did not appear in print, however, until 1958, and then only in very condensed and sketchy form.[2]

In the Yale lecture, entitled "In What Sense is Intuitionistic Logic Constructive?," Gödel began by distinguishing two kinds of objections raised by intuitionists against classical mathematics: first, against the use of "so-called impredicative definitions"; second, against "the law of excluded middle and related theorems of the propositional calculus." Contrary to appearances, Gödel argued, it is the first of these objections that carries greater weight. Indeed,

---

[2] In reply to a graduate student's inquiry many years later Gödel stated, "There were several reasons why I did not publish [those results] then. One was that my interest shifted to other problems, another that there was not too much interest in Hilbert's Program at that time" [379].

he remarked, "it can be shown quite generally that, as long as no impredicative definitions come in, every proof of classical mathematics becomes a correct intuitionistic proof if existence and disjunction are [re]defined" in an appropriate way.

That fact — a consequence, in Gödel's view, of the vagueness of the intuitionists' (necessarily) informal notion of "derivation" — cast doubt on the purity of the intuitionists' commitment to constructive methods. Gödel proposed instead to redefine the primitive notions of intuitionistic logic "in terms of strictly constructive ones." Accordingly, he stipulated, first, that all primitive functions were to be calculable and all primitive relations definable; second, that existential assertions were to be regarded "only as abbreviations for actual constructions"; and, third, that the negation of a universal proposition was to be understood as asserting the existence (in the sense just specified) of a counterexample.

He went on to outline how that could be done, "not for intuitionistic logic as a whole, but for its applications in . . . number theory," where impredicative definitions are not employed.

As the "concrete objects" of his system Σ Gödel took the natural numbers together with the functions of all finite types (functions of functions, and so on) that can be defined from them by means of either explicit or recursive definitions. He then illustrated by example how one could show that the axioms and rules of inference of intuitionistic logic are theorems of the system Σ, and he concluded by mentioning several applications of his interpretation. Among them were a proof of the consistency with intuitionistic logic of the principle $\neg(\forall A)(A \vee \neg A)$ (despite the inconsistency of the assumption $(\exists A)\neg(A \vee \neg A)$) and the reduction of the consistency of classical number theory to that of Σ.

◇　◇　◇

By the summer of 1941 Kurt and Adele were ready for an extended vacation. They left Princeton in early July and traveled to Brooklin, Maine, a resort community near Bar Harbor, probably on the recommendation of Veblen, who had often summered there. For the next two months they lodged at the Mountain Ash Inn, where the pleasant climate and delicious food greatly appealed to them. It was, Gödel wrote to Veblen, "really one of the most attractive places I have ever seen in my life." He was particularly struck by the lilac-colored flowers, which, he later recalled, were the same as those he had seen blooming profusely at Marienbad twenty years before [380]. Toward the end of their stay, the Gödels also visited Kleene at the home of his parents in nearby Hope, Maine [381].

No doubt the cool, clear air was one of the things that Gödel appreciated most about his visit to Maine. Indeed, it may have been a primary reason for his going there, for on their return he and Adele moved into an apartment at 3 Chambers Terrace — the first of three moves made, by his own admission, in an effort to escape "the bad air from the central heating" [382]. The effort, however, was apparently to no avail, for barely a month after the move Morgenstern complained that Gödel continually brought up the subject of "the gases": He was convinced that the heating unit emitted "smoke gases", and he had gotten rid of a bed they had just bought because after a few days he could no longer bear the smell of its wood and polish [383].

Such behavior led some of Gödel's colleagues to become seriously alarmed. Aydelotte, in particular, was worried enough to contact Gödel's physician, Dr. Max Gruenthal. "I am naturally a good deal concerned about Dr. Goedel's condition," he wrote,

> and should be most grateful if you would give me your general opinion of his case, ... especially whether you think it is all right for him to go ahead with his work ... and whether there is anything we could do here at the Institute to ease the mental tension from which he is evidently suffering.
>
> I should like also especially to know whether you consider that there is any danger of his malady taking a violent form which might involve his doing injury either to himself or to other[s] ... [384].

Dr. Gruenthal replied that while he could not give a detailed report on Gödel's condition without his consent, he could assure Aydelotte that he saw no "acute danger of his malady taking a violent form." But how, he asked, had it become evident that Gödel was "suffering from a 'mental tension'"?

Aydelotte then explained that "the evidence we have had here of Dr. Goedel's difficulties comes from the fact that he thinks the radiators and ice box in his apartment give off some kind of poison gas. He has accordingly had them removed, which makes the apartment a pretty uncomfortable place in the winter time." However, he continued, "[He] seems to have no such distrust of the heating plant at the Institute and ... carries on his work here very successfully" [385].

Since there were indications that Gödel's condition seemed to be improving, Aydelotte decided to do nothing further beyond urging him to see his doctor more frequently; and, as Gruenthal had predicted, no violent actions ensued. But Gödel's obsession with fresh air and his fears about his refrigerator remained conspicuous throughout much of the rest of his life, and it seems certain that doubts about his mental stability lay behind the institute's reluctance to grant him a permanent position.[3]

In fact, the institute continued to support Gödel on a year-to-year basis until 1946. The impetus finally to appoint him as a permanent member appears to have come from von Neumann, among whose papers at the Library of Congress is a cryptic note from Veblen, dated 5 December 1945, containing the remark "Many thanks for your action and letter about Gödel. I took it up with Aydelotte, and I think there is a chance that we may be able to put that business through at once." What "that business" was is made clear by the minutes of a meeting that took place at the IAS thirteen days later: "The Director recommended that Professor Kurt Gödel be made a permanent member with a stipend of $6000, plus a contribution to [his] retiring allowance to be matched by him sufficient to give him a pension of $1500 at the age of 65." There was, however, the proviso "that the terms of the appointment ... be such that if for reasons of physical [or mental?] disability Professor Gödel should at any time be unable to perform his duties he should be retired on a pension of $1500" [387].

◇   ◇   ◇

From the spring of 1941 until the fall of 1946 Gödel did no more lecturing. Instead, he devoted himself to further attempts to prove the independence of the Axiom of Choice and the Continuum Hypothesis, and to renewed study of the life and work of Leibniz.

Within a year after the Brown lecture Morgenstern reported that Gödel was "making good progress with his proof for the independence of the Continuum Hypothesis" and thought he might have it ready "in a few months" [388]. Ultimately that optimism proved unfounded, but for a time, at least, it spurred him on: The (few) dated entries in volumes 14 and 15 of his *Arbeitshefte* leave no doubt that during 1942 he concentrated most of his efforts on the independence questions.

---

[3] As one example, in June of 1941 Aydelotte included Gödel's name on a list of scholars who had been accepted by the IAS for the coming year but who were "still seeking posts." The list was sent to the Emergency Committee, which provided no further support for Gödel but did forward information about him to the chairman of the mathematics department at the University of Wyoming [386].

In the summer of that year he and Adele returned to Maine, this time to Blue Hill House (now the Blue Hill Inn) in the community of that same name. And there too he continued to work intensively [389].

Years later Mrs. Louise Frederick, the manager of Blue Hill House at the time of Gödel's visit, shared her vivid memories of that occasion [390]. Throughout his stay, she recalled, Gödel was "unremittingly taciturn and dour," his countenance that "of a man lost in thought."

> During the daylight hours [he] spent most of his time in his room . . ., [where] he could remain out of sight all day long. Adele did the beds herself, and did not even allow the staff to enter the room. [He] did most of his thinking on long walks . . . at night, leaving the inn at sunset and returning after midnight. He walked with his hands behind his back, leaning forward, looking down, . . . usually . . . along the Parker Point Road, a narrow road through the pine forest along the coast, abutting some of the wealthiest homes in the area.

By then, of course, the United States was at war, so Gödel's nocturnal perambulations aroused suspicions. According to Mrs. Frederick, many residents thought "this scowling man with a thick German accent walking alone at night along the shore . . . [must be] a German spy, trying to signal ships and submarines in the bay"![4]

Gödel himself left no record of his impressions of Blue Hill, probably because he was so absorbed in his work. Apparently he hardly socialized with the other guests there at all: Frederick recalled that she never saw the Gödels in the inn's common room, and "although [they both] came to the dining room for meals, they virtually never ate." They left Maine at the end of the summer, never to return. Afterward Gödel twice wrote to Frederick, "accusing her of stealing his trunk key."

Much later, Gödel did offer an assessment of his accomplishments that year. Contrary to persistent rumors, he had not, he said, succeeded in establishing the independence of the Axiom of Choice with respect to full Zermelo-Fraenkel set theory. He had only obtained "certain partial results, namely, proofs for the independence of the axioms of constructibility and choice in type theory," and, he declared, "on the basis of my very incomplete memoranda from that time (i.e., 1942) I could, without difficulty, reconstruct only the first of those two proofs" [391].

---

[4] On the contrary, Gödel once told his colleague Atle Selberg that after the war he had volunteered for duty as a civil defense aircraft spotter.

◇ ◇ ◇

On their return to Princeton the Gödels moved again, this time to an apartment at 108 Stockton Street, on the grounds of what had been the Hun preparatory school. Among their neighbors there, by coincidence, was George Brown, who had returned from Boston to accept a position as a research associate. He had recently married, and his wife, Bobbie, had been engaged to draft the illustrations for von Neumann and Morgenstern's *Theory of Games and Economic Behavior*, then in preparation.

According to the Browns [392], the Gödels were "not very social." Indeed, more than once they noted that Kurt "avoided leaving [his] apartment when certain foreign visitors came to town" — apparently because he feared they might try to kill him[5]. The Gödels did invite the Browns in, but because Kurt "insisted that the screens be left out of the windows" in order for him to breathe properly, the rooms in their apartment were opened to dust, drafts, and insects — a circumstance that discouraged Mrs. Brown from wanting to go there.

Nevertheless, the two couples became fairly well acquainted during the few months they lived nearby. Mrs. Brown (now Mrs. Dorothy Paris) recalled that Adele was extremely lonely, in part because of the wartime interruption in communications between the United States and Austria, and was "rueful that [she] had no child." Since Brown herself was then both busy and pregnant, there was ample reason for Adele to be envious of her. But when the baby arrived Adele "was most attentive and generous of her time and help" [394].

There were some aspects of Adele's behavior, though, that struck the Browns as "more than a bit strange." There was, for example, her fascination with the Browns' pet kittens, which were tailless (presumably part Manx) and possessed an unusual body structure. On seeing them Adele at once conceived a desire to get such a pair of kittens for herself. Manx cats, however, were not then available, so she had to settle for the ordinary kind — after which Mrs. Brown had a hard time persuading her not to amputate their tails! Only by stressing how essential tails are for balance in normal cats was Adele finally talked out of the idea [395].

Over the years the Gödels had a number of other pets. Those mentioned in the family correspondence include a cocker spaniel — unfortunately hit

---

[5] Brown stated that he had "deliberately banished from memory any particular names, usually those of well-known mathematicians." According to Paul Erdös [393], one such was the topologist Eduard Čech, who had been a professor in Brno from 1923 to 1939.

by a car and killed after only about a year — a pair of lovebirds, and a dog named Penny, acquired in 1953 and still "fresh and lively as ever" twelve years later. They were a source of enjoyment for both Kurt and Adele and no doubt helped to combat her loneliness, as well as to provide her with living creatures on which she might lavish her affection, in lieu of the child for which she longed.

Why the Gödels remained childless is a matter for speculation. To Louise Frederick it seemed obvious that Kurt simply "didn't have the strength to make a baby!," but Adele's age (she was nearly forty at the time of her second marriage) was likely an important factor. In addition, there are indications that the two practiced contraception. Rudolf Gödel thought his brother might have been averse to having children because he was concerned about the prevalence of cancer in Adele's family [396], while Adele once told a friend that a doctor had advised her that Kurt's mental troubles might be hereditary. She also confided that during the war the two of them had supported a foreign orphan through one of the international relief organizations. After the war they were given the opportunity to adopt the girl, sight unseen, but Kurt had objected that only a blood relative should bear the Gödel name [397].

Whatever the real reason may have been, Adele was forced to sublimate her maternal feelings. She did so in several ways besides those already mentioned. A talented seamstress, she is said, for example, to have made a child's dress each day of the war for donation to the relief effort. Her contributions were so highly valued that after the war the Viennese gave her a bust of her father (who had died, out of touch with her, during the war years) as a token of appreciation [398].

◇ ◇ ◇

The frustration stemming from his inability to better the results he had obtained during that "Summer of '42" eventually led Gödel to abandon work on the independence problem. He turned instead to philosophy, and just at that juncture — fortuitously, so it would seem — Professor Paul Arthur Schilpp, editor of the distinguished *Library of Living Philosophers* series, invited him to contribute an essay on Russell's mathematical logic to a forthcoming volume on *The Philosophy of Bertrand Russell.*

When Schilpp tendered the invitation in November of 1942 he was probably unaware of Gödel's failed collaboration with Heyting eight years earlier. Had he known of it, his pleasure at securing Gödel's acceptance might have been tempered by doubts about what lay ahead. In fact, however, Gödel delivered a draft manuscript just six months later.

A protracted editorial correspondence then ensued, reminiscent of Gödel's exchanges with Neugebauer. As always, Gödel was exacting and could not be hurried. He promised Schilpp that he would submit a revised manuscript before leaving for vacation at the Hotel St. Charles in Seaside Heights, New Jersey (to which he returned each of the following two summers as well); but he did not in fact submit the final version until 28 September. The reasons for the delay, as he explained to Schilpp, were that his own health had been bad, his wife had just undergone surgery, and in the midst of her hospitalization he had had to move to another apartment (at 120 Alexander Street, where they would remain until their purchase of a home six years later) [399]. In addition, earlier that summer he had had to undergo his examination for American military service.

The format for the *Library of Living Philosophers* series called for the subject of each volume to reply to the essays in it. Gödel looked forward to Russell's response, but by the time he submitted his revised commentary Russell had already completed his replies to the other essays and felt he had "no [further] leisure" to devote to the volume. Gödel did his best to persuade him otherwise, but in the end Russell contributed only a brief remark about Gödel's article, in which he admitted that it had been some "eighteen years since [he had] last worked on mathematical logic" and that it would therefore "have taken [him] a long time to form a critical estimate of Dr. Gödel's opinions."

In fact, Russell had spent some time in Princeton that very spring, while Gödel was in the process of composing his critique. He had lectured at the IAS and Gödel had been in the audience; but according to Morgenstern, Russell was then so out of touch with mathematics that he had never heard of such luminaries as von Neumann and Siegel, both of whom were then at the institute.

Russell mentioned his stay in Princeton in the second volume of his *Autobiography*, where he remarked that he "used to go to [Einstein's] house once a week" for discussions with him, Gödel, and Pauli[6]. There is no evidence, however, that the conversation ever turned to Gödel's essay; indeed, Russell characterized the discussions as "in some ways disappointing" since "although all three of them were Jews and exiles and, in intention, cosmopolitans ... they all had a German bias toward metaphysics." Moreover, he said, "Gödel turned out to be an unadulterated Platonist" [401].

---

[6] In the draft of a letter to Kenneth Blackwell, curator of the Russell archives at McMaster University, Gödel later said that he recalled only one such occasion; he also disputed Russell's other claims about him [400].

Of course, Gödel was not a Jew. He *had* become a Platonist; but, as he stressed in his essay, Russell too had made "pronouncedly realistic" assertions. In his *Introduction to Mathematical Philosophy*, for example, Russell had declared that "Logic is concerned with the real world just as truly as zoology, though with its more abstract and general features."

Elsewhere, Gödel noted, Russell had "compared the axioms of logic and mathematics with the laws of nature, and logical evidence with sense perception" — a parallel that he himself wholeheartedly endorsed. Indeed, Gödel asserted, "Classes and concepts may ... also be conceived as real objects," and the assumption of their existence "is quite as legitimate as the assumption of physical bodies ... [since] they are in the same sense necessary to obtain a satisfactory system of mathematics as physical bodies are necessary for a satisfactory theory of our sense perceptions." Moreover, Gödel suggested, "axioms need not necessarily be evident in themselves." Rather, "their justification lies (exactly as in physics) in the fact that they make it possible for ... 'sense perceptions' to be deduced" [402]. (Compare Russell: "We tend to believe the premises because we can see that their consequences are true, instead of believing the consequences because we know the premises are true" [403].) Three years later Gödel would make the same point again, even more forcefully, in his essay "What is Cantor's Continuum Problem?"

Taken as a whole, Gödel's 1944 essay is an intricately interwoven tapestry (what Hermann Weyl, in his review *1946*, described as "a delicate pattern of partly disconnected, partly interrelated, critical remarks and suggestions") that served as much to promulgate his own philosophical views as it did to analyze Russell's. A detailed discussion of its contents would be out of place here (see, e.g., *Parsons 1990* for further commentary), but two further points are worth noting.

In line with the emphasis in his Yale lecture on the problematic role of impredicative definitions, Gödel devoted considerable space in his 1944 essay to a critique of Russell's "vicious-circle principle," according to which "no totality can contain members definable only in terms of [that] totality." Under the strictest interpretation, Gödel argued, that principle would rule out impredicative definitions altogether, thereby rendering impossible both "the derivation of mathematics from logic" (a principal aim of *Principia Mathematica*) and "a good deal of modern mathematics itself" — a circumstance that Gödel took "rather as a proof that the ... principle is false than that classical mathematics is."

Many years later the vicious-circle principle and Gödel's remarks about it were taken up in a book by Charles Chihara [404], one of whose chapters, entitled "Gödel's Ontological Platonism," contains some unwittingly apt remarks. At the outset Chihara distinguishes "ontological" Platonists such as Gödel ("who believe actually that there are such objects as sets") from what he calls "mythological" Platonists (who "hold that mathematicians construct their systems *as if* they were describing existing objects").[7] He then argues that "even if we ... agree with Gödel that no ... nominalistic reduction of mathematics is possible, [that] by itself does not commit us to ontological Platonism," for "one is not tempted to say that we ought to believe in the existence of ghosts just because some theory of ghosts ... demands the[ir] existence." Furthermore, he protests, "Gödel's reasoning seems to lead to a massive population explosion of our ontology: if we use mathematical intuitions to postulate mathematical objects, it would seem that we could use 'theological intuitions' to postulate ... objects like angels." Ironically, though, Gödel *did* believe in the existence of ghosts — quite apart from any theory postulating their existence — and it seems unlikely that he would have had qualms about the validity of theological intuitions, either.

Later on in the chapter Chihara pursues the related question of how "a systematic philosopher" might attempt to axiomatize the notion of "a perfect being ... call[ed] 'God'". He imagines that such a philosopher might think it impossible that "a perfect being could be ... created by some other thing" and that he might "express this 'insight' by saying that *God is a necessary being*," an axiom he could say had "forced itself on him as being true" (a reference to one of Gödel's statements about set theory in his paper *1964*). Even so, Chihara suggests, the philosopher might not be a believer but merely one who found that the necessary existence of God "forced itself upon him as being *true to* [that particular] *concept of God*" [405]. Here again the suggestion is ironic since at that very time Gödel was attempting just the sort of axiomatization Chihara had envisioned — ostensibly as a purely formal exercise in modal logic.

The final paragraph of Gödel's essay on Russell's logic also merits special attention. There Gödel deemed the "incomplete understanding of the foundations" to have been the reason why "mathematical logic has up to now remained so far behind the expectations of Peano and others" — especially Leibniz's vision that logic "would facilitate theoretical mathematics to the same

---

[7] Gödel did not broach the idea of "as if" Platonism in his Russell essay, but he had in fact studied Hans Vaihinger's *Die Philosophie des Als Ob* (The Philosophy of the "As If") and taken detailed notes on it.

extent as the decimal system of numbers has facilitated numerical computations." Nevertheless he saw "no need to give up hope," for in his view the
Characteristica universalis that Leibniz had outlined was not "a utopian project"
but a calculus that, "if we are to believe his [own] words," Leibniz "had [al-
ready] developed ... to a large extent."

Gödel went on to assert that Leibniz had intentionally withheld publication
of his *Calculus ratiocinator* because he "was waiting ... till the seed could fall
on fertile ground." Privately, though, Gödel harbored graver suspicions: As
already noted, while at Notre Dame he had told Menger that he believed
publication of some of Leibniz's works had been suppressed by a hostile conspiracy.

That the study of Leibniz was the primary focus of Gödel's attention during
the years 1944–45 is attested both by IAS *Bulletin* nos. 11 and 12 and by entries
in Morgenstern's diary. To Morgenstern Gödel reiterated his conviction that
Leibniz had been "systematically sabotaged by his editors," and he also made
several extraordinary claims about what Leibniz had done. Morgenstern was
particularly surprised — and not a little dubious — when Gödel told him he
had found various declarations by Leibniz about the scientific importance of
developing a theory of games; and his doubts turned to outright disbelief when
Gödel claimed further that Leibniz had already discovered the antinomies of
set theory ("cloaked in the language of concepts, but exactly the same") and
had anticipated both Helmholtz's resonance theory of hearing and the law of
conservation of energy [406].

Morgenstern discussed Gödel's "fantasies" with Menger, who was equally
skeptical of them. But he also told Menger of a strange incident that had taken
place:

> Gödel took him one day into the Princeton University Library and gath-
> ered together an abundance ... of material [for comparison]: *firstly* books
> and articles which [had] appeared during or shortly after Leibniz's life
> containing exact references to [his] writings. ... *Secondly* those very col-
> lections or series [to which the references had been made]. ... In some
> cases neither in the cited pages nor elsewhere was any writing of Leibniz
> to be found, whereas in other cases the series broke off just before the
> cited [passages] ... or [else] the volumes containing the cited writings
> never appeared [407].

Morgenstern agreed with Menger that Gödel was too much alone and that
regular teaching duties would be good for him. Yet however absurd many of
Gödel's ideas seemed to Morgenstern, the library demonstration struck him as

inexplicable and "highly astonishing" [408]; and when relations with European institutions were reestablished after the war, Morgenstern collaborated with Gödel in attempts to have copies of Leibniz's manuscripts brought to the United States (see chapter IX).

◇ ◇ ◇

Communications between persons in America and Austria were restored some years before that. Mail was still sometimes lost, however, and censorship remained in effect throughout the Allied occupation. For a time, therefore, Gödel and his mother numbered their letters to each other.

Their correspondence resumed with her letter of 9 June 1945, to which Gödel replied on 7 September, following his return from vacation at the seashore. As was to be expected, the news from Vienna was a mixture of good and bad. Marianne herself had survived unharmed, both in Brno (from which she had escaped prior to the Czechs' expulsion of the Sudeten Germans) and Vienna (where she and Rudolf had lived through the Allied bombings). But part of the villa had been confiscated by the Czech state, many of Gödel's other relatives had been forced to resettle in Germany, and his godfather, Redlich, had perished in a Nazi gas chamber.

Reestablishing contact with Adele's family proved more difficult. From Kurt's brother she learned that her father had died during the interim, and not long after word came of the deaths of three other family members as well. Her mother was reported to be safe, but Adele worried a great deal about how well she was being cared for, especially in view of the perilous economic conditions that prevailed in Vienna during the immediate postwar period. She desired desperately to revisit Vienna and assess the situation for herself, but until the spring of 1947 she was not able to do so. In the meantime she and Kurt did all they could to render assistance to their families by sending food, clothing, and other staples to them via the Red Cross, CARE, and other organizations.

How effective their efforts were — and indeed, just how bad life in Vienna really was then for the Gödel and Porkert families — is difficult to judge, in part because only Gödel's side of the correspondence has been preserved, and also because detailed references to the extent of the privations in Vienna were rigorously censored. The letters do, however, provide insight into Kurt and Adele's own lives.

The picture that emerges from them is one of contrast between his growing contentment with their life in Princeton and her mounting unhappiness. He, for example, was pleased with their Alexander Street apartment, which was located near the train station (a plus, in his estimation), "directly opposite the most elegant hotel in town" (the Princeton Inn). It occupied the entire upper floor of the building and had windows on all sides, which made it possible to endure the "sweltering heat" of the summers in Princeton. Adele, however, "complained about everything": about hygienic conditions, about the physical condition of the apartment, and about their not living in a better neighborhood. Above all, despite their proximity to New York City, she hated the idea of living in a small town [409].

The contrast extended also — at least ostensibly — to matters of health. In the second of his letters to his mother, written 22 January 1946, Gödel described his own health as "quite good." He conceded that his stomach was "somewhat worse than in Vienna," but added reassuringly that walking to and from the institute gave him adequate exercise. He worried only that he might be eating too much — a matter that was a source of "constant disagreement" between him and Adele. Nevertheless he reported that it was she who had lost weight, mostly from the strain of worrying about her family. (Late in 1945 she also underwent an appendectomy, and the following summer all her remaining lower teeth were extracted and replaced by a denture.)

With his physician brother he was more honest: Whenever he ate "somewhat more than usual (... 'the usual' ... [being] not much)", especially of highly spiced foods, meat that was too rare, or coffee or tea that was too strong, he experienced mild stomach pains "high up in the right rear." In addition he suffered from "persistent constipation," for which he "constantly" resorted to laxatives such as milk of magnesia, ExLax, and Imbricol. Indeed, in 1946 he began to keep a daily record of his laxative consumption — a record that he maintained for the next thirty years and that occupies five folders in his *Nachlaß.* Consequently he was always underweight. "In the last few years," he confessed, "I have *never* weighed more than *54 kg*" (about 120 pounds) [410].

Gödel thought that a return to Vienna for a few months might help to lessen Adele's dissatisfaction with conditions in Princeton. He declared, however, that he was happy in America and would not consider going back to Vienna to live, even if he were offered a position there. The land and the people in Princeton were "ten times more congenial," he said, and the "prompt functioning of government officials in America" made life there "$10 \times 10 \times ...$" better than in Austria [411].

Nonetheless he was not uncritical of America's foreign and domestic policies. To his mother's opinion that it was fortunate Austria's fate had been left in American hands he added the qualification "Roosevelt's America." He regarded Roosevelt's death as "one of the most depressing events of the 20th century" and thought its circumstances and timing "suspicious." Perhaps, he declared, the president *did* die of natural causes; but if so it seemed that a "secret power" opposed the furtherance of Roosevelt's plans. In any case, he averred, the deleterious effects of the Republican landslide in the 1946 elections were already becoming evident in everyday affairs — movies, for example, had become "decidedly worse"!

More rationally, Gödel also objected to the growing secrecy surrounding scientific work ("principally on account of the atom bombs") [412]. Such secrecy was manifest, he believed, even in arrangements for the upcoming Princeton Bicentennial Conference on Problems of Mathematics, at which he had been invited to speak.

On its face the charge seems absurd since the conference was a large one (more than ninety participants) attended by mathematicians from eight foreign countries. It is true, though, that no proceedings of the conference ever appeared. The only publication chronicling the event as a whole was a small pamphlet of limited circulation (*Princeton University 1947*), which contained brief overviews of the nine sessions,[8] without technical details.

The session on logic took place on the first day of the conference (17 December 1946) and focused on the single broad topic of decision problems [413]. It was chaired by Church, with Tarski serving as discussion leader and J. C. C. McKinsey as reporter. As reported in the official summary, Tarski spoke first, apparently at some length. His still unpublished talk "surveyed the status of the decision problem" in various areas within logic and "pointed out open, important problems." More specifically, according to a letter he wrote to Gödel a week before the session, he aimed to draw attention to the contrast between the relatively undecidable statements of arithmetic and the apparently absolutely undecidable statements of set theory and to pose two questions: first, whether the distinction between relative and absolute undecidability could be made precise, and second, "whether, on the basis of some adequate definition of those notions, it will be possible to show that a number-theoretical problem can be undecidable only in a relative sense."

Gödel's own much briefer remarks (*1946*) began with a reference to Tarski, who had, Gödel said, justly "stressed . . . the great importance of the concept

---

[8] On algebra, algebraic geometry, differential geometry, mathematical logic, topology, "new fields" (mostly areas of applied mathematics), mathematical probability, analysis, and analysis in the large.

of general recursiveness," the first instance in which it had proved possible to give "an absolute definition of an interesting epistemological notion" (a definition, that is, "not dependent on the formalism chosen"). That the diagonal procedure did not lead outside that concept was, Gödel declared, "a kind of miracle" that should "encourage [us] to expect the same thing to be possible ... in other cases," too. He proposed two possible candidates, the second of which — that of definability in terms of ordinals — he had studied in sufficient detail to venture a few "more definite" assertions.[9]

He recognized that some might object to taking all the ordinals as primitive symbols of a language, since "it has some plausibility that all things conceivable by us are denumerable." Nonetheless, he argued, there was "much to be said in favor" of the idea; in particular, though the formal notion of ordinal definability might not capture the informal one of "comprehensibility by our mind," it did, he believed, provide "an adequate formulation in an absolute sense ... of [a set's] 'being formed according to a law' ... [rather than] by a random choice of ... elements." In addition, he predicted that the sets definable in terms of ordinals would satisfy all the axioms of set theory, including the Axiom of Choice, and thereby provide "another, and probably simpler, proof" of the latter's (relative) consistency. He doubted, however, that they could also be proved to satisfy the Continuum Hypothesis.

Gödel's proposal to countenance the use of non-denumerable languages reportedly "led to a spirited discussion" between him and Church, "centered on ... what could reasonably be called a 'proof' and when [one] could 'reasonably' doubt a proof" [414]. Sadly, however, no detailed record of that debate is extant.

Whatever impression Gödel's remarks made at the time, they were subsequently forgotten or ignored for nearly twenty years. A typescript of his talk was prepared and a copy of it sent to McKinsey for inclusion in a proposed volume of conference proceedings, but the volume never materialized and Gödel apparently made no attempt to publicize his ideas elsewhere. Perhaps he still felt constrained by the "secrecy" that he thought had surrounded the conference. In any case, the typescript appears to have circulated only among a small group of conference participants. It did not appear in print until 1965, after Kleene brought the piece to the attention of Martin Davis and suggested that he include it in his anthology *The Undecidable.*

---

[9] His first candidate was that of demonstrability on the basis of stronger and stronger axioms of infinity, for which he though it "not impossible that ... some completeness theorem would hold."

Two years before that Paul J. Cohen had at last proved the independence of the Continuum Hypothesis and thereby sparked a resurgence of interest in set theory. There followed a period of intense research effort, in the course of which the concept of ordinal definable set was repeatedly rediscovered. Among those who did so were John Myhill and Dana Scott (jointly) and a group in Czechoslovakia led by Petr Vopěnka.[10]

Each of those efforts had been carried out independently, and none of the results obtained had been published prior to 1965. Indeed, not until 1971 — a quarter century after the conference in Princeton — did detailed proofs of Gödel's conjectures finally appear in print [415]: First, Myhill and Scott formally defined the class OD of ordinal-definable sets (as $\{x:$ For some ordinal α, x is definable over $R(\alpha)$ by a formula of set theory with one free variable and no parameters$\}$) and showed that it was absolute (in the sense that definitions over OD by means of first-order formulas do not lead outside the class) and possessed a definable well-ordering. They then used those facts to show that the class HOD of *hereditarily* ordinal-definable sets (those in OD all of whose ancestors under the $\in$-relation are also in OD) satisfies all the axioms of ZFC, and thus that the Axiom of Choice is relatively consistent with the axioms of ZF set theory. Second, Kenneth McAloon used Cohen's methods to show that $2^{\aleph_0} \neq \aleph_1$ is consistent with the axioms of ZF together with the assumption that every set is ordinal definable.

Gödel himself never returned to the subject of ordinal definability. Until Cohen's breakthrough his attention had turned increasingly away from set theory back to his earlier interests in physics and philosophy.

---

[10] Later it was discovered that Emil Post had done so, too, some years before. He died in 1954, unaware that he had once again been anticipated by Gödel.

IX
Philosophy and Cosmology

(1946–1951)

Time present and time past
Are both perhaps present in time future,
And time future contained in time past.
If all time is eternally present
All time is unredeemable.

— T. S. Eliot, "Burnt Norton" [416]

ALTHOUGH GÖDEL'S APPEARANCE at the Princeton Bicentennial Conference was the most conspicuous event in his life during 1946, there is little to suggest that his brief remarks there reflect work to which he had devoted a great deal of time and effort; the ideas are not fully worked out, the typescript appears to have been composed on the basis of a single preliminary draft, and the topics in question take up little space in his workbooks. It seems rather that those remarks were a digression from two more important endeavors to which he had committed himself earlier that year: a request from Lester R. Ford, editor of the *American Mathematical Monthly*, for an expository article on the continuum problem, and a second invitation from Schilpp, this time for a contribution to a volume on Einstein.

As envisioned by Ford, the *Monthly* article was to be a "formula piece", one of a series of "What Is ...?" papers whose purpose was to expound, "in as simple, elementary and popular a way" as possible, certain "small aspect[s] of higher mathematics." Accordingly, he advised that it should "not be a long paper and its writing ought not to take a great deal of time" [417].

Gödel, however, took the commission much more seriously. For him the paper was not merely a popular exposition of a topic to which he had made an important contribution — though it was certainly a superb example of that — but another opportunity to propound his philosophical views. Its

173

composition therefore demanded the same meticulous care and attention to detail (for example, thirty-five footnotes in the space of ten pages) that he was wont to devote to his research contributions.

Ford had hoped to have the paper by July of 1946, four months after Gödel agreed to write it. But when his editorship ended the following December he still had not received it. Indeed, Gödel did not finally submit the manuscript until near the end of May the following year. He had produced a typed draft by the middle of August, but, as usual, he had subsequently "found some insertions desirable."

In its format, "What is Cantor's Continuum Problem?" followed the model for the series of which it was a part. In the first two sections Gödel reviewed the concept of cardinal number, stated Cantor's Continuum Hypothesis in its historically correct form ("every infinite subset of the continuum has either the power of the set of integers or of the whole continuum"), and gave a succinct summary of what little was then known about the power of the continuum. In the last two sections he then attempted to account for the "scarcity of results." To some extent, he granted, the latter might be ascribed to "purely mathematical difficulties," but he thought that there were "deeper reasons" as well, involving difficulties that called for "a more profound analysis" of the meanings of such terms as "set" and "one-to-one correspondence."

After briefly disposing of the "negative attitude[s] towards Cantor's set theory" expressed by Brouwer, Weyl, and Poincaré, Gödel maintained that regardless of one's philosophical standpoint, Cantor's continuum problem retained a precise meaning — namely, whether the power of the continuum could be formally deduced from a given set of axioms. As in the Brown lecture, he conjectured that the question would turn out to be undecidable on the basis of the axioms so far considered. Yet he also insisted that "even if [some]one should succeed in proving [its] undemonstrability" (as Cohen later did), that would "by no means settle the question" for someone like himself who believed that "the concepts and axioms of classical set theory ... describe some well-determined reality." For such a person, a proof of undecidability would show only that the presently accepted axioms did not provide "a complete description" of that reality. By augmenting them, a solution might still be obtained.

But how to do so? How were new axioms — statements expressing *true* properties of the "well-determined reality" of sets — to be found? Gödel suggested that "a more profound understanding of the concepts underlying logic and mathematics" — entities that in his view constituted a Platonic universe of their own — might "enable us to recognize" certain "hitherto unknown" axioms.

Such concepts would serve to clarify the very *nature* of sets. Thus, he pointed out, those who understood sets to exist only "in the sense of extensions of definable properties" might regard his Axiom of Constructibility as a true statement that ought to be adjoined to the axioms of ZFC. If so, then the proofs he had already given in his monograph *1940* would establish not just the relative *consistency* of the Continuum Hypothesis, but its *truth*.

If, however, one conceived of sets "in the sense of arbitrary multitudes" (as he himself apparently did), then the question would remain open; in that case, he conjectured, the Continuum Hypothesis would turn out to be false, because it seemed to him to have certain "paradoxical" consequences.

In line with his remarks in *1946*, Gödel suggested that to settle the continuum problem in this broader sense it might be fruitful to consider certain strong "axioms of infinity" (principles asserting the existence of sets having very large cardinalities), since the "weak" large cardinal axioms studied up to that time had already been shown to have consequences "far outside the domain of very great transfinite numbers." Moreover, he asserted that even if some new axiom "had no intrinsic necessity at all," its truth might come to be accepted inductively, on the basis of its " 'verifiable' consequences" (those "demonstrable without the new axiom, whose proofs by means of the new axiom are considerably simpler and easier to discover"). Indeed, he declared, "There might exist axioms so abundant in their verifiable consequences, shedding so much light upon a whole discipline, and furnishing such powerful methods for solving given problems ... that they would have to be assumed ... in the same sense as any well-established physical theory" [418].

In a letter to his mother written just as his essay was about to appear, Gödel described it as "an exposé of no very great significance" — a characterization that clearly belied the effort he had put into it. In fact, the essay was a manifesto for mathematical Platonism, and as such was to prove quite influential.

◇ ◇ ◇

Toward the end of May 1946, while Gödel was still finishing his draft for the *Monthly*, Schilpp paid a brief visit to the IAS and tendered his invitation. Unlike Ford, Schilpp did not initially propose a topic for Gödel's contribution (as he had for the Russell volume).[1] He was probably aware of Gödel's close

---

[1] Some months later he suggested "The Realistic Standpoint in Physics and Mathematics." But Gödel had other ideas.

friendship with Einstein and may have expected that he would offer to write something in the way of a personal tribute or memoir.

By Gödel's own account, he had first met Einstein in 1933; they were introduced by Paul Oppenheim, a fellow refugee who lived in Princeton and was a good friend not only of Einstein but of many other prominent intellectuals, including Morgenstern, Russell, Hans Reichenbach, and Carl Hempel [419]. But Gödel did not come to know Einstein well until about 1942. The attraction between them was puzzling to many, for, as Ernst Straus has recalled, Einstein and Gödel "were very different in almost every personal way": Einstein was "gregarious, happy, full of laughter and common sense" while Gödel was "extremely solemn, very serious, quite solitary, and distrustful of common sense as a means of arriving at the truth." Yet "for some reason they understood each other well and appreciated each other enormously" [420]. In later years they were often seen together, especially during their noontime strolls to and from the institute — occasions when they engaged primarily in discussions of philosophy, physics, and politics [421].

Of course, one reason for the attraction between the two was that both possessed intellects of the highest order. Another, noted by several informants, was that Einstein recognized Gödel's need for someone to look after him and was willing to serve as his protector (as Veblen had during Gödel's early years in Princeton and as Morgenstern would in the years after Einstein's death). Gödel himself, however, offered a different explanation: In reply to a query from Carl Seelig, one of Einstein's biographers, he wrote that he had "often thought about why Einstein took pleasure in conversing with me" and had concluded that it was partly because "I frequently held an opinion counter to his and made no attempt to conceal my disagreement." In particular, he said, he had always been very skeptical about the idea of a unified field theory [422].

Despite their differences, the two shared certain attitudes toward the world and their work. Both rejected the idea of indeterminacy or chaos in physics.[2] Both "went directly and wholeheartedly to ... questions at the very center of things" [424]. Both, in their later years, came to be regarded as outsiders, engaged, as Freeman Dyson has put it, in "unfashionable pursuits." And both, as young men, had to choose between mathematics and physics. They

---

[2]Einstein's opposition to quantum mechanics and his faith in rational inquiry, as expressed, for example, in his famous dictum "God may be subtle, but He isn't malicious," are well known. As for Gödel, though he discussed quantum mechanics with von Neumann and was apparently interested in its underlying logic, Straus has spoken of his "interesting axiom," according to which "nothing that happens" in our world "is due to accident or stupidity" [423].

did so, of course, in opposite directions: Gödel turned away from physics (so Morgenstern reported) because he found it logically incoherent, whereas Einstein, confronted with "many beautiful questions" in mathematics that all seemed equally important, could "never ... decide which were central and which peripheral," and so turned to physics, where he had a profound sense for what the important problems were [425].

It may be that the ultimate basis for the bond between the two great men was their complementary outlook. But in any case their friendship had little to do with the topic Gödel ultimately chose to write about for Schilpp: the relationship between relativity theory and idealistic philosophy. The latter grew instead out of Gödel's interest in philosophers such as Kant, who (though in a different sense than Einstein) had denied the objectivity of time. In particular, the idea that the passage of time might be a subjective phenomenon led Gödel to consider "whether [the concept of absolute time] ... is a necessary property of all possible cosmological solutions" of the field equations of general relativity theory [426]. He discovered that it was not, in a very strong sense; and that surprising result led him to undertake further investigations, which culminated in far more than the brief essay (*1949b*) he finally submitted to Schilpp.

The excursion into relativity theory began in August 1946 when Gödel informed Schilpp that he would contribute an essay of three to five pages to be entitled "The theory of relativity and Kant." Evidently he took up the question almost immediately and quickly became absorbed by it, for just a month later he wrote his mother that he was so deeply involved in his work that he found it hard to summon the concentration for writing letters. Nevertheless, despite the intensity of his efforts, he subsequently discovered mistakes in what he had done and so wrote Schilpp in February to request an extension of the deadline for submission.

◇   ◇   ◇

Among the factors responsible for Gödel's false start were, perhaps, the preparations for Adele's long-awaited trip to Europe. It had taken months for her to book passage and obtain the necessary documents — in part, Gödel thought, because of her occasionally rude outbursts to Austrian officials. Those matters had necessitated several trips to New York and Washington, and as the time wore on her worries about her family became almost hysterical (to the point that Gödel thought the trip necessary in order to restore her mental well-being, despite the strain it placed on their finances [427]).

In addition, there was trouble with the Vienna apartment that the two of them still owned and rented out: Either as a result of the war or through their tenants' neglect it had suffered damage, and efforts to solve the problem at long range had proved unsuccessful.

Adele finally left for Vienna in May of 1947. She remained there for the next seven months, looking after her mother, visiting other family members and friends, and doing what she could to settle the apartment dispute without resort to litigation. In stark contrast to the isolation she had endured in Princeton, her life was suddenly hectic, filled with many responsibilities.

◇ ◇ ◇

For Kurt, on the other hand, the routine of daily life during Adele's absence reverted to what it had been during his prewar visits to Princeton; he was lonely, but free of distractions. For the first two months that Adele was away Morgenstern went out of his way to keep Gödel company. The two frequently went out to dinner together, and on those occasions Gödel apparently ate well, in contrast to the meager diet he followed at home. In July, however, Morgenstern himself traveled to Vienna — there, among other things, to visit with Gödel's mother and brother and apprise them of Kurt's fame, of which, until then, they had remained completely unaware. Consequently, for several weeks, while Einstein was away taking a rest cure, Gödel was left to fend for himself.

He did so, typically, by burying himself in his research. In late August he wrote his mother that his work had occupied him so much that he had abandoned plans to leave Princeton for a vacation at the shore; and a month later he announced that he had finally finished the manuscript for Schilpp. But then, as usual, he found it necessary to make revisions, which further delayed its submission. Nevertheless, he told his mother that it would be "the editor" who would determine when his essay appeared [428].

In fact, Schilpp was doing everything in his power to coax the manuscript from him. He pleaded, chided, complained, and cajoled, all to no avail. Gödel did not finally submit his short essay until March of 1949, having in the meantime drafted and rejected a total of five much longer manuscripts.

Those rejected drafts remained unpublished until long after Gödel's death (two of them are included in volume III of his *Collected Works*); but from them, together with his letters to his mother and remarks he made to Morgenstern, it is possible to reconstruct in some detail how his cosmological discoveries evolved. In particular, an entry in Morgenstern's diary makes clear what Gödel had achieved at the time he first told his mother of having completed his manuscript: He had, Morgenstern reported, found "a world in which simultaneity can not be defined," a discovery that was of great interest to Einstein and von Neumann [429]. More specifically, internal evidence in the first two of Gödel's five manuscripts suggests that he recognized early on that in cosmological models satisfying Einstein's field equations "the existence of a 'natural' cosmic time ... depends upon the non-rotation" of the matter in them relative to the so-called compass of inertia [430].

It seems then that by September of 1947 Gödel had succeeded in construct-ing a "rotating universe" in which there was no privileged notion of universal time that could, without arbitrariness, be regarded as absolute throughout the cosmos. The subsequent revisions he undertook during the fall of 1947 were presumably in the nature of detailed calculations, since in his November letter to his mother he remarked that the philosophical investigations he had origi-nally undertaken had "led to purely mathematical results" that he intended to publish separately.[3]

$$\diamond \quad \diamond \quad \diamond$$

In December Gödel's research was interrupted by two events: his wife's return, and their citizenship hearings, his on the 5th, hers one week later.[4]

The hearings took place in Trenton, and the account of Gödel's behavior on that occasion has become the most famous of the anecdotes about him. The story is told once again by Morgenstern, who served that day both as Gödel's chauffeur[5] and, with Einstein, as one of his two witnesses. (Morgenstern served in the same capacities for Adele, but Einstein did not.)

Like all applicants for citizenship, Gödel expected to be questioned on matters concerning the American system of government; and being the person he was, he had prepared for the examination much more thoroughly than was customary or necessary. As the day approached he became visibly agitated, until finally he confided to Morgenstern that he had found an inconsistency in the Constitution.

Morgenstern was much amused, but realized that Gödel was deadly serious about the matter and that his bid for citizenship might be jeopardized if he were

---

[3]In fact, he published two technical accounts of his cosmological work: the papers *1949a*, addressed to an audience of physicists, and *1952*, for mathematicians.

[4]The hearings had originally been scheduled together, but Adele's return was delayed by unrest in France, so she did not arrive back in time to accompany her husband.

[5]Gödel apparently never drove a car in America, though he had done so years before in Europe. Perhaps it was just as well, to judge from a tale that Adele later told one of her neighbors: She herself, she said, had ridden with him only once. During the ride he became so absorbed in his thoughts that they nearly crashed, and after that she insisted that he give up driving.

allowed to expound upon his "discovery". Morgenstern therefore discussed the problem with Einstein, and the two agreed that they must work together to prevent such a thing from happening.

On the appointed day Morgenstern picked up his two passengers, and as they rode the short distance to Trenton Einstein endeavored to distract Gödel. On entering the car, he turned to him and exclaimed, "Well, are you ready for your next-to-last test?"

"What do you mean, 'next-to-last'?"

"Very simple. The last will be when you step into your grave."

After that macabre beginning, Einstein went on to regale his two friends with jokes and stories. He told, for example, of an autograph hunter who had recently accosted him. Such individuals, he said, were the last of the cannibals: They sought to appropriate the spirits of those they consumed. Gödel was probably not much calmed by such humor, but during the ride he did keep quiet about his legalistic concerns.

When the trio arrived at the courthouse they found several other applicants in line ahead of them, so it appeared that Einstein and Morgenstern might have to continue their distractive efforts for some time. As it turned out, however, the judge, Philip Forman, was the same one who had administered the oath of citizenship to Einstein some years before, and on seeing him there he ushered the three into his chambers at once.

For a while Gödel was all but forgotten, but eventually Forman turned to him and asked, "Do you think a dictatorship like that in Germany could ever arise in the United States?"

It was just the opening Gödel had been waiting for. He answered affirmatively and began to explain how the Constitution might permit such a thing to happen. Fortunately, Forman quickly understood what he had set in motion and stepped in to rescue the situation. "You needn't go into that," he interrupted, and went on to more mundane questions [431].

Surprisingly, Gödel himself gave the hearings only passing mention in his correspondence with his mother. He provided a much fuller account of the later ceremony (on 2 April) at which he and Adele took their oaths of citizenship. Forman, whom Gödel described as "an especially sympathetic person," again presided, and after the oath-taking he spoke "for about an hour on past and present circumstances in the United States." In its "simplicity and naturalness," Gödel declared, what he said "had quite a telling effect," so that "one went home with the impression that American citizenship, in contrast to most others, really meant something" [432]. For unknown reasons, at the time his citizenship was granted Gödel also legally dropped his middle name — a matter he discreetly neglected to report to his mother.

◇   ◇   ◇

With Adele's return, life in the Gödel househeld settled back into its usual routine. On Sundays, for example, he and Adele would arise around noon. She would prepare a meal, and after they had eaten he would make the weekly rental payment. He would then spend the rest of the afternoon reading the *New York Times.* He subscribed only to the Sunday edition, but found even that too much to absorb [433]. On weekdays he worked at his office at the institute while Adele went shopping and busied herself with housekeeping chores. In the evenings they sometimes went to see a movie[6] — their only regular source of entertainment — but otherwise their interactions with the community around them were very limited.

It was important to Gödel's health and well-being that intrusions to his daily routine be minimized and that Adele be on hand to supervise his diet. It is not surprising therefore that when Morgenstern had dinner with the Gödels early in the new year Kurt was looking much better. He had regained the weight he had lost during Adele's absence, though he was still seriously underweight for a man of his height (his citizenship certificate listed his height as 5'7" and his weight as 110 pounds), and he told Morgenstern that he was eager to get back to work in logic. As for his cosmological work, he intended "to publish his proof but leave the galactic statistics for others to work out" [434].

Morgenstern did not elaborate further, but from one of Gödel's later publications on the subject it is clear what the significance of such "galactic statistics" was. In his paper *1952* Gödel stated that "*a directly observable necessary and sufficient criterion for the rotation of an expanding spatially homogeneous and finite universe*" was that "*for sufficiently great distances there must be more galaxies in one half of the sky than in the other half.*" Observational statistics could thus be employed to test whether the universe we live in is one of the sort Gödel had envisioned; whether, that is, the notion of time as an absolute quantity really *is* a chimera.

A later entry by Morgenstern indicates that the models Gödel had obtained up to that time were not expanding, and so could not correspond to our own universe, with its observed red shift for distant stellar objects. They did, however, suffice to demonstrate the abstract *possibility* of the nonexistence of

---

[6] Gödel was especially fond of Disney films, particularly *Snow White*, which he saw on at least three occasions, and *Bambi.* He also enjoyed an occasional thriller such as "*M,*" but he told his mother that he "hated comedies." (FC 43, 22 August 1948)

an absolute world time. It would seem then that, as the year 1948 began, Gödel did not intend to pursue the mathematical construction of rotating *expanding* models. Rather, he preferred to cease theorizing and leave the question of the actual structure of our universe to physicists. (In the paper he ultimately submitted to Schilpp he noted that expanding rotating solutions did exist and that "in such universes an absolute time also might[7] fail to exist." But in any case, he argued, "the mere compatibility with the laws of nature of worlds in which there is no distinguished absolute time ... throws some light on the meaning of time also in those worlds in which an absolute time *can* be defined.")

Nevertheless, Gödel was not content to leave the gathering of statistics entirely to others, for in his *Nachlaß* are two bound notebooks in which he recorded angular orientations of galaxies — a circumstance that suggests he did hope to find physical confirmation for his theories after all. Indeed, it seems he was firmly convinced that such confirmation would occur, for many years later, not long before he died, he called Freeman Dyson to inquire about the latest observational evidence for rotation; and when Dyson informed him there was none, he was unwilling to accept the conclusion [435].

In late February Gödel promised Morgenstern that he would "soon" give him a copy of his Kant-Einstein essay. But then something quite unexpected happened. In the same letter in which he told his mother of the citizenship ceremony, Gödel apologized for a two-month delay in his correspondence with her. He had intended to write long before, he said, but for several weeks he had been beset by a "problem" that had driven everything else out of his mind. He would have preferred not to work on it uninterruptedly, but that had proved impossible. Even when he listened to the radio or went to the movies he did so "with only half an ear." At last, though, just a few days before, he had "settled the matter enough to be able to sleep well again."

Gödel did not say what the problem was that had so obsessed him, but the answer is provided by Morgenstern's diary entry for 12 May (two days after Gödel's letter). There Morgenstern reported that he had spoken with Gödel at length about his cosmological work and had learned that he was making "good progress" with it. In particular, he added, "Now in his universe one can travel into the past."

---

[7]From what has been said above about the connection between rotation and temporal absoluteness, Gödel's use of the word "might" in this passage may seem problematic — especially since the paper *1949a*, in which he pointed out that connection, appeared *before* that in the Schilpp volume. The apparent discrepancy is resolved in a footnote, where Gödel imposed the additional requirement that "absolute time should agree in direction with the times of all possible observers."

It was that revolutionary discovery — that in the universes he had already constructed there were "closed timelike lines," which, at least in theory, would allow one to revisit one's own past — that Gödel described in his paper *1949a*. The existence of such lines (which are *not* present in the expanding universes he considered in his paper *1952*) showed that in those universes there was not only no "natural" notion of absolute time, there was none at all. As Gödel put it in the fourth of his unpublished manuscripts, "in whatever way one may [try to] introduce an absolute 'before,' there always exist either temporally incomparable events or cyclically ordered events."

In a footnote, later partly deleted, Gödel acknowledged that "This state of affairs seems to imply an absurdity." What if, for example, one were to revisit a point in one's past and attempt to alter what had happened then? He noted, however, that such inconsistencies

> presuppose not only the practical feasibility of . . . [such a] trip into the past ([for which] velocities very close to the speed of light would be necessary . . .) but also certain decisions on the part of the traveler, whose possibility one concludes only from [some] vague conviction of the freedom of the will.

But given such a conviction, "practically the same inconsistencies . . . can be derived from the assumption of strict causality" — a tenet to which he himself was firmly committed. In matters of causation, therefore, he believed one must carefully distinguish what is logically possible from what is temporally permissible [436]. Schilpp had hoped to have his volume ready for presentation to Einstein on the occasion of Einstein's seventieth birthday (14 March 1949). Gödel, however, did not finish his essay until the month before, and even then he delayed sending it off — apparently because he was still ruminating on some of the ideas in it. Finally, Schilpp extracted Gödel's promise that he would give a copy of the essay to Einstein at the gala birthday celebration that was to be held in his honor at the Princeton Inn on 19 March. Gödel did so, and the long-awaited manuscript reached Schilpp shortly thereafter.

How much Gödel had told Einstein about his results before that is unknown. In his published remarks in the volume containing Gödel's essay, Einstein acknowledged that the possibility of closed time-like lines was one that had already "disturbed" him at the time he developed the general theory of relativity. Having failed to clarify the question himself, he hailed Gödel's discovery as "an important contribution." Nevertheless, he opined that it would "be interesting to weigh" whether Gödel's solutions might not, after all, "be excluded on physical grounds" [437].

◇ ◇ ◇

On 7 May Gödel lectured on his discoveries at the IAS. The talk was well attended,[8] though to judge from Morgenstern's account of the event, few of Gödel's colleagues had any idea what he had been working on. (Most, in fact, were astounded to discover the depth of his knowledge of physics.) He spoke for an hour and a half, "in good form" according to Morgenstern, but apparently over the heads of most of his audience [438].

Afterward Gödel's discoveries were the talk of the institute. As might be expected, the initial reaction was a mixture of surprise, high praise, and lack of full understanding. Later, however — although Gödel published details of his calculations in his papers *1949a* and *1952* — his universes were largely forgotten. Only recently has there been a resurgence of interest and appreciation.

In several respects Gödel's cosmological results were accorded much the same reception as his incompleteness theorems. Both discoveries upset firmly held preconceptions. Both came from an unexpected quarter. Both were motivated by philosophical issues outside the concerns of many in the scientific community. Both had an air of paradox about them that fostered dubiety. And both appeared to be theoretical curiosities, of little apparent relevance to mainstream work in physics or mathematics.

Like the incompleteness theorems, the correctness of Gödel's cosmological results was also challenged in print, in a paper by S. Chandrasekhar and J. P. Wright that appeared in 1961 in the *Proceedings of the National Academy of Sciences* [439]. Chandrasekhar, who had been in the audience for Gödel's talk at the IAS, was one of the most eminent and respected figures in modern physics. Accordingly, his opinions carried a great deal of weight — so much so that Gödel's conclusions about the possibility of time travel in relativistic universes were for some time thereafter "treated as doubtful in the ... philosophical literature" [440].

It is hard to fathom why Gödel was not consulted while an article critical of his work was under review, especially since he had by then been a member of the National Academy of Sciences for some eight years. Had he been made

---

[8] In addition to Einstein and Morgenstern those present included Veblen, Robert Oppenheimer, S. Chandrasekhar, S.S. Chern, and Martin Schwarzschild, but not von Neumann, Weyl, or Siegel.

aware of Chandrasekhar and Wright's claims he would surely have pointed out that they were mistaken in asserting that he had assumed certain lines in his universes to be geodesics.

The article was finally brought to his attention in 1969 by Professor Howard Stein of Case Western Reserve University, who had recognized the error and had submitted a paper [441] defending Gödel's results. The editors, however, were reluctant to accept Stein's criticisms, so he determined to settle the issue once and for all by writing to Gödel himself.

When Gödel quickly confirmed that he had "never said, or meant to say" what Chandrasekhar and Wright had imputed to him, Stein's paper was promptly accepted and published; and with that, the alleged refutation of Gödel's results was recognized for what it was: a simple misunderstanding.

◇ ◇ ◇

For some time after his 1949 lecture at the institute Gödel endeavored to extend his cosmological results still further. But early that summer Adele became infatuated with a house that had come on the market, and her insistence that they buy it forced him to interrupt his work.

In letters to his mother written after negotiations for the sale were well under way [442] Gödel described the house and the "vexations" its purchase had entailed. Located at 129 (later 145) Linden Lane, near the edge of town, it was a small one-story dwelling of "sturdy cinderblock" construction. (Figure 14) It had been built just three years before and possessed "all the modern comforts," including automatic oil heating and an air conditioner that pumped cool air into the rooms from the basement. The living room seemed "awfully large" to Gödel (it could, he declared, easily accommodate a dancing party for fifty persons), but it lacked the entrance vestibule common in European homes. It was paneled in part with wood and had a wood-burning fireplace in one wall. A short hallway led from the living room to two small bedrooms and bath, and a small fully paneled room opened off the kitchen. The house was surrounded by a large yard in which a few young trees had been planted, and there was also a garden plot nearby.

Gödel deemed the house to be "just the right size" for the two of them. Its relatively secluded location appealed to his reclusive nature, and its situation "in one of the higher-lying parts of Princeton" significantly cooler in the sum-

FIGURE 14. The Gödel home, 145 Linden Lane, Princeton, New Jersey, ca. 1950.

mer than their apartment had been.[9] Unfortunately, though, its price ($12,500 plus closing costs[10]) was more than they could afford. Nevertheless, Adele was determined that they should buy it "at any cost."

In the end, Gödel borrowed the entire amount: Three-quarters of the cost was covered by a mortgage and the rest by a salary advance from Robert Oppenheimer, who had become IAS director two years before. It was an arrangement that Gödel admitted "appeared somewhat shaky," but he reassured his mother that by virtue of Oppenheimer's approval the transaction "had the institute's backing." In addition, he said, an "expert" from the institute had inspected the property and declared it worth the price.

Whether Gödel really believed what he told his mother is questionable. Certainly Morgenstern saw the matter very differently. In his diary entry for 16 July he reported that Gödel "had a bad time behind him." The house Adele had compelled him to buy was not only worth much less than he had

---

[9] Gödel would later become renowned for wearing sweaters and earmuffs in the middle of summer, but during his early years in Princeton he often complained about the stifling heat.

[10] Equivalent to about $80,500 today — far below its present market value.

paid for it but was near the edge of town, beyond convenient walking distance to the IAS and to stores, and was "not in a good neighborhood."[11] The negotiations concerning it had been stressful and time consuming and had interrupted Gödel's research at a time when it had been progressing well.

By then Morgenstern's perspective was no longer that of a bachelor, for he himself had married the previous year. But his opinion of Adele remained unchanged. She was, he granted, a good cook, albeit in the heavy German style. In other respects, however, she was unpleasant to be around. She was coarse and loud, monopolized conversation, displayed "astonishingly bad taste" in the furnishings she chose, and after ten years in America still spoke English very badly. On one occasion, during a dinner at the Gödels', Morgenstern's wife Dorothy was shocked and "roused to indignation" by the way Adele spoke about Kurt and "terrorized" him [443]. And despite her husband's real or imagined sensitivity to "bad air," Adele was a smoker.

As an economist Morgenstern was presumably well aware of property values in Princeton, and his descriptions of Adele's behavior are in accord with those of other observers. She was, by all accounts, a sharp-tongued, strong-willed woman who could be rude and irascible. Gödel himself admitted that on occasion she "displayed symptoms of clinical hysteria" and that she often formed "quite wrong impressions" of other people. In particular, he said, she sometimes took "mortal offense" at remarks others would take little notice of and frequently imagined there was enmity toward her when none in fact existed [444].

Indisputably, Adele made a bad impression on many of those — especially women — who came in contact with her; and given Gödel's aversion to personal conflict of any kind it is easy to believe that she may have bullied him, at times to an embarrassing extent. Still, there is little reason to think that she was quite the millstone Morgenstern made her out to be. For Gödel, after all, had no social pretensions and scant interest in social interaction. So his wife's coarseness, however much it may have contributed to her own isolation, cannot have had much impact on his relationships with colleagues; he would have been an outsider in any case. Her demands did sometimes keep him from his work, but he accomplished a great deal nonetheless — probably much more than he would have without her to care for him and to force him, at times, to make decisions. And if her taste tended toward kitsch, so did his.

---

[11] By Princeton standards; in fact it was not far from the Princeton High School, an academically prestigious institution whose building and "luxurious" grounds reminded Gödel of a castle (FC 54, 18 October 1949).

For example, he described a pink flamingo lawn statue that she placed in the garden outside his window as "awfully charming" (*furchtbar herzig*) [445].

The Gödels purchased the house on Linden Lane on 3 August and moved into it about a month later. Adele was thrilled to have a home of her own, and she immediately threw all her energies into fixing up the rooms and planting flowers in the garden. The far end of the living room, next to the kitchen, was converted into a dining room, and Gödel chose the small room off the kitchen for his study. They did not need to buy furniture, he told his mother, because they already had an oversupply.

The following January, after they were well settled in to their new abode, Gödel expressed his hope that the house would never become a burden; and in fact, it served them well for the rest of their married life, especially after Adele made a number of improvements to the property. Over the years she built a small sun porch on the back of the house, paneled the basement, and created a miniature *Heuriger*[12] in the back yard. When Kurt's study eventually proved too small she exchanged rooms with him and had glass-enclosed bookcases built into the room that had been hers. And after a time she converted the room off the kitchen into a rustic *Bauernstube*.

According to Elizabeth Glinka, a neighbor who in later years became Adele's companion and nurse, the Gödel home was for the most part furnished simply, in a contemporary style. There were, however, two extravagances: Adele loved oriental rugs and chandeliers, of which, at the time Glinka began visiting the house, four hung in the combined living/dining room, two from the ceiling and two from the wall.

To outsiders the house no doubt seemed quite modest and its furnishings somewhat eccentric. But it was not intended to appeal to visitors, of whom in any case there were very few. It served rather as a quiet haven where Gödel could pursue his investigations undisturbed. After a few years the evergreens Adele had planted grew to screen the house from the view of passersby, and in 1957, to prevent a house from being built next door, the Gödels purchased the adjoining lot to the north.

What had seemed to Morgenstern a poor buy thus turned out, in the long run, to be a sound investment. The price of the house may have been inflated and the arrangements for its purchase temporarily disruptive to Gödel's intellectual endeavors, but overall the move made life more stable for both him and Adele. While she devoted herself to the affairs of the house, he was

---

[12] An outdoor café where new wine is served, for which the Viennese suburb of Grinzing is especially famous.

left free to pursue his research. Her complaints grew less frequent, and his preoccupation with "bad air" subsided markedly.

◇ ◇ ◇

Gödel must have resumed his cosmological investigations almost immediately after details of the house contract were settled, for on 2 September 1949 he submitted a final addition to one of the footnotes in the Schilpp paper. About that time he also accepted an invitation to speak on his results in relativity theory at the International Congress of Mathematicians (the first to be held after the war), which was to take place in Cambridge, Massachusetts, in August of the next year.

Nevertheless, some months earlier Gödel had told Morgenstern that he was eager to curtail his work in physics in order to resume his studies of Leibniz. Leibniz's manuscripts in Hanover had somehow escaped damage during the war, and Gödel and Morgenstern conceived the idea of requesting that they be microfilmed for deposition in the Princeton University library. Morgenstern began to look into the matter in May of 1949, but almost immediately his inquiries ran into difficulties that served once again to fuel Gödel's conspiracy theories.

Problems first arose when Morgenstern attempted to locate a copy of the critical catalog of Leibniz's manuscripts, prepared early in the century by the editors of the interacademy edition of Leibniz's works [446]. Reportedly, the only copy of the catalog in the United States had been deposited at the National Academy of Sciences in 1908. Yet when Morgenstern wrote the academy about it, they could offer no clue as to its whereabouts. Subsequent inquiries to the Smithsonian Institution and the Library of Congress proved equally fruitless, and in the end the mystery of the catalog's disappearance never was explained.

Negotiations concerning the microfilming continued for the next several years. They were carried on mostly by Morgenstern, but with Gödel's keen interest and assistance. Early on, permission to copy the manuscripts was granted by the Niedersächische Landesbibliothek in Hanover, but the amount of material (700,000 to 800,000 pages by one estimate) posed a funding problem. Morgenstern approached the Rockefeller Foundation for support, not only for the photocopying itself but for a travel grant for Gödel, who, he said, was eager to visit the archives in Hanover and Paris.[13] But various complications arose. First it was reported that a firm in Germany planned to undertake

---

[13] Indeed, in 1951 Gödel went so far as to obtain a passport, though he never used it.

the filming as a commercial enterprise; later it turned out that those plans had been given up because of the costs involved; and then, in the autumn of 1951, Morgenstern and Gödel learned that others were pursuing aims similar to their own.

In September of that year Professor Paul Schrecker of the University of Pennsylvania reported that all the "apparatus" for editing Leibniz's papers was now in the Russian occupation zone, and the following month the Library of Congress advised that the authorities in Hanover were no longer interested in having the Leibniz documents filmed. Subsequently, however, in June of 1952, Morgenstern himself visited Hanover and once again obtained permission for the filming, as well as a revised estimate of the number of pages involved, which he reported as 300,00–400,000.

In the meantime the Rockefeller Foundation had also learned of Schrecker's efforts. To avoid further conflict, Morgenstern proposed that the two competing projects be united; but the foundation deemed Schrecker's plans to be further advanced. In May of 1953 the American Philosophical Association withdrew its support from Morgenstern and Gödel's endeavor, and not long afterward, with Rockefeller assistance, Schrecker at last succeeded in having a copy of the Leibniz materials deposited at the library of the University of Pennsylvania. To what extent Gödel subsequently made use of it is unknown.

◇ ◇ ◇

By the spring of 1950 the intensification of the Cold War and the nuclear arms race had begun to temper Gödel's respect for America. In a letter to his mother written on 30 July, about a month after the start of the Korean conflict, he lamented such perilous times. He cited the resumption of sugar hoarding as one of several signs that the United States was returning to a wartime economy, and blamed Truman for fomenting anti-Communist hysteria. There was, he declared, strong opposition to the administration's policies, but critics were kept muzzled.

That was the era of Senator Joseph McCarthy and his persecution of intellectuals — including Oppenheimer, the institute's director — so the threat to civil liberties was clear and present. Nevertheless, Gödel did not realize how pervasive the government's surveillance of political opinions had become. He knew that mail between America and Austria continued to be censored by the occupation authorities, for he advised his mother that an article about Einstein she had tried to send him had been removed as a "forbidden enclosure"; but he surely did not suspect that extracts from two of his letters to her would be passed on to U.S. Army intelligence and the FBI [447].

The first of those extracts, taken from his letter of 1 November 1950, also involved Einstein. As translated by the censors, it read:

> Einstein warned the world not to try to attain peace by rearmament and intimidating the adversaries. He said that this procedure would lead to war and not to peace, and he was quite right. And the fact is well known that the other procedure (trying to come to an agreement in an amicable way) wasn't even attempted by America, but [was] refused from the first. It isn't the one and only question who started matters, and for the most part it would be difficult to establish. But one thing is certain: under the slogan "democracy" America is waging war for an absolutely unpopular regime and under the name of a "police action" for the U.N., and does things to which even the U.N. doesn't agree [448].

The second, from his subsequent letter of 8 January 1951, reported that

> The political situation developed wonderfully here during the holidays, and you only hear of defense of the homeland, compulsory military service, increase of taxes, increase of prices, etc. I think, even in the blackest (or brownest) Hitler Germany, things were not that bad. People who talk such nonsense again over there as they did in Hitler times are probably in the minority, and I hope the Germans will not be so silly as to have themselves used as cannon-fodder against the Russians. I am under the impression that America with her madness will soon be isolated [449].

The excerpts were forwarded to the FBI "for information purposes" because they were deemed to reflect "a pro-Russian attitude on the part of the subject." But since the Bureau could find nothing else in its files on Gödel save his registration with the German consulate, shortly after the war broke out, as a German alien of military age, no investigation of him was undertaken. When he applied for a passport a few months later he was subjected to a routine loyalty investigation, but the document was issued with little delay. (By then he had retreated somewhat from his comparison of America's policies with those of Hitler. In a letter to his mother written in March of 1951 he acknowledged that "in many respects" the situation was "of course not comparable with that of Germany in 1933–39"; but with regard to "the building up of armaments and the suppression of every pacifist stirring," he thought it "would soon surpass" the German example.)

◇ ◇ ◇

On 31 August 1950 Gödel delivered his address to the International Congress of Mathematicians (ICM). There were, he reported, several hundred people in the audience, and he received much applause both before and after his talk. The congress was quite well attended (about twenty-five hundred participants in all), but he belittled its international character; nine-tenths of those in attendance, he said, were either Americans or European émigrés who had resided in the United States for ten years or more. The reason, he suspected, was that the political regime in the United States desired "to build a Chinese wall around the country" to ensure that Americans would remain ignorant of world opinion and that those outside the country would not become aware of what was happening within it [450]. (In fact, the restrictive U.S. visa policies then in effect — a reflection of the anti-Communist hysteria that afflicted the nation at that time — were widely criticized both at home and abroad.)

Morgenstern was among those who attended the congress, and while there he once again heard rumors that Gödel had proved the independence of the Continuum Hypothesis. Gödel readily confirmed that he had obtained some partial results — one of many discoveries, so Morgenstern thought, that lay hidden behind the shorthand in his writing desk — but to write everything up would, he said, be a great deal of trouble.

Four months later, however, he told Morgenstern that he had changed his mind: As soon as he finished revising his ICM address for publication he intended to begin writing out his independence proofs. After that he would publish a paper on Diophantine equations, extending his undecidability results; and there were several other manuscripts that he was eager to publish as well. The reason, he said simply, was that he had come to realize that "one does not live long enough to do everything."

Morgenstern guessed that Gödel's about-face had been brought about by the untimely death of Abraham Wald, his former colleague from Menger's colloquium, who had been killed on 13 December in a plane crash in India. But Gödel may have experienced more direct intimations of his own mortality, for within a few weeks he was faced with a life-threatening health crisis.

X
Recognition and Reclusion

(1951–1961)

WITH HINDSIGHT THE chronic digestive troubles Gödel had described in 1946, especially the discomfort he experienced whenever he drank strong coffee or ate acidic or highly spiced foods, can be seen as warning signs of an incipient ulcer problem. But at the time his doctors did not recognize them as such. The very obsessiveness of Gödel's concern for his diet and bowel habits made it easy to dismiss his complaints as those of a hypochondriac, particularly since the dietary restrictions he imposed upon himself were effective in alleviating the symptoms.

Although the diet had caused him to become severely underweight, Gödel had otherwise enjoyed generally good health, and as the year 1951 began there was nothing to suggest that he was under undue emotional stress or that he had experienced any symptoms foreshadowing a resurgence of his stomach trouble. But then suddenly, early in February, he began to hemorrhage. He was rushed to the hospital, where X-rays confirmed the presence of a duodenal ulcer. Blood transfusions and intravenous feedings were administered at once, but even so his survival was in doubt for several days.

When Morgenstern first visited Gödel his condition was still precarious. The prospects for his recovery were all the more uncertain because of his conviction that he was about to die. (He even gave Morgenstern an oral testament.) Nevertheless, Morgenstern believed that his friend would eventually pull through, and a week later, despite his fears, Gödel had begun to show some signs of recovery.

Not long afterward Gödel's condition improved enough that he was allowed to return home, and there his depression quickly subsided. Adele's solicitous care was surely the principal reason for the change in his outlook, but his recovery was also spurred by a development that took place behind the scenes: Shortly after the worst of the crisis had passed Oppenheimer asked Morgenstern what he thought could be done to boost Gödel's spirits, and Morgenstern told him forthrightly that the best thing would surely be to promote Gödel to professor without further delay.

Both agreed that the action was long overdue, but they knew, too, that some of Gödel's IAS colleagues continued to oppose the idea. In particular, C. L. Siegel had declared some months before that "one crazy man on the faculty" (himself) was enough [451]. So Oppenheimer doubted that he could effect a promotion for Gödel within a short time.

Then an alternative occurred to him. At that time Oppenheimer was serving on the selection committee for the first Einstein Award, a prestigious honor that had been endowed only the year before by Admiral Lewis L. Strauss, one of the institute's trustees. The award consisted of a gold medallion and a check for $15,000 and was to be presented once every three years on Einstein's birthday (14 March). To bestow it on Gödel would be especially timely, not only because of his close friendship with Einstein and his own recent contribution to relativity theory, but because his achievements had until then received little notice outside the mathematical community. And of course, the extra money would also help Gödel to pay the medical bills he had incurred.

Oppenheimer had little doubt that the other committee members — von Neumann, Weyl, and Einstein himself — would agree to such a proposal. But he was concerned about matters of propriety. As IAS director he was reluctant to nominate anyone from the institute. Weyl was unlikely to take the initiative, and for von Neumann or Einstein to promote Gödel's candidacy might, Oppenheimer thought, smack of cronyism. On the other hand, there could hardly be any question as to Gödel's worthiness. The real problem was that the committee had already recommended that the award be given to Julian Schwinger of Harvard [452], a brilliant young mathematical physicist who had made fundamental contributions to several areas of research and would later share a Nobel Prize for his work in quantum electrodynamics. No public announcement of Schwinger's selection had been made, as the identity of the recipient was to be kept confidential until the day of the awards ceremony; but Admiral Strauss had been informed, and it would be awkward to retract the recommendation.

By mid-February a solution had been found: The award would be made jointly to both Schwinger and Gödel, each of whom would receive a medal and check. No record of the negotiations involved has been preserved [453], but the decision was evidently reached swiftly, for even before Gödel's release from the hospital Einstein had dropped hints to him that an honor was forthcoming.

Official notification of the award reached Gödel shortly after his return home, and, as Oppenheimer had hoped, he was both surprised and excited by the news. He did not look forward to the awards ceremony itself — he confided to his mother that he hated such formal occasions — but he was elated that his work was at last to be accorded a measure of public recognition.

The presentation of the awards took place during a luncheon at the Princeton Inn. As he had for their citizenship hearings, Morgenstern served as the Gödels' chauffeur to and from the event; and, as before, he had misgivings about what might occur. This time it was not Kurt's behavior that concerned him, but Adele's, for in the weeks preceding the luncheon she had caused no little trouble with the arrangements. Among other things, she had insisted on being seated beside her husband throughout the entire proceedings — hardly an unreasonable demand, but, in the view of the organizers, an act of effrontery that had disrupted the official protocol. To Morgenstern's relief, however, she comported herself very well on the appointed day.

The luncheon was a rather intimate affair, hosted by Admiral Strauss and attended by some seventy-five invited guests. The atmosphere was festive yet unpretentious. Oppenheimer delivered a brief tribute to Schwinger and his work, and von Neumann did the same for Gödel, whose achievement in modern logic he described as "a landmark which will remain visible far in space and time." Einstein then personally presented the medals, quipping to Gödel as he did so, "And here my dear friend for you. — And you don't need it!" [454]. In reality, of course, Einstein knew full well that Gödel *did* need such a token of esteem. Apart from invitations to lecture it was the first academic honor that had ever been bestowed upon him.[1]

Morgenstern hoped that the extensive press coverage the Einstein award received would prompt Princeton University to award Gödel an honorary doctorate as well. And indeed, within three months Gödel did receive a Litt.D. — from Yale, on the occasion of that university's 250th anniversary. The next year brought an honorary Doctor of Science from Harvard. Princeton, however, perhaps in part because of its lingering rivalry with the IAS, continued to ignore the genius within its midst.

That such recognition meant a great deal to Gödel and played a role in restoring both his physical and mental health is indicated not least by the fact that the strict diet he was then obliged to follow did not deter him from

---

[1] In contrast to Schwinger, who, though twelve years younger, had already been elected to the National Academy of Sciences.

going to New Haven to accept the Yale degree. On the contrary, far from worrying that travel might adversely affect his recovery, he had even begun to contemplate a sojourn in Europe, both to visit the Leibniz archives and to see his mother again after their eleven-year separation.

In his correspondence with her and his brother, Gödel first broached the idea of vacationing with them at a spa. The most appealing destination, he said, would be Velden am Wörther See, where there was a sanatorium. Alternatively, he considered renting a hotel room in Vienna. In either case, he stressed that it would be essential for him to maintain his dietary regimen. He lived, he said, "principally on butter" (about a quarter pound a day), supplemented by eggs (three per day plus the whites of two more, often in the form of egg nog), milk, puréed potatoes, and baby food. No other fat was permitted, no soups or anything sharp or sour, and no fresh fruit. All bread must be toasted. As for meat, he ate very little; a chicken would last him a week and would be all that he required [455].

While the three of them contemplated the various possibilities, the summer swiftly passed by. Toward the end of it Kurt and Adele vacationed once more at the New Jersey shore, this time at Asbury. They stayed first at the Hotel Carlton, but soon found a less expensive small apartment that they could rent. The landlady was very nice, Gödel reported, and the apartment had a wood-paneled room like that in their home. Unfortunately, however, the night vapors there proved to be "so bad that it completely counteracted the tonic effect of the sea air." Because of that they finally moved to the Hotel Monterey, a *sehr sympathisches* establishment that featured an enormous outdoor swimming pool. To Gödel it seemed "fabulously elegant," reminiscent of the Hofburg in Vienna. But because of its age it was, he reassured his mother, still "affordable for non-millionaires" [456].

To Gödel's regret the summer was a cool one and the salutary effects of their seaside excursion less than he had expected. Only after his return to Princeton did he begin to feel somewhat rejuvenated. As late as mid-September he still held out hope of going to Vienna, but eventually, to his mother's great dismay and disappointment, he decided not to go that year after all. The main reason was that he had fallen behind in preparing an important lecture that he was to deliver to the American Mathematical Society on 26 December.

◇　◇　◇

The address in question was the Gibbs Lecture, named for the great nineteenth-century American mathematical physicist Josiah Willard Gibbs. Initiated in 1924, the lecture was a centerpiece of the society's annual meeting, and the invitation to present it was a tribute reserved for the most eminent practitioners of pure and applied mathematics. Previous lecturers had included such luminaries as G. H. Hardy, Vannevar Bush, and Theodore von Kármán, as well as three other IAS faculty (Einstein, von Neumann and Weyl). Gödel, however, was the first (and has remained the only) logician to be so honored.[2]

He had accepted the invitation shortly before his illness, and by his own account had devoted much of the time since his recovery to preparing the text of his speech. Its title, "Some Basic Theorems on the Foundations of Mathematics and Their Philosophical Implications," only hinted at the controversial nature of its subject: the significance of the incompleteness theorems for debates on the nature of mathematics and the limitations of human intelligence.

Gödel began his talk by acknowledging that the theorems themselves had by then become fairly widely known and that, especially through the work of Turing, they had also "taken on a much more satisfactory form than they had had originally." He believed, however, that their philosophical consequences had never been adequately discussed.

As a first consequence Gödel observed that whatever philosophical position one may take toward mathematics, "the phenomenon of . . . [its] inexhaustibility" will necessarily arise in some form. That is so because the first incompleteness theorem shows that *"whatever well-defined system of axioms and rules of inference may be chosen, there always exist diophantine problems . . . which are undecidable by th[o]se rules, provided only that no false propositions . . . are derivable."* More explicitly, the second incompleteness theorem establishes that one cannot consistently assert both that the axioms and rules of a given formal system are correct *and* that they embrace all of mathematics; for the perception that the axioms are correct will entail the perception of their consistency, "a mathematical insight not [formally] derivable" from the axioms themselves.

Gödel distinguished the totality of all *objectively* true mathematical propositions from the totality of those that are formally demonstrable on the basis of *evidently* true axioms. The incompleteness theorems did not, he said, preclude the existence of "a finite rule producing all . . . evident axioms." But if there were such a rule we would be unable to recognize it as such, since "we could never know with mathematical certainty that all the propositions it produce[d

---

[2]His achievements, within their sphere, were certainly on a par with those of Gibbs, and in other respects, too, there were parallels between the two men. Like Gödel, Gibbs was a loner who rarely ventured far from his intellectual haven; and he too made his greatest contributions before he obtained a paid position, while he lived on the inheritance left him by his father.

were] correct; ... the human mind (in the realm of pure mathematics) ... [would thus be] equivalent to a finite machine that ... is unable to understand completely its own functioning." Because the theorems it produced would be subject to the limitations of the first incompleteness theorem, "there would exist ... diophantine problems ... undecidable ... by *any* mathematical proof the human mind can conceive."

Gödel certainly did not find such a conclusion attractive. Nevertheless he did not claim, as others later would, that the incompleteness theorems *refuted* the mechanistic view of mind. He asserted only that a certain "disjunctive conclusion" must hold — namely, that "*Either ... the human mind ... infinitely surpasses the powers of any finite machine, or else there exist absolutely unsolvable diophantine problems.*"

There followed, he believed, a corresponding disjunction of philosophical alternatives: Either "the working of the human mind cannot be reduced to the working of the brain, which to all appearances is a finite machine," or else "mathematical objects and facts ... exist objectively and independently of our mental acts and decisions." Those alternatives were not, of course, mutually exclusive. Indeed, he was firmly convinced of the truth of both.

With regard to the first possibility, Gödel remarked only that "some of the leading men in brain and nerve physiology" had "very decidedly" denied that psychical and nervous processes could be explained in purely mechanistic terms. He devoted the rest of the lecture to consideration of the second alternative, which, he claimed, was "supported by modern developments in the foundations of mathematics." In particular, he argued against the view that mathematics is only our own creation, and especially against the contention that mathematical propositions "express ... solely certain aspects of syntactical (or linguistic) conventions" and are therefore "void of content."

At first, Gödel tried to assert that a creator necessarily knows all the properties of his creatures — a view tinged with theological overtones related to ideas Anselm had invoked in his putative proof of the existence of God (which Gödel would later attempt to formalize); but he soon backed away from full-fledged advocacy of that position. To be sure, he conceded, we build machines whose behavior we cannot predict in every detail. Thus "if mathematics were our own free creation, ignorance as to the objects we created ... might still occur," as a result, say, of "the practical difficulty of too complicated computations." But if so, then "at least in principle, although perhaps not in practice," such ignorance would have to disappear as greater conceptual clarity was attained. Yet despite having provided "an insurmountable degree of exactness," research in the foundations of mathematics had, he avowed, contributed "practically nothing" to the solution of outstanding mathematical problems.

The reasoning seems strangely contorted and double edged, especially in view of Gödel's contention that progress in settling open questions about sets was to be sought through clarification and extension of the axioms concerning them. One has the feeling that he sensed, but was unable fully to confront, that there was a conflict between some of his most deeply held beliefs.

He went on to offer other arguments as well. He maintained, for example, that because a mathematician cannot "create the validity of ... theorems ... at his will," mathematical activity "shows very little of the freedom a creator should enjoy." On the contrary, he argued, "what any theorem does is ... to restrict [that] freedom," and whatever restricts the freedom of creation "must evidently exist independently" of it. He made no attempt to justify the reification involved, which very likely reflected his own theological convictions that God's creativity is unfettered, whereas man's is not.

He agreed that "a mathematical proposition says nothing about the physical or psychical reality existing in space and time, because it is true already owing to the meaning of the terms occurring in it." But he rejected the contention that "the meaning of th[ose] terms ... [is] something man-made, consisting merely in semantical conventions." Instead, he reaffirmed his Platonistic view that "concepts form an objective reality of their own, which we cannot create or change, but only perceive and describe." The meaning of mathematical statements thus inheres in what they say about relations among concepts.

Gödel had intended to publish his lecture in the *Bulletin of the American Mathematical Society*, and toward that end he spent much of the next two years revising it [457]. The resulting manuscript was found in his *Nachlaß*, but the textual alterations are so extensive that it is impossible to ascertain exactly what he said in the lecture. In addition to routine insertions, deletions and changes of wording, several whole pages are crossed out and two series of interpolations have been added, the first consisting of insertions to the body of the text, the second of footnotes. The sequence in which the interpolations occur in the text is chaotic, and the interrelations among them are byzantine: There are insertions within insertions, insertions to the footnotes, footnotes to insertions, and even footnotes to other footnotes. The situation finally became so confusing that Gödel himself found it necessary to draw up a concordance between the interpolation numbers and their locations.

In the end he never did submit the manuscript for publication; it appeared for the first time in 1995, in the third volume of his *Collected Works.* Probably he recognized that, however strong his own convictions about its subject matter, his arguments lacked persuasiveness. As he strove to refine them his attempts to refute mathematical conventionalism gradually became the focus of his intellectual endeavors, until finally he abandoned work on the Gibbs Lecture and turned to the writing of a separate essay entitled "Is Mathematics Syntax of Language?"

◊   ◊   ◊

The view that mathematics *is* syntax of language had been advocated especially by Rudolf Carnap and was one of the points on which Gödel had circumspectly disagreed with the Vienna Circle. In the quarter-century that had elapsed since his association with the Circle his views on the subject had matured and crystallized, but until the Gibbs Lecture he had always refrained from direct criticism of the conventionalist position.

In that regard it is notable that Carnap came to the institute in the fall of 1952 and remained there until the spring of 1954. The extent of his contacts with Gödel during that time is largely undocumented, but the two old friends must surely have crossed paths regularly and exchanged views on issues of mutual concern. Carnap's book *Logical Foundations of Probability* had appeared not long before, and Gödel evidently studied it at some length, for he remarked to Morgenstern that he considered Carnap's approach to probability the "most acceptable" of all that had so far been proposed [458]. It does not seem, though, that he discussed his critique of conventionalism with Carnap much, if at all, nor does Carnap's presence in Princeton appear to have been the direct impetus for its composition.

Rather, it was once again Schilpp who, on 15 May 1953, wrote Gödel to solicit a contribution for his forthcoming volume *The Philosophy of Rudolf Carnap.* Schilpp suggested that Gödel submit an essay of twenty-five to forty pages, to be entitled "Carnap and the Ontology of Mathematics." Gödel replied that he did not have the time to prepare such a lengthy article. Instead he offered to write a brief note, to be entitled "Some Observations About the Nominalistic View on the Nature of Mathematics." Parts of the text, he told Schilpp, he might take verbatim from his Gibbs Lecture.

Schilpp willingly accepted Gödel's counterproposal. After all, he replied, it was not the quantity but the quality of the writing that mattered. He had no doubt that whatever Gödel wrote would be of great interest, and perhaps, despite his earlier experiences, he thought that the composition of such a short piece would not take Gödel very long.

He should have known better. For although Gödel began work right away, he was, as always, a perfectionist, and he found it much harder to express his views than he had expected. As he struggled to find a satisfactory formulation of his thoughts the "few pages" he had envisioned grew into a lengthy manuscript that he revised and rewrote repeatedly. In the end he generated six distinct versions, two of which were eventually published in volume III of his *Collected Works*. His efforts extended until 1959, nearly five years past Schilpp's original deadline for receipt of the paper. Finally, on 3 February of that year Gödel advised Schilpp that since it was too late for Carnap to reply to his criticisms publication of them would be neither fair nor "conducive to an elucidation of the situation" [459].

In the interim he had promised several times to send a manuscript within a few weeks or months; but various circumstances had interfered. Toward the end of 1954 he experienced a brief episode of depression, during which he became convinced he was suffering from serious heart trouble and was about to die [460]. Not long after that his attention was diverted by a further request from Schilpp, who inquired whether he possessed a German translation of the article he had prepared for the Einstein volume. He did not, and thought for a time of preparing one. In the end, he left the translation to the editor of the German edition (Hans Hartmann), but he could not forgo the opportunity to provide a few additions to the footnotes. The following year Adele was hospitalized briefly with sciatica, and then some "unusual and important faculty matters" arose — presumably the contentious appointments of Alexander Koyré and George Kennan — that Gödel felt demanded his attention [461].

◇ ◇ ◇

His involvement with such internal political matters was a reflection of his changed status. For on 1 July 1953 — the same year that he was elected to membership in the National Academy of Sciences — the IAS had finally appointed him as a professor.

The delay in Gödel's promotion has puzzled many. Some have even regarded it as scandalous. (Von Neumann, for example, is said to have remarked, "How can any of us be called professor when Gödel is not?" [462].) Several of his colleagues, though, have suggested two principal reasons why he was not made a member of the faculty sooner. First, there was ongoing concern about his mental stability. Second, there were many who doubted that he would want to take on the administrative duties expected of faculty and who worried that his indecision and legalistic habit of mind might impede the conduct of institute business.

The former objection was voiced especially by Weyl and Siegel, both of whom left the IAS in 1951. Their opposition was surely the primary obstacle to Gödel's advancement. The other qualms were less publicly articulated but were probably more widely shared, and their substance is affirmed by Gödel's own testimony. He once admitted, for example, that apprehension that he might be expected to serve as an officer or committee member of the Association for Symbolic Logic had deterred him from joining that organization until ten years after its founding [463].

As it turned out, Gödel's mental problems did not affect his performance as a faculty member any more than they did his research. They simply formed an undercurrent to his existence, an omnipresent force that now and again surfaced temporarily but that for the most part remained submerged and subdued. And though he more than once complained of the burden his faculty duties imposed (he told his mother that he "often thought with regret back to the beautiful time when he didn't have the honor of being a professor" [464]), he nonetheless took a keen interest in institute affairs. Immediately following his promotion, for example, he began to raise questions about the adequacy of the institute's endowment and the security of its investments [465]; and within a short time, partly because he deliberated *so* much when asked to make decisions, his colleagues ceded virtually complete authority to him to determine which logicians would be offered visiting IAS memberships.

Deane Montgomery, who became a member of the IAS faculty in 1948 and continued beyond Gödel's retirement, described Gödel as having been a very good faculty member *if* one were willing to be patient with him and to put up with long telephone conversations, sometimes at inconvenient hours[3]; but he added that Gödel's tenacious concern for minor matters could become obstructive. Atle Selberg, another long-time colleague, agreed. He noted in particular how difficult it was for Gödel to make up his mind about candidates for membership.[4] And Armand Borel, who joined the faculty four years after Gödel, has written that despite his apparent remoteness, Gödel "acquit[ted]

---

[3] Gödel's fondness for telephone conversations, his difficulty in ending them, and his insensitivity to others' sleeping habits have been attested by a number of informants.

[4] Gödel himself acknowledged the problem in his letters to his mother. It was easiest, he said, "simply to say 'yes' to them all," which was "nearly always the end result" (FC 97 and 101, 31 October 1953 and 19 March 1954).

himself well of some of the school business." However, "in more difficult affairs" Borel sometimes "found the logic of Aristotle's successor ... quite baffling" [466].

Ironically, Gödel's elevation to professor coincided with his withdrawal from public participation in mathematical activities: After the Gibbs Lecture he never again addressed a mathematical audience nor attended a meeting of any mathematical society; he seldom went to lectures at the institute and never conducted any seminars there[5]; and though he continued to engage in mathematical research, all his publications after 1952 were revisions of earlier work.[6]

In part his professional isolation reflected the continuing shift of his interests away from mathematics toward philosophy; in part the intrusion of personal and family concerns; and, not least, the deaths within a little over five years of Einstein, von Neumann, and Veblen — the three close colleagues who had looked after him since his arrival in Princeton.

Of the three, it was Einstein's death on 18 April 1955 that had the greatest impact on him. In a letter to his mother written just three months earlier, Gödel mentioned that he often visited Einstein, whom he described as "the personi-fication of friendliness." He reported that Einstein was suffering from anemia and was unable to go out, but he said nothing about the aortic aneurysm that Einstein's doctors had discovered some six years earlier.

Those who lived within Einstein's household had known for some time that the aneurysm was growing and might burst at any moment. With others, how-ever, including Gödel, Einstein had been very discreet about personal matters. Fully aware of the time bomb ticking away within himself, he remained calm and stoic to the end. He continued to pursue his scientific and pacifist endeav-ors until the day the aneurysm ruptured, and he was alert and communicative until his last hours [467].

Had he understood the true nature of his friend's affliction Gödel would surely have attempted to visit Einstein during the four days that elapsed be-tween his collapse and death. As it was, however, his demise came as a

---

[5] When approached about doing so in 1967 he firmly declined. "I have never in my life conducted a seminar," he declared, and "It is a little late to start at 59. I am not good at such things anyway. I never go to lectures because I have diff[iculty] in following them even if I am well acquainted with the subj[ect] matter" (IAS memo, GN 020934.6).

[6] His *Dialectica* paper of 1958 presented results that, though previously unpublished, had been obtained nearly two decades before. Late in his life he also submitted a paper in which he proposed axioms "leading to the probable conclusion" that the power of the continuum is $\aleph_2$; but when errors were found he withdrew it (see chapter XII).

complete surprise; Gödel did not learn of the aneurysm until one of Einstein's doctors told him what the cause of death had been.

The suddenness of the loss left Gödel shocked and disturbed. He sought in vain to find a hidden meaning in its timing ("Isn't it curious," he remarked to his mother, "that Einstein's death followed scarcely 14 days after the twenty-fifth anniversary of the Institute's founding?" [468]), and for several days his sleep and appetite were affected. Shortly afterward he helped Bruria Kaufman, Einstein's last collaborator, to arrange the scientific papers Einstein had left in his office in Fuld Hall [469], and some months later he attended a concert at the institute given in Einstein's memory — the first time, he declared, that he had allowed himself "to endure two hours of Bach, Händel and the like."[7]

By then it was clear that von Neumann, too, was mortally ill. In August of 1955 bone cancer had unexpectedly been discovered in his shoulder, and though he had attempted to conceal his condition and maintain his usual, amazingly hectic schedule of work, by November he was confined to a wheelchair. The following January he reentered Walter Reed hospital, where he remained, except for a few brief appearances, until his death in February of 1957.

During and after the war von Neumann had often been away from the institute, serving as a collaborator in the design of the EDVAC computer at the University of Pennsylvania and as an adviser to various military and governmental agencies, including the Manhattan Project. His contacts with Gödel were consequently much less frequent than they had been in earlier years. Nevertheless, the two remained good friends and saw each other often enough that they felt no need to communicate by letter.

What they may have discussed on the occasions when they did meet is a matter for speculation. After 1946 von Neumann's activity at the IAS centered on the Electronic Computer Project, an endeavor he initiated and to which he contributed many ideas, especially concerning the logical organization of computers, that are now recognized as fundamental [470]. At about the same time he also began to develop a theory of self-reproducing automata. Both efforts involved aspects of formal logic that bore obvious connections to Gödel's earlier work, so one might expect Gödel to have taken more than a passing interest in them.

---

[7](FC 119, 18 December 1955.) Though he attended opera performances in New York with some frequency, Gödel otherwise cared little for classical music. He claimed that Bach and Wagner made him "nervous" and once asked his mother why good music should have to be "tragic." He preferred popular songs like "O mein Papa," "Harbor Lights," and "The Wheel of Fortune."

There is no direct evidence, however, that he did. Gödel was surely well aware of the computer project, for it was a matter of considerable contention among the IAS faculty.[8] But the first several years of the project were those in which he had turned away from logic to pursue his interests in philosophy and cosmology; besides, he had never displayed much interest in the practical applications of recursion theory. Von Neumann probably kept him informed of the direction of his research efforts, but beyond that it seems unlikely that Gödel paid much attention to developments in computer science.

If he did not, then his last contact with von Neumann is all the more remarkable. For in March of 1956 Gödel sent him a letter in which, after expressing the hope that the latest achievements in medicine might lead to his complete recovery, he broached a mathematical problem for his consideration. "Evidently," he noted, "one can easily construct a Turing machine that is capable of deciding, for each formula F of the predicate calculus and each natural number n, whether F has a proof of length n (where length is measured as the number of symbols in it)." Given such a machine, one can then define a function ψ of F and n and another function ϕ of n alone by taking $\psi(F, n)$ to be the number of steps required for the machine to make its determination and $\phi(n)$ to be the maximum value of $\psi(F, n)$ over all formulas F. Gödel then asked: "How rapidly does $\phi(n)$ grow for an optimal machine?" If there were a machine for which $\phi(n)$ grew no faster than n or n^2, "the consequences would," he declared, "be of the greatest significance"; in particular, it would mean that "despite the unsolvability of the decision problem, mathematicians' reasoning about yes-or-no questions could be completely mechanized" [472].

Gödel's recognition of the question's significance demonstrates once more the sureness of his instinct for what was most fundamental; for his query to von Neumann is the first known statement of what has since come to be called the "$P = NP$" problem[9] — the central unsolved problem in theoretical computer science [473]. One can only wonder what von Neumann might have said about it, for by then he was too ill to respond.

---

[8] As "the first ... venture of the Institute outside the realm of purely theoretical work," it was considered "out of place" even by "faculty members who had a high regard for the endeavour in itself" [471]. After von Neumann's death the project was phased out and the computer given to Princeton University.

[9] The equation serves as an acronym for the modern formulation of the question: Are the problems solvable by a deterministic algorithm in polynomial time the same as those solvable by a non-deterministic one in polynomial time?

◇ ◇ ◇

The void in Gödel's life created by the deaths of his two closest professional associates could never be fully overcome, of course. But soon afterward he did establish close contacts with two other mathematical colleagues with whom he corresponded until a few years before his own death.

One of those was Georg Kreisel, a younger logician who was a resident scholar at the IAS from 1955 to 1957. Kreisel had done important work in proof theory and intuitionistic logic and was interested in cultivating Gödel's acquaintance. He often went to Princeton expressly to see Gödel, and for a number of years the two carried on a prolific correspondence. They exchanged views on a wide spectrum of logical questions, including the extension of some of Gödel's own results.[10]

The other individual was Paul Bernays, with whom Gödel had lost contact after 1942. A Swiss citizen, Bernays had moved to Zürich in 1934 after having been dismissed from Göttingen as a "non-Aryan." Thereafter he had clung to rather low rungs on the academic ladder at the ETH, serving first as a temporary instructor and then, after 1945, as an "extraordinary" (that is, associate) professor with a half-time appointment. During the spring semester of 1956 he served as a visiting professor at the University of Pennsylvania, and while there he took advantage of his proximity to Princeton to visit Gödel and renew their lapsed friendship.

Then age sixty-seven, Bernays had broad interests in logic and philosophy. Since the 1930s his work had had many points of contact with Gödel's. It was he who had first presented a fully detailed proof of the second incompleteness theorem and who, later, had proved a syntactically formalized version of the completeness theorem. It was his much improved version of von Neumann's axiomatization of set theory that Gödel had adapted in his 1940 monograph. And it was his interest in Gödel's "functional interpretation" of intuitionistic logic (of which he had been unaware until that spring of 1956) that finally induced Gödel to publish the ideas he had presented in his 1941 lectures at Yale and the IAS.

The article in question (*Gödel 1958*) appeared in *Dialectica*, a journal that Bernays, Ferdinand Gonseth, and Karl Popper had founded in 1947, in a *Festschrift* issue honoring Bernays's seventieth birthday. Relatively brief, it was Gödel's only publication in German since his emigration, and not surprisingly

---

[10] It has come to light recently that one of the results for which Kreisel is best known, the so-called no-counterexample interpretation, had been partly anticipated by Gödel years before, in his lecture to the Zilsel circle.

betrayed some loss of facility in his native language. Nonetheless it was a seminal contribution, both technically and philosophically, and as such was a fitting tribute to a colleague who, from that time forward, was to serve as one of Gödel's principal sounding boards on mathematical and philosophical issues. (Others who served in that capacity included Georg Kreisel, Hao Wang, William Boone, and Gaisi Takeuti.)

The content of the paper — essentially a précis of what Gödel had expounded more fully in his lectures seventeen years before — requires little further elaboration. In it Gödel outlined his notion of a computable function of finite type and stressed how it could be applied to provide a constructive, though not finitary, proof of the consistency of classical number theory. He also noted that "starting from the same basic idea" it was possible to construct much stronger systems, which he thought might be employed in a similar fashion to prove the consistency of analysis.

Privately, he had already suggested that possibility to Bernays [474], and also to Kreisel, who was strongly interested in such matters. Kreisel soon succeeded in extending Gödel's interpretation to analysis, and shortly afterward he lectured on his results at the Summer Institute for Symbolic Logic, held at Cornell in June of 1957. It was there that Gödel's notion of functional interpretations first came to the attention of logicians at large, whose interest was immediate and widespread.

During the next few years Kreisel maintained an active correspondence with both Gödel and Bernays, and also with Clifford Spector, a young logician of great promise who was pursuing similar directions in his own research. Spector made further progress toward the desired consistency proof, and when, at Gödel's invitation, Bernays came to the IAS for the 1959–60 academic year, Spector took the opportunity to discuss his results with the two of them.

The following year Spector himself came to the IAS and there prepared an account of his work. His achievements greatly impressed Gödel, who wrote a glowing job recommendation for him in which he described Spector as "probably *the* best logician of his age group in this country" [475]. Sadly, however, Spector never had the chance to take up a new academic position; for the following summer, just as his stay at the institute was drawing to a close, he suddenly died of an acute form of leukemia. It was left for Kreisel to edit the manuscript of his paper, which was published posthumously with a postscript by Gödel [476].

◇ ◇ ◇

The foregoing account has focused on the external events in Gödel's life during the 1950s. But professional concerns were no longer the only matters that occupied his attention. Other sources, especially his correspondence with his mother, reveal that, in marked contrast to his behavior during the years leading up to World War II, after the war personal and political worries often interfered with his work and his peace of mind.

He was concerned especially about the situation in his former homeland, where the Allied occupation was finally coming to an end. He worried about the welfare of his and Adele's relatives in Austria and Germany. And he continued to seek compensation from the Czech regime for the confiscation of the family villa in Brno.[11]

At times he became very pessimistic. In 1953, for example, he declared that "we live in a world in which 99% of all beautiful things are destroyed in the bud" and asserted that "certain forces" were at work that were "directly submerging the good" [477]. Yet he also contrasted the situation in Europe with that in America. In the United States, he said, one felt "surrounded by genuinely good and helpful people"; those in public service seemed actually to be interested in helping the citizenry rather than in making their lives miserable [478].

Six months earlier his opinion of government officials had been less charitable: In reporting on the ceremony at Harvard where he received his honorary Doctor of Science he said that he had been "thrust quite undeservedly into the most highly bellicose company" because Robert A. Lovett (then Secretary of Defense) and John Foster Dulles ("the architect of the peace treaty with Japan," which Gödel viewed as a step toward war with Russia) were among the other honorary degree recipients [479]. The about-face is explained by the fact that an election had taken place in the meantime and Truman was about to be replaced as president.

Indeed, in October of 1952 Gödel wrote his mother that his preoccupation with politics during the preceding two months had left him little time for anything else. He blamed Truman for fomenting war hysteria, for creating the climate for McCarthy, for increasing taxes, and for prolonging the war in Korea. Like most other intellectuals, he looked forward eagerly to a change of administration. But his failure to support Stevenson set him apart from most of his friends. After the election, for example, Einstein quipped to Ernst Straus,

---

[11] In the end his mother received about 60,000 Austrian shillings as recompense (the equivalent then of about $2,400), an amount far below the value of the property.

"You know, Gödel has really gone completely crazy — he voted for Eisenhower!" [480]

Gödel held out great expectations for the new president and seems to have approved of his subsequent performance in office. Within a year of his inauguration Eisenhower had, Gödel wrote, negotiated a cease-fire in Korea, reduced the military budget by $3 billion, and brought inflation to a standstill [481]. Gödel was disappointed that Eisenhower did not extend clemency to the Rosenbergs, but he laid blame for their execution primarily on their defense team, which, he believed, had "intentionally" made "unbelievable mistakes" in order to give the Communists "new proof of American barbarity" [482].

In foreign affairs Gödel regarded Konrad Adenauer as "a second Brüning" and feared for a time that a second Hitler might follow. He was also pessimistic about the situation in the Middle East, where he correctly foresaw that the intransigence on both sides would lead to a local war with worldwide repercussions. But in contrast to his fears during the Truman administration he no longer feared that a third World War was imminent, partly because of his faith in Eisenhower's leadership (someone like him, he declared, came to power "only once every few hundred years") and partly because he thought the horror of atomic weapons would act as a deterrent. In addition, he believed it was "too early" for another World War, since the first two had been separated by twenty-five years and only seventeen had elapsed since the start of the second [483].

Besides politics, Gödel pursued a variety of other outside interests. He had no hobbies, but he read omnivorously, occasionally attended the opera or visited museums in New York (he had at least a passing interest in modern art), and, for diversion, watched plays and variety programs on television. He exercised by doing calisthenics, and when he had the opportunity he enjoyed swimming in the ocean. On his visits to the New Jersey shore he also liked to exhibit his skill at the boardwalk game of Skee-ball; he boasted to his mother of the prizes he had won there for Adele.

In his letters to his mother and brother, as nowhere else, Gödel appears as a human being caught up in the affairs of daily life.[12] Among the other topics that figure prominently in his correspondence with them, apart from discussions of diet and health, are German literature and Austro-German history. Gödel was especially fond of some of the nineteenth-century German roman-

---

[12] Prior to the 1950s the correspondence focused on conditions in Austria and on Kurt and Adele's efforts to help their relatives there. After Austria's economic stabilization, however, the letters are concerned primarily with day-to-day affairs.

tics: He mentioned in particular the poetry of Rudolf Baumbach and Theodor Fontane's novel *Effi Briest.* He also liked fairy tales (those by Eduard Mörike, for example) and declared that only in such stories was the world represented as it should be [484]. He was disdainful toward much of Goethe's writings and found Shakespeare "hard to get into." Among more modern writers he esteemed the "bittersweet love stories" of the feuilletonist Raoul Auernheimer, as well as the stories of Gogol and Zweig and some of those by Kafka.

As to history, both Gödel and his mother had an abiding interest in the Mayerling affair, the suicide/murder of Crown Prince Rudolf and his mistress Mary Vetsera at a hunting lodge outside Vienna in 1889. Gödel read each new account of the incident avidly, but always with skepticism; for he suspected, probably with good reason, that details of the affair had never been fully revealed. He maintained a similar interest in the life of Ludwig II of Bavaria and was equally skeptical of the received views about him. He doubted, for example, that Ludwig was actually mad and, despite the extravagance of his castles, thought that his taste in art was probably not as bad as it was reputed to be. He believed that Ludwig had been committed for political reasons and had not drowned himself, as historians claimed, but had been murdered because of his opposition to the Prussian regime [485].

Another subject that Gödel felt free to discuss with his mother was religion.[13] His own beliefs are revealed most fully in a series of four letters that he wrote to her during the summer and fall of 1961. In them he reiterated both his faith in the power of rational inquiry and his contempt for the attitudes of contemporary philosophers toward religious questions. "Ninety per cent" of them, he said, "see their principal task as that of beating religion out of men's heads." He freely acknowledged that at present "we are far from being able to provide a scientific basis for the theological world view." Nevertheless, he believed it was "quite unjustified" to maintain that "just in the sphere of religion nothing can be achieved through our understanding." On the contrary, he thought it might be possible "purely rationally" to reconcile the theological *Weltanschauung* "with all known facts" [486].

What prompted Gödel to discuss his religious views was his mother's query whether he believed in an afterlife — a question to which his answer was

---

[13] Although each of them had married an individual who was at least nominally Catholic, they shared an antipathy toward Austria's state religion — an attitude based on opposition to religious dogmatism and reinforced by their experience of Austrian clerical fascism. (In Europe, Gödel declared, religion was a matter not of conviction but of allegiance to "manifestly harmful" political parties.)

unequivocally affirmative. Science, he asserted, has shown that "the greatest regularity and order prevail in all things" and so has confirmed that the world is rationally organized. Hence there must be another world beyond the present; for "what sense would there be," he asked, in creating a being such as man, with all his potential for development, if he were allowed to achieve "only one in a thousand" of the things of which he is capable? To be sure, he conceded, the nature of that other world is something about which we can only conjecture. But since "we one day found ourselves in this world without knowing why or whence," the same thing could also happen again "in the same way." Such an expectation is not, he argued, in conflict with modern cosmological theories, for though they posit that the universe had a definite beginning and will "in all probability also have an end," they leave room for the creation of a new heaven and earth thereafter, as prophesied in the book of Revelation [487].

One might ask: If we are to enter into the next world in the "same way" as this, with no greater awareness of why we are there or whence we have come, what is the point?

Gödel's answer was that he did not, in fact, believe that we would begin our second life in a comparable state of ignorance. On the contrary, he expected not only that we would be born into the next world with "latent recollections" of this one, but that the new abode would be an everlasting intellectual paradise in which "everything of importance" would be perceived "with the same unerring certainty as that $2 \times 2 = 4$" [488].

But why should two worlds be necessary? If such a state of perfect knowledge is to be attained in the hereafter, why not here and now?

Gödel addressed those questions in the second of his theological letters, by invoking an argument akin to the so-called anthropic principle. If we were able to look "deeply enough within ourselves," he reasoned, each of us could conclude that "among all possible beings, I am the one with just this combination of characteristics" — including the propensity to make mistakes and the capacity to learn from them. Hence "if God had created in our place beings who did not need to learn anything," they would simply not be *us*[489].

In essence, the argument begs the question by asserting that God created us as we are because otherwise we wouldn't *be* what we are, and so *we* wouldn't exist at all. Of course, Gödel was not defending a scholarly thesis, but writing informally to his mother; still, the discussion seems rather naive for one so philosophically sophisticated.

Overall, the theological view put forward in these letters combines certain traditional elements of Christian faith (man's fallibility, yet ultimate perfectability; the promise of eternal life) with an abiding rationalistic optimism characteristic of the previous century. In particular, Gödel's belief that science has affirmed "the greatest regularity in all things" — especially his claim in the last of the four letters that all of science is based on the premise that "everything has a cause" — flies in the face of the quantum-mechanical world view.

◇   ◇   ◇

Gödel's relationship to his mother, as conveyed in his letters to her, was that of a modest, respectful, and loving son who, despite the years spent apart, remained concerned about her welfare, interested in her activities, and nostalgic about their past times together. He wrote to her about once a month, usually at some length, each time enclosing a monetary contribution toward her support.

The tone of the letters is relaxed and deferential, but there were occasional points of contention. In particular, Gödel frequently had to reassure his mother about his diet; from time to time he found it necessary to defend Adele against Marianne's criticisms; above all, he was at pains to explain his continuing failure to return to Vienna to visit.

Adele's own trips to Europe exacerbated the tensions. Marianne objected to her son's being left alone in Princeton, and she suspected her daughter-in-law of using what money there was for travel to visit her own relatives. Gödel chided his mother for making such accusations, but the excuses he gave for not coming along with his wife (for example, that he had nightmares about not being allowed to return to the United States) were not very convincing. How sincere he was about them is hard to judge.

Marianne was particularly incensed by a sudden trip Adele made to Vienna in the spring of 1953. The sojourn was occasioned by Adele's concern for her sister Liesl, whom she had been led to believe was gravely ill. On arrival, however, she discovered that the situation was not as it had been portrayed to her, and it was not long before conflict erupted between her and her family. She wrote back that she would not have come had she known the true state of affairs and declared that she now realized her family had "always regarded her as a milk cow."

In the letter in which he confided that news to his mother Gödel expressed the hope that the family strife might actually have a beneficial effect, since he believed that in the years prior to their marriage Adele had developed a pathologically exaggerated dependence on her relatives that had aggravated her nervous disposition and been responsible for the "abnormal attitude" she was wont to display toward people she was with. He did not deny that his wife had "great faults" (attributable in large part, he thought, to her "morbid mental state"), but he maintained that "as far as life in Princeton is concerned," she was "in general quite normal" [490].

Three years later, in 1956, family problems again impelled Adele to return to Vienna on short notice. She went then to rescue her eighty-eight-year-old mother, who was living alone and not being properly cared for. She wasted little time in selling her mother's apartment and bringing her back to share their home in Princeton. Before leaving the city, however, she took time to visit her mother-in-law and managed at last to set matters right between them. Evidently Marianne recognized that Adele's concern for her mother was justified and that Mrs. Porkert was no longer capable of managing her own affairs.

Adele's mother knew no English, and at the time she moved in with the Gödels she had also begun to exhibit some signs of mental confusion. She had trouble finding her way around the house, often going into the wrong room, and occasionally acted as though she were still living in her apartment on Langegasse. Nevertheless she blended unobtrusively into the household. Adele cared for her devotedly, and Kurt seems to have been quite fond of her. She remained with them until her death from heart disease on 22 March 1959.

◇ ◇ ◇

The year before that Gödel finally extended an invitation to his own mother to come to Princeton for a visit. At first he suggested she might travel with Morgenstern's sister, who was just then planning a visit of her own. In the end, however, Rudolf decided to accompany her. The two arrived in mid-April and stayed until the first week of May. Because of Mrs. Porkert (who by then was confined to a wheelchair) there was no room for them to stay in the house on Linden Lane, so Gödel offered instead to find accommodation for them in a nearby hotel or apartment. He told his mother not to worry about the cost, as he would of course pay for her lodging; but he did not extend the same offer to his brother.

As it turned out the weather was bad during most of their visit. Nonetheless the reunion was a joyous occasion for all concerned. Marianne was impressed

with Princeton and enjoyed visiting with the Morgensterns and going on excursions to New Hope and Washington Crossing. Most of all, of course, she
was happy to see her son, his home, and the place he worked, and to reassure
herself about his diet and state of health.

The visit proved so pleasurable, in fact, that Marianne planned to return the
following year as well; but the expenses associated with Mrs. Porkert's final
illness made it impossible for her to do so. She came instead in the spring
of 1960, and every two years thereafter — a total of four times altogether —
until her own death in the summer of 1966.

XI
New Light on the Continuum Problem
(1961–1968)

MATHEMATICAL LOGIC CAME of age in America during the 1950s. It was then that comprehensive textbooks on modern logic first appeared in English — most notably, Kleene's *Introduction to Metamathematics*; then that the development of electronic computers began to spark increasing interest in applications of recursion theory; and then that the successes of A. I. Maltsev, Leon Henkin, Abraham Robinson, and Alfred Tarski in applying Gödel's completeness and compactness theorems to problems in algebra fostered the creation of the new discipline of model theory.

Gödel himself played no direct role in any of those developments; but he kept abreast of important advances,[1] and because his work was of seminal importance to so much of logic, the growing interest in the field at last brought his theorems to the attention of a wider audience: Between 1962 and 1967 three different English translations of the incompleteness paper appeared, as well as a popular exposition of the incompleteness theorems by Ernest Nagel and James Newman.[2]

The earliest of those translations, by Professor Bernard Meltzer of the University of Edinburgh, was published without Gödel's authorization. Its pub-

---

[1] Among the results that most excited and pleased him were Richard Friedberg's solution of Post's Problem, William Boone's proof of the unsolvability of the word problem for groups, and Dana Scott's discovery that the existence of a measurable cardinal is incompatible with the Axiom of Constructibility.

[2] Their article "Gödel's Proof" was first published in 1956 in *Scientific American*. Shortly thereafter it was reprinted as a chapter in Newman's anthology *The World of Mathematics*, and two years later an expanded version was issued as a separate monograph.

lisher, the British firm of Oliver and Boyd, had requested his permission, but he had failed to respond in a timely fashion. After waiting six months they had therefore gone ahead and obtained permission from the editors of the journal in which the article had originally appeared.

The Meltzer translation was seriously deficient in many respects and received a devastating review in the *Journal of Symbolic Logic* [491]. Gödel complained to the publisher of the American edition (Basic Books) that "The introduction [by the philosopher R. B. Braithwaite] treats some questions very inadequately and contains several wrong or misleading statements." He worried that the dedication might be mistaken as his own and feared that readers might think he had cooperated in the book's production. He questioned "the whole idea of making a paper written for specialists available to a large public" and declared that "if . . . carried out at all" such an endeavor "should at least be done in the right manner." He could do little to rectify the damage that had already been done, but he demanded that in any subsequent printings of the book the full title of his paper be given on the cover, the dedication be deleted, and a note by him be inserted after the preface [492].

Fortunately, the Meltzer translation was soon supplanted by a better one prepared by Elliott Mendelson for Martin Davis's anthology *The Undecidable*; but it, too, was not brought to Gödel's attention until almost the last minute, and the new translation was still not wholly to his liking.

Davis had previously obtained Gödel's permission to include translations of some other papers of his in the anthology, but he had inadvertently failed to mention *1931a*. Gödel first became aware of the intention to include it only in 1964, just a year before the volume was scheduled to come out, and while he did not object to the idea of having the incompleteness paper appear in the collection, he found the translation of it "not quite so good" as he had expected. "Some improvements," he thought, "would certainly be desirable." He asked Davis whether the translation could perhaps be replaced by another then in preparation [493]; but when informed that there was not time enough to consider substituting another text, he declared that Mendelson's translation was "on the whole very good" and agreed to its publication.[3]

The translation Gödel favored was that by Jean van Heijenoort, an erstwhile Trotskyite revolutionary who in later life became the most eminent historian of mathematical logic [494]. Van Heijenoort was not only a meticulous scholar but an accomplished linguist. Trained in both mathematics and philosophy

---

[3] Afterward he would regret his compliance, for the published volume was marred throughout by sloppy typography and numerous misprints.

— he was a member of the mathematics faculty of New York University and taught part time in the philosophy department at Columbia — he spoke English, German, French, Spanish, and Russian fluently and could read several other languages, including Latin and ancient Greek. He had developed an interest in logic after delving into *Principia Mathematica* in the New York Public Library, and in 1957, "spurred on by the inspirational encounters he had" during a landmark conference at Cornell (the Summer Institute of the Association for Symbolic Logic), he had begun a systematic survey of the literature of modern logic [495].

Two years later Professor Willard van Orman Quine of Harvard learned of van Heijenoort's work and forthwith invited him to undertake the editing of a source book in mathematical logic, one of a series of such source texts that Harvard University Press had issued in the history of the sciences. The resulting work, *From Frege to Gödel: A Source Book in Mathematical Logic, 1879–1931*, was an exemplar of its genre. Beginning with Frege's *Begriffsschrift* and ending with Gödel's incompleteness paper, the volume contained English texts of forty-six papers, translated from French, German, Italian, Russian, Latin, and Dutch. (A second volume, covering the period 1932–60, was planned but never produced, perhaps because preparation of the first involved so much more time and effort than had been expected.) Each paper was prefaced by an incisive introductory note, and references to works cited in the extensive bibliography at the end of the volume were made uniform. Great care was taken to ensure the accuracy of the translations and citations and to adopt editorial apparatus that would help rather than hinder the reader.

In the end van Heijenoort devoted seven years to the project. It brought him into contact with some of the greatest figures in the field, including Russell and Gödel, and secured his international reputation once and for all.

In the preface to the volume van Heijenoort noted that Gödel was one of four authors who had personally read and approved the translations of his works. He had done so, van Heijenoort remarked, "after introducing a number of changes" in the text of *1931a* — a statement that barely hinted at the protracted negotiations that had taken place over details of the translation.

Privately van Heijenoort declared that Gödel was the most doggedly fastidious individual he had ever known. Between 10 May 1961 and 5 June 1966 the two exchanged a total of seventy letters and met twice in Gödel's office in order to resolve questions concerning subtleties in the meanings and usage of German and English words. The potential for conflict between them was high, for Gödel was concerned that van Heijenoort was not a native speaker

of English,[4] while van Heijenoort had his own strong and sometimes idiosyn-
cratic opinions about linguistic matters. With Gödel, however, van Heijenoort
was always patient, deferential, and accommodating; and though both men
were perfectionists, each of them respected that trait in the other. Accordingly, their interchanges remained thoroughly cordial.

◇   ◇   ◇

Gödel's efforts in overseeing translations and reprints of his earlier works
took up a substantial amount of his time during the 1960s — time that he
would have preferred to devote to the study of philosophy. The translation
work interfered in particular with his studies of Husserl's writings, to which
he had begun to devote himself in 1959 after abandoning work on the essay
for Schilpp's Carnap volume.

Details of those studies can be gleaned from Gödel's shorthand notes, from
Hao Wang's published recollections of his conversations with Gödel (*Wang
1987*), and from a shorthand manuscript entitled "The Modern Development
of the Foundations of Mathematics in the Light of Philosophy," found among
Gödel's papers after his death and published posthumously in volume III of
his *Collected Works*. The stimulus for the latter seems to have been Gödel's
election to membership in the American Philosophical Society, for filed with
the manuscript was a letter from the society's executive secretary inviting him,
as a new member, to deliver a talk at an upcoming meeting [496]. So far as is
known Gödel never replied to the invitation, but he probably composed the
paper in response to it.

The stated purpose of the essay was "to describe ... the development of
foundational research in mathematics since the turn of the century, and to fit
it into a general schema of possible world-views ... [classified] according to
the degree ... of their affinity or diffinity with metaphysics." In his scheme
Gödel placed skepticism, materialism, and positivism on the left, spiritualism,
idealism, and theology on the right, declaring that it was "a familiar fact" that
since the Renaissance the development of philosophy had "by and large gone
from right to left."

Gödel noted that "mathematics, by its nature as an apriori science," had
"long withstood" that leftward trend. But, he charged, "around the turn of
the century" the significance of the antinomies of set theory had been "exaggerated by skeptics and empiricists" and "employed as a pretext for [a]

---

[4] Ironically, the English grammar Gödel owned and cited as authoritative, Poutsma's *A Grammar of Late Modern English*, was written by a Dutchman.

leftward upheaval." In fact, Gödel asserted, the antinomies "did not appear within mathematics, but near its outermost boundary toward philosophy" and had subsequently "been resolved in a manner that is completely satisfactory and, for everyone who understands the theory, nearly obvious." Nonetheless, in deference to the prevailing *Zeitgeist,* Hilbert had created a "curious hermaphroditic . . . formalism . . . which sought to do justice both to the spirit of the time and to the nature of mathematics."

Hilbert's program had foundered on the incompleteness theorems, in the wake of which Gödel saw only two possibilities: "One must either give up the old rightward aspects of mathematics or attempt to uphold them in contradiction to the spirit of the time." The latter course, whereby "the certainty of mathematics is to be secured not by . . . the manipulation of physical symbols, but rather by cultivating . . . knowledge of the abstract concepts themselves" was in Gödel's view "undoubtedly worth the effort." What was needed was "a clarification of meaning that does not consist in giving definitions"; and that, Gödel believed, might be provided by Husserl's phenomenology.

Gödel argued that there was "no objective reason" to reject phenomenology, and he offered two reasons why it might be expected to lead to progress in foundational questions. First, it was in accord with the way children developed understanding of concepts by passing through succesively higher "states of consciousness." Second, "in the systematic establishment of the axioms of mathematics, new axioms, . . . [not derivable] by formal logic from those previously established, [had] again and again become evident."

The claims have a familiar ring. Gödel had advanced similar ideas, though without reference to Husserl, in his Princeton Bicentennial remarks, in his essay on Cantor's continuum problem (which he would revise and strengthen for a 1964 reprint), and in his Gibbs Lecture. But the tone of the "modern development" essay is far more polemical.

Perhaps that is why he never published it or told anyone else about it. His reluctance to give public expression to views that might be considered iconoclastic was of long standing, so his lashing out against the pernicious effects of the *Zeitgeist* was very likely an act of private catharsis. The essay was not among the items he later listed as candidates for publication, and the text exists only in a shorthand draft that shows little evidence of rewriting.

◇   ◇   ◇

Some months before, Gödel had declined an invitation from Harvard to deliver the prestigious William James Lectures, and had thereby passed up an important opportunity to promulgate his philosophical views. Presumably he did so on grounds of health, for at about that same time he wrote Bernays that he had felt "very bad the whole past year."

How bad his condition really was is hard to say. The previous May, during Rudolf and Marianne's second visit, Morgenstern had reported that Gödel looked unwell and undernourished. He wore an alarm wrist watch to remind him when to take various pills and carried a box of baking soda around with him in a pouch. For some time, too, he had been making weekly visits to a psychiatrist. But the sessions had done little to counter his anorexia or lessen his fixation on his bowel habits. Shortly after his mother's return home, for example, he wrote to thank her for the enema tubes she had given him. He would soon try them, he said, but in the meantime he had found that taking milk of magnesia on an empty stomach was almost as effective [497].

Despite his chronic malnutrition, there is little to suggest that Gödel's physical condition was worse than usual or that Adele had become much concerned about it. On the contrary, she was frequently away on long trips — to Canada in 1963, and to Europe in each of the years 1959–61 and 1965–66 — and, despite her husband's phobia about air quality, she continued to smoke in their home.[5]

In the end it is impossible to separate Gödel's genuine health problems from his hypochondria and his desire to maintain his privacy in the face of growing public recognition. He resisted anything that would take him away from his work, and frail health, whether real or imagined, was an effective excuse for avoiding unwanted commitments.

Still, he could not avoid all intrusions. In particular, he could not ignore an internal dispute that arose at the IAS in 1962 — an affair that he characterized as "a great row" [498].

The conflict erupted when the mathematics faculty at the IAS proposed to offer a faculty appointment to John Milnor, a differential topologist at Princeton. Milnor was an eminent figure in the field, so no opposition was expected. But the proposal precipitated a heated debate concerning the propriety of Princeton's educational institutions vying with each other for faculty.

According to Armand Borel [499], the mathematicians both at Princeton and at the IAS believed that offers to a faculty member should be based solely on merit and that individuals receiving such offers should be free to choose among them. The chairman of the institute's board of trustees, however, maintained

---

[5] Gödel never complained about either aspect of his wife's behavior. He believed that travel was good for her, and he continued to reassure his mother that he was well taken care of during her absences.

that years before, during Flexner's tenure as director, an agreement had been entered into that prohibited the institute from making any offers to colleagues at Princeton.

The chairman's opinion was communicated to the mathematics faculty by director Oppenheimer prior to the vote recommending Milnor's appointment — an action that Atle Selberg, then the executive officer of the School of Mathematics, deemed highly improper. To prevent further interference by the administration in faculty affairs, Selberg thenceforth excluded Oppenheimer from the meetings of the mathematics faculty; but Gödel regarded that, too, as improper, and so ceased to attend the meetings himself. From then on Selberg had to solicit Gödel's vote by proxy [500].

After much discussion the IAS faculty finally passed a resolution urging that they "be free to extend professorial appointments to faculty ... [at] Princeton University, with due regard to the interest of science and scholarship and the welfare of both institutions"; the vote was fourteen to four with two abstentions. The trustees, however, disregarded the faculty's plea. At their meeting in April of 1962 they declared that the IAS would continue to abide by the supposed earlier agreement. Accordingly, they refused to consider Milnor's nomination [501].

The trustees' rebuff of the faculty kindled resentment that smoldered for years afterward. Yet, as it turned out, the issue that created the furor was shortly resolved in a manner very much in accord with the faculty's wishes; for before the trustees met again the following April Oppenheimer secured the assent of President Goheen of Princeton to the adoption of a new policy, whereby either institution, after mutual consultation, might offer a position to a faculty member at the other. It was not possible for Milnor's nomination to be reconsidered, however, because in the meantime all available positions in mathematics had been filled.

The new policy was finally placed into effect in 1969, when another opening arose. Once again the mathematicians nominated a Princeton topologist (Michael Atiyah), and this time the appointment was approved [502]. (Milnor had accepted a position at MIT the year before, but in 1970 he too finally came to the IAS.)

◇ ◇ ◇

The appointment controversy disturbed and distracted Gödel; and just as it began to die down another, most unexpected, event occurred that captured his full attention. In the spring of 1963 Paul J. Cohen, a young mathematician at Stanford, proved at last that the Axiom of Choice was independent of ZF set theory and the Continuum Hypothesis independent of ZFC.

Gödel had of course conjectured both results long before and had attempted to prove them. But in the two decades that had elapsed since then, little progress had been made. The most important discovery during that period was a negative result found by the English logician John Shepherdson: In 1953 he showed that the method of inner models that Gödel had used to establish the relative consistency of the Axiom of Choice and the Generalized Continuum Hypothesis could *not* be used to establish their independence. Specifically, he proved that for no formula $\Phi(x)$ of ZF could one prove in ZF that the class of sets that satisfy Φ form a model of ZF $+ V \neq L$ [503]. Hence to establish the independence of any statement implied by the Axiom of Constructibility, a new method would have to be devised.

Such a development had not seemed imminent, and Cohen's breakthrough was all the more surprising because he was not a logician (although, from another point of view, he was by that very fact not steeped in methods that had failed to solve the problem). He had received his Ph.D. from the University of Chicago in 1958 in harmonic analysis, had gone on to solve an important problem in Banach algebras, and had then spent the years 1959–61 at the IAS, where he spoke with some logicians, particularly Solomon Feferman, concerning outstanding problems in their area of specialization. But at that time he had not had any contact with Gödel [504].

After leaving the IAS Cohen went to Stanford, where he continued to discuss logic with Feferman and Kreisel. Feferman referred him to several important papers in the field, and Cohen used Feferman as a sounding board for various ideas he had about how one might prove the independence of the Axiom of Choice. In the course of those musings Cohen independently rediscovered some of Shepherdson's results — in particular, the fact that there must be a *minimal* standard model of ZF (a countable model that is a submodel of every other standard model) [505].

Feferman had tried to use nonstandard models of ZF (those in which the $\in$-relation is not well founded) as a means of demonstrating the independence of AC, but Cohen made "a firm decision ... to consider only standard models" [506]. Cohen saw that the Skolem Paradox — the fact that ZF set theory, though it asserts the existence of uncountable sets, must (if consistent) admit countable models — could be exploited to construct countable *extensions* of a given model (for example, the minimal one) in which all the ordinals (and hence all the constructible sets in the model) were left unchanged, while new subsets of the integers were adjoined.

Cohen's method, which he called "forcing," is, as he noted, somewhat analogous to transcendental field extensions in algebra. The basic idea is that what passes for $\mathcal{P}(\omega)$ (the set of all subsets of the integers) *within* a countable standard model M of ZF is, if viewed from *outside* that model, actually a countable set; it appears uncountable within the model because the one-to-one correspondence between it and ω is not itself a set within the model. Hence there must be uncountably many subsets of ω that lie outside the model, one of which (hopefully) might be adjoined to M so as to obtain a new model $N \supset M$.

The trick is to pick the right subset of ω to add. On the one hand, the axioms of ZF guarantee that any subset definable within M must already belong to M itself. On the other, if the resulting structure N is to be a model of ZF with the same ordinals as M, not just any subset $a \notin M$ will do; from the point of view of M the subset to be adjoined must (as Cohen put it) in some sense be "generic," that is, it must not provide outside information about M.

Cohen succeeded in giving a precise technical meaning to the notion of being generic over M. To demonstrate the independence of $V = L$ from ZFC, for example, he defined a new subset a by considering, for each $n \in \omega$, the pair of formulas $n \in a$ and $n \notin a$ (in the *forcing language* of M, containing names for each element of M together with the new constant symbol a), and called a finite, consistent set of such formulas a *condition*. He then showed how to construct a compatible infinite sequence P_n of conditions, ordered by inclusion, such that for each sentence A of the forcing language either A or its negation would be forced to be true in the eventual model N by some P_n (and so by all its successors).[6] The set a was then determined by the limiting sequence P (the union of all the P_n's).

Like all path-breaking results in mathematics Cohen's proofs were immediately subjected to intense critical scrutiny — in the course of which, as is very often the case, a few errors were discovered. Those were soon corrected, but Cohen nonetheless found the process of validation upsetting. He feared that some hidden, irreparable mistake might suddenly come to light and that, even if correct, his proofs were likely to encounter resistance among logicians. To quell the criticism he therefore wrote Gödel seeking his "stamp of approval" [507].

In response, Gödel advised Cohen to put aside any feelings of dismay. "You have," he declared, "achieved the most important progress in set theory since its axiomatization." The proofs, in his opinion, were "in all essential re-

---

[6]The forcing relation itself was defined by a complicated induction.

spects ... *the* best possible" and in reading them he had experienced "delight" comparable to "seeing a really good play."

With utter disregard for his own past practice, Gödel urged Cohen to publish his proofs without delay. "If I were in your place," he solemnly advised, "I [would] ... not now try to improve the results, ... [for that] is an infinite process." At the same time, however, he offered some suggestions on how Cohen might alter the manuscript to improve the exposition [508].

Throughout the summer a substantial amount of correspondence was exchanged between the two, almost all of it on editorial matters. Gödel had published his consistency results in the *Proceedings of the National Academy of Sciences*, and he offered to submit Cohen's work there also. (Cohen could not do so himself since he was not a member of the academy.) Cohen accepted eagerly, but it was not long before he became frustrated by Gödel's continued deliberation over fine points. Because of length restrictions in the journal he readily agreed to Gödel's suggestion that the paper be broken into two parts, and eventually, in an effort to avoid further delays, he gave Gödel carte blanche to make alterations to the text as he saw fit. Gödel finally sent off part I of the paper on 27 September, and part II arrived in the hands of the editors exactly two months later.

The method of forcing immediately became the focus of an intense flurry of research activity in set theory that extended well into the 1970s. In the course of just a few years Cohen's technique was refined, extended, and simplified in various ways and used to demonstrate the independence from ZFC of a great many statements. Several particularly important results were obtained by Robert M. Solovay, another self-taught logician, who, together with Dana Scott, also recast Cohen's forcing method as the method of Boolean-valued models (in which the truth value of a statement is taken to be an element of a complete Boolean algebra).

For his pioneering work Cohen was made a full professor at Stanford, and at the International Congress of Mathematicians held in Moscow in 1966 he was awarded a Fields Medal, the highest honor bestowed by the mathematical community, and one awarded to no other logician before or since.

Alonzo Church delivered the official tribute to Cohen on that occasion, and in preparing his remarks he consulted with Gödel concerning the relation between his and Cohen's work. He was especially concerned to establish, once and for all, whether there was any truth to the still-recurrent rumors that Gödel had proved the independence of the Axiom of Choice and the Continuum Hypothesis in the 1940s but had chosen to keep the proofs to himself.

Gödel replied to Church's inquiry on 29 September 1966. "As far as your mention of my result of 1942 is concerned," he wrote, "I can ... reconstruct the independence proof ... only of $V = L$ (in type theory including the axiom of choice)." At the time he had thought that that proof "could be extended to an independence proof of the axiom of choice" as well, but he had "never worked [it] out in full detail."

In a draft of that same letter, and also in a letter to Wolfgang Rautenberg, portions of which were later published, he went further: He had, he said, been quite doubtful that his method would yield an independence proof for the Continuum Hypothesis. He did not say what the method was, but he did remark that it had more in common with Scott and Solovay's approach than with Cohen's [509].

As it happened, Cohen's breakthrough occurred just after Gödel had finished preparing a revised version of his article "What is Cantor's Continuum Problem?," which, together with his essay "Russell's Mathematical Logic," was to appear in an anthology of papers on the philosophy of mathematics (*Benacerraf and Putnam 1964*). Most uncharacteristically, Gödel allowed his Russell essay to be reprinted there unchanged, except for the addition of one introductory footnote; but because he wished to take account of subsequent developments in set theory, he made numerous alterations to the text and footnotes of his 1947 paper and added a supplement in which he summarized some of the most important advances that had occurred prior to 1963.

Cohen's work came to Gödel's attention just in time for him to add a brief postscript about it as well. It did not, however, cause him to retract any of the views he had earlier put forward about the Continuum Hypothesis. In particular, he continued to maintain that there was "good reason for suspecting" not only that the Continuum Hypothesis would turn out to be false, but that work on the question of its independence would lead to the discovery of new axioms from which it could be disproved.[7]

In the supplement Gödel had reiterated his belief that even if the Continuum Hypothesis were proved to be formally undecidable within ZFC, the question of its truth would nevertheless remain meaningful. "By a proof of undecidability," he asserted, "a question loses its meaning only if the system of axioms under consideration is interpreted as a hypothetico-deductive system, i.e., if the meanings of the primitive terms are left undetermined." In geometry, he pointed out, the truth or falsity of Euclid's fifth postulate "retains

---

[7] *Gödel 1964*, pp. 266–268. Such axioms have yet to be proposed and, to many who have studied the question, now seem unlikely [510].

its meaning if the primitive terms are taken ... as referring to the behavior of rigid bodies, rays of light, etc." So too, he believed, the objects of set theory were well-defined conceptual entities which, "despite their remoteness from sense experience," were perceptible via mathematical intuition. Hence "new mathematical intuitions leading to a decision of such problems as [the] continuum hypothesis" were "perfectly possible" [511].

In that regard Gödel thought that intuitions concerning "orders of growth" of sequences of integers were particularly promising candidates for investigation. In one of his letters to Cohen [512] he opined that "Once the continuum hypothesis is dropped the key problem concerning the structure of the continuum ... is ... the question of whether there exists a set of sequences of integers of power $\aleph_1$ which for any given sequence of integers contains one major[iz]ing it from a certain point on." The attempt to formulate axioms about such sequences and to use them to settle the Continuum Hypothesis would engage Gödel's attention for much of the rest of his life.

$$\diamond \quad \diamond \quad \diamond$$

Throughout the rest of the '60s translations of Gödel's works continued to proliferate: The anthologies of Davis and van Heijenoort appeared in 1965 and 1967, followed soon afterward by translations of various papers into Italian, French, and Russian. Gödel made no attempt to supervise any of the latter, but when Bernays informed him, late in 1965, that *Dialectica* was planning to publish an English translation of his 1958 article he took an active interest in the matter. Unhappy with the draft translation that had been submitted for consideration and perhaps recalling his experiences with the translations of the incompleteness paper, he decided to revise it himself — and thereby became caught up in an ever-expanding endeavor of the sort he had warned Cohen against [513].

During the same period Gödel received numerous invitations to participate in scholarly convocations, all of which he politely but firmly refused. In 1966, for example, he declined to attend two events organized in honor of his sixtieth birthday (a celebration at the University of Vienna and a conference at Ohio State University), and the following year he turned down Cohen's invitation to come to California to take part in a major symposium on axiomatic set theory. In each case he stated that he was no longer willing to travel far because of his poor health. (He did, however, go to Massachusetts in June of 1967 to receive an honorary Doctor of Science from Amherst.)

In 1966, too, he refused both an honorary professorship offered him by the University of Vienna and membership in the Austrian Academy of Sciences, and he balked as well at Austria's attempt to award him a national medal for art and science.[8] Ostensibly he feared that acceptance of foreign honors might compromise his United States citizenship [514]. More likely, however, his demurral reflected a lingering disdain for his former homeland; for he made no similar objection when the Royal Society (U.K.) elected him a foreign member in 1968, nor did he do so in 1972 when he became a corresponding member both of the British Academy and of the Institut de France (Académie des Sciences Morales et Politiques).

◇ ◇ ◇

Had Gödel chosen to participate in the Austrian recognition ceremonies he could have celebrated his sixtieth birthday with his mother, whom he had not seen since the spring of 1964. As it was, however, they only spoke by telephone, for by then she herself was too frail to travel.

Marianne's health had begun to decline shortly after her last visit to Princeton. In July of 1964 she had broken her arm in a fall, and while recovering from that she suffered a heart attack. For several months thereafter she required continual bed rest, and early the following year she was diagnosed with angina [515]. Gödel was manifestly concerned about his mother's condition, and, in his own hypochondriacal way, he did what he could to combat her depression. He tried to convince her that angina was actually a nervous affliction and that her recuperation would not require an unforeseeable time; he told her that he himself had begun taking nitroglycerin long before; and he empathized with her preoccupation with thoughts of death, which, he assured her, would haunt anyone confronted with serious heart trouble. He also arranged to share the cost of her medical treatment with his brother Rudolf. Yet he never considered going to see her, even though Adele sojourned in Europe both in 1965 and 1966.

In one of his last letters to his mother, written on 18 June 1966, Gödel mentioned that Adele, then traveling in Italy, planned to spend a few weeks in Vienna before returning home. She intended to arrive around the 26th, not long after Marianne's own return from a visit to the Salzkammergut.

As it turned out Adele was still in Vienna at the time of her mother-in-law's death, which occurred suddenly on 23 July. Adele attended the funeral in Kurt's stead, and there, as a result of inclement weather during the interment

---

[8] The University of Vienna did eventually succeed in awarding Gödel an honorary doctorate by doing so posthumously.

ceremony, contracted a lingering case of bronchitis that Gödel afterward invoked in defense of his own absence from the services. ("Why," he asked his brother, "ought I to have stood in the rain for an hour by an open grave?" [516])

Gödel left disposition of his mother's estate to his brother. He doubted, he said, that he had any right to claim a share of the inheritance, for he felt Rudolf had done much more for their mother than he had. He told Rudolf that he had discussed the matter with Adele, from whom he believed nothing should be kept secret, and that she too wished to have nothing to do with details of the settlement. He did suggest, though, that if Rudolf thought it would not conflict with their mother's wishes he might send Adele a piece of her jewelry as a remembrance of her.

That Marianne's death should have affected Rudolf more deeply than it did his brother is not surprising, for Rudolf had lived with his mother for many years and had no other companion to turn to after her demise. There is no question that Kurt loved his mother, but his grief for her was tempered by their long separation.

At age eighty-six Marianne had led a long and active life. She remained mentally alert until her very last days. And by dying when she did she was spared having to witness Gödel's own incipient deterioration.

XII
Withdrawal
(1969–1978)

> The paranoic is logical. Indeed, he is strikingly meticulously logical.
>
> —Yehuda Fried and Joseph Agassi,
> *Paranoia: A Study in Diagnosis*

DURING THE FINAL decade of his life Gödel withdrew more and more from the world around him. He continued to work, not only on revisions to earlier papers, but on some new results as well, including two that he regarded as especially important: the determination of the 'true' power of the continuum, and a formalization of Anselm's 'ontological argument' for the existence of God. (As far back as 1940, in private discussions with Carnap, Gödel had maintained that a precise system of postulates could be set up for such allegedly metaphysical concepts as "God" and "the soul.") But, increasingly, his thoughts were directed inward. After his mother's death he maintained only a sporadic correspondence with his brother, and his circle of close friends, never large, gradually narrowed. Eventually only Oskar Morgenstern was left.

Though he did not officially retire from the IAS until 1976, Gödel had begun to lose interest in institute affairs well before then. By 1964 almost all the professors who had been at the institute when he first arrived had died, and the interests of their successors were far removed from logic. Gödel's contacts with his colleagues, which had never been extensive, became even less frequent the following year, when his office was relocated into the institute's newly constructed library building. Situated at the rear of the structure on a charming site overlooking the institute's woods and pond, the new office was quiet, out of the way, and close to the stacks in which Gödel was wont to spend so much of his time. But it was also isolated from the offices of all the other mathematics faculty.

Nevertheless, Gödel could not help being distracted, both by internal and external events. Some, such as the escalating conflict in Vietnam or the recurrent and sometimes bitter controversies over the appointment of new IAS faculty members, added to his depression, while others, like the space program, engaged his keen interest. He persevered with his research efforts, but as time went on he lost touch with what other logicians were doing. There were a few exceptions, notably Abraham Robinson and Hao Wang, with whom his contacts actually increased during the years 1968–74; and his correspondence with Paul Bernays continued until 1975. But with most of his other colleagues his communications lapsed or ceased altogether.

Among the latter was Georg Kreisel, who had come to know Gödel well in the mid-'50s and had maintained intensive contact with him for some years afterward. In the obituary memoir he wrote for the Royal Society Kreisel noted that "superficially . . . the changes [in Gödel's behavior] appeared minor" up to the time their contacts ceased: "His mind remained nimble" into the early '70s, and in some respects he was "less formidable" than before; "only his exquisite sense of direction had obviously gone." Still, "several [of those closest to him] were already alarmed at the end of the sixties" [517].

Kreisel himself withdrew, declaring that he found Gödel's "efforts not to show his depressions" to be too painful to watch. Had he remained on the scene he would have witnessed a series of psychotic crises, intermittent at first, which became chronic by the fall of 1975. Like the episodes in 1936 and 1954, they were characterized by hypochondria and paranoia. But in contrast to those earlier times Adele was no longer capable of serving as a bulwark against the currents of irrationality; for by then she too was seriously ill, and the burdens of caring for her and tending to household affairs devolved increasingly on Gödel himself.

It is difficult to assess the full extent of Adele's disabilities or just when they began, but from Gödel's letters to his mother it seems that she suffered either an episode of heat exhaustion or a mild stroke during her trip to Italy in the summer of 1965. Whatever it was, it forced her to cut short her itinerary and fly directly home from Naples. (Already the year before Gödel had remarked how overweight she had become, and after her return from Italy Adele began to be treated for high blood pressure.)

By September she had recovered, and Gödel deemed that "on the whole" her visit to Ischia had done her "a lot of good," though he acknowledged that she seemed to have lost some of her spirit of adventure. But when Morgenstern saw her the following spring, on the occasion of Gödel's sixtieth birthday, he thought she looked quite bad [518].

In 1968 Adele was hospitalized briefly for tests, and two years after that Morgenstern reported that Gödel had become depressed about her condition. Morgenstern gave no indication what the nature of her illness was, but in shorthand drafts of letters that Gödel sent to his brother between 1970 and 1972 [519] (the last known correspondence between them) Adele's afflictions are variously described as hypertension, purpura, bowel trouble, gall bladder disease, arthritis, and bursitis. Walking had become especially hard for her — the result, according to Gödel, of neuritis, but more likely of another stroke, whose effects were exacerbated by her corpulence and her failure afterward to persevere with physical therapy.

◇ ◇ ◇

Kurt's own decline is chronicled in Morgenstern's diaries. By the spring of 1968 Morgenstern had already begun to express alarm at Gödel's emaciated appearance: Never before, Morgenstern thought, had he looked so gaunt. Gödel had once again convinced himself that he suffered from heart trouble, and Morgenstern marveled that he was still alive [520]. As before, the root problem was apparently psychosomatic.

When Gödel wrote to Bernays the following Christmas he confirmed that his state of health had been "rather bad . . . in recent months." Yet his hypochondriacal preoccupations had not prevented him from carrying on meticulous logical work, for the bulk of the letter was concerned with his continuing struggle to achieve precision in the English revision of his *Dialectica* paper [521]. Nor had his anxieties blunted his concern for the welfare of others. Early in the new year Morgenstern underwent surgery for prostate cancer, and during his convalescence Gödel visited him several times. On those occasions Morgenstern found Gödel's demeanor "especially charming" [522], and in the months that followed Gödel continued to display solicitous concern for all of Morgenstern's family. He maintained an especially keen interest in the intellectual development of Morgenstern's son Carl, then a precocious Princeton undergraduate, whom he encouraged in his mathematical studies.

Throughout 1969 there was little apparent change in Gödel's condition or his work regimen. In addition to his work on the *Dialectica* revision he deliberated further on issues he had raised five years before in the postscript to his 1934 lectures. There he had stressed his belief that Turing's results placed no bounds on the power of human reason, but only on the potentialities of formal systems [523], "whose essence ... is that reasoning [within them] is completely replaced by mechanical operations on formulas." He had failed to mention, however, that Turing himself had presented an argument purporting to show that mental procedures can *not* transcend mechanical ones. Indeed, in his 1936 paper Turing stressed that "*the justification* [for the definitions of how his machines were to operate] *lies in the fact that the human memory is necessarily limited.*"

Turing's claim troubled Gödel, but at the time he wrote the postscript he probably felt he could offer no definitive refutation of it. Nevertheless he continued to ponder the matter until, in early December of 1969, he announced to Morgenstern that he had found an error in Turing's proof, a discovery he believed might have momentous philosophical consequences [524]. What Turing had disregarded, Gödel asserted, was "that *mind, in its use, is not static, but constantly developing.* ... Therefore, although at each stage of the mind's development the number of its possible [distinguishable] states is finite there is no reason why this number should not converge to infinity in the course of its development" [525]. Our capacity for greater understanding is thus potentially unlimited — extending, in Gödel's view, even into the afterlife, as he had already maintained in the third of his "theological" letters to his mother.

Three days before Christmas Morgenstern saw Gödel once again, and on that occasion the two had a long discussion about economics, a subject on which Morgenstern found Gödel's ideas "as good as they were precise." Morgenstern remained concerned about Gödel's appearance but found his mental energy undiminished. He considered Gödel by far the most stimulating of his friends and said that though Gödel kept up with the news solely by reading the Sunday *New York Times*, he did so with such thoroughness that he was always well informed [526].

A month later, however, Gödel's health suddenly worsened. On 23 January 1970 Morgenstern took him to the hospital, convinced that he was suffering from genuine heart trouble of some sort, as well as diabetes. Morgenstern hoped then that the nature of Gödel's health problems would at last be thoroughly investigated; but, much to his surprise, Gödel was home again just four days later, uncharacteristically full of praise for his doctor [527].

What the diagnosis was is unclear, but for a short time Gödel's condition did seem to improve. The main difficulty, as always, was to persuade him to eat. Morgenstern kept in touch by telephone and, on 2 February, reported that though Gödel was somewhat depressed, he was "certainly better." But

Morgenstern noted too that Gödel had convinced himself that he needed digitalis, even though the medication his doctor had prescribed[1] had been effective.

Four days later Gödel began to exhibit signs of full-blown paranoia. His doctors, he declared, were lying; his medications were not being correctly identified; even the descriptions of them in reference works were wrong. Morgenstern was deeply troubled by his friend's behavior, but he believed that it was to some extent an act, a state of mind from which Gödel would in time recover [528]. And so when, after a few more days, Gödel seemed once again to be better, Morgenstern predicted that everything would soon be in order.

Gödel himself, however, was afraid he was about to die — so much so that he asked Morgenstern to arrange for the posthumous publication of seven articles that he felt were "nearly ready" for printing. Besides the *Dialectica* revision and the note criticizing Turing's argument the list included "a work on the consequences of his result of 1931" [529], his formal analysis of the ontological proof for the existence of God, the revised text of the Gibbs Lecture, an essay on Carnap (presumably one of the six drafts of the article intended for Schilpp's volume, though which one is unclear), and, most important, his "proof that the 'true' power of the continuum is $\aleph_2$" [530]. There were also, Gödel said, "many, many memoranda" in his shorthand notebooks, mostly of a philosophical nature [531].

About some of the items Gödel spoke in greater detail. With regard to the continuum problem he asserted that his proof was based on some quite reasonable additions to the axioms of set theory. He was convinced that the proof was correct, and Morgenstern urged him to publish it without delay. Morgenstern doubted, though, that his urgings would be heeded, for he knew full well how long Gödel was wont to deliberate over his intellectual creations.

Between 13 February and 8 March Morgenstern mentioned Gödel only twice in his diary: On the 15th he reported having discussed his condition with Paul Oppenheim, who agreed that Gödel was improving but was reluctant to admit it; and early the next month Morgenstern met Dana Scott, who told him that Gödel had also entrusted some of his manuscripts to him.

On 10 March Morgenstern declared that despite Gödel's belief that he was about to die, his virtual refusal to eat, and his continued obsession with digitalis, his mind was nonetheless "of the greatest clarity and as sharp as ever." Two days later, however, Gödel telephoned him, and it was at once clear that he was again deeply disturbed. He felt, he said, as though he were under

---

[1] Isordil, used in the treatment of angina.

a hypnotic spell, compelled to do the opposite of what he knew to be right. He recognized that he was very weak, but attributed his lack of energy not to malnutrition, but to the *freedom* he had enjoyed at the IAS, where he had had no lecturing responsibilities or other duties to toughen him and renew his strength. Not without some reason, he also feared that his doctor would either commit him to an asylum or refuse to continue treating him.

When Morgenstern visited Gödel at his home the next afternoon he was profoundly shocked both by his appearance and his behavior. Gödel looked, he said, "like a living corpse." He no longer trusted his doctor, and seemed to be having hallucinations. He had suddenly taken up smoking. And, in contrast to their earlier meetings, he now spoke of little but his illness. All at once, he declared that he was not respected in Princeton (a contention that was not altogether absurd, given that Morgenstern himself had made three unsuccessful attempts to get Princeton University to award Gödel an honorary doctorate); and when Morgenstern countered that he and von Neumann, at least, had always held him in the highest regard, Gödel replied that if Morgenstern were a true friend, he would bring him cyanide [532].

It seemed to Morgenstern that Gödel could not long survive unless he were placed in a home and fed intravenously. He recalled Adele's tale of how, long before the war when Gödel had been overcome by fear of poisoning, she had fed him, spoonful by spoonful, until she had brought his weight up from forty-eight kilograms to sixty-four. But given her own present state of health Morgenstern doubted she could save him again [533].

Somehow, through it all, Gödel hung on. A month later his hallucinations still persisted, and in desperation he began calling one doctor after another, seeking help only to reject it. He was severely emaciated and his behavior was manifestly psychotic. Yet he was not placed back in a hospital. Some days, indeed, he even went to his office at the institute.

At the height of the crisis, on 11 April, Rudolf Gödel arrived from Vienna. He had probably learned of the gravity of his brother's condition from Morgenstern, for Adele had not summoned him and did not welcome his presence (perhaps rightly, for despite his medical training he seemed unsure what to do). He, too, thought that his brother ought to be fed intravenously, but his efforts to intervene were ineffectual.

On the 14th Gödel telephoned Morgenstern again. Now convinced that he was about to be placed under guardianship, he reported that on each of the preceding four nights someone had stolen into his room and secretly given him injections; and, once more, he appealed to Morgenstern to help him commit suicide.

◇ ◇ ◇

At that critical juncture, with tantalizing abruptness, Gödel's name disappears from Morgenstern's diary. Certainly Morgenstern remained deeply concerned about his friend, but his own professional activities took him away from Princeton for extended periods. He did not mention Gödel again until August, and then he could only marvel at the miraculous recovery that had taken place: During the intervening four months Gödel had gained eighteen pounds, and his mind was once again "crystal clear" [534]. What brought about the change remains unclear, but a remark in one of Gödel's letters to Tarski indicates that potent psychoactive drugs may have been administered [535].

It was during that interim, apparently, that Gödel wrote up his results on the power of the continuum, for on 19 May Tarski returned the copy of a manuscript on the subject that Gödel had sent to him. Its curiously qualified title ("Some Considerations Leading to the Probable Conclusion That the True Power of the Continuum is $\aleph_2$") suggests that Gödel harbored lingering doubts about either the correctness of his arguments or the reasonableness of the axioms on which they were based and wished to obtain Tarski's imprimatur before submitting the paper to the *Proceedings of the National Academy of Sciences.*

In any case, serious errors in Gödel's proofs were soon discovered, though it is unclear whether they had come to light by the time Tarski returned the manuscript. In his cover letter Tarski said that he would write more about the paper shortly, but no follow-up letter from him is preserved. Nor does anything in Morgenstern's diaries suggest that difficulties had arisen. In the entry for 4 August there is, however, an interesting discrepancy; for there Morgenstern reported that Gödel had sent off his highly important result that the power of the continuum was $\leq \aleph_1$.

It may be that Morgenstern simply committed a slip of the pen. Perhaps he intended to write $\aleph_2$ rather than $\aleph_1$, or $\nleq$ instead of $\leq$. But in either case the result would still have been weaker than the equality that, according to Morgenstern's entry for 10 February, Gödel had earlier claimed to have proved. On the other hand, if Morgenstern meant what he wrote, then, by Cantor's theorem, Gödel would in fact have established the *truth* of the Continuum Hypothesis.[2]

The latter possibility is of particular interest in light of Gödel's subsequent actions; for at some point he added the notation "I. Fassung" (first version) to

---

[2]Morgenstern, perhaps lacking a sufficiently deep understanding of the subject matter, seems not to have noticed the discrepancy.

the text he had sent to Tarski and then composed a second text labeled "II. Fassung," bearing the doubly underlined injunction "nur für mich geschrieben" (written just for me), that he entitled "A Proof of Cantor's Continuum Hypothesis from a Highly Plausible Axiom About Orders of Growth."

Later still, apparently that same year, Gödel drafted a letter to Tarski [536] that he labeled "III. Fassung", in which he acknowledged that his original manuscript was "no good." He made no mention of the second version, but his comments on the mistakes in his proofs make it clear that he had discovered errors in that manuscript as well.

In the third version Gödel came to no definite conclusion as to the value of $2^{\aleph_0}$. Nevertheless, he declared that he still believed the Continuum Hypothesis to be false. He was obviously embarrassed at having circulated an incorrect proof, but he continued to work on the problem.

In September of 1972 Morgenstern reported that Gödel had once again found new axioms that implied that $2^{\aleph_0} = \aleph_2$. But this time he was in no hurry to publish his discovery. He was still working on details of the proof two years later. Finally, in August of 1975, Morgenstern reported that he had spoken to Gödel once more about the matter and had persuaded him to go ahead and submit his result to the *Proceedings*. Gödel remarked, though, that he no longer saw any point in doing so, for by the time it appeared he would have retired from the IAS and no further publication would be expected of him!

A month later Morgenstern wrote that Gödel was convinced at last that his proof concerning the power of the continuum was correct. There was, however, one final surprise: The power was "*not* $\aleph_2$, but rather 'different from $\aleph_1$'" [537].

◇ ◇ ◇

In fact, Gödel never did publish anything more on the subject; his three drafts appeared posthumously, in volume III of his *Collected Works*. Subsequently, though, a number of other mathematicians, notably Erik Ellentuck, Robert M. Solovay, and Gaisi Takeuti, carried out investigations that helped to clarify the relationships among the ideas contained in Gödel's manuscripts [538]. Their results may be summarized as follows.

Of the four axioms in the manuscript Gödel sent to Tarski, the first two, taken together, are equivalent to a schema of statements $A(\omega_n, \omega_n)$ (for each $n < \omega$) that have come to be known as the "square axioms." More generally, for $n \geq m$ the "rectangular axiom" $A(\omega_n, \omega_m)$ asserts that among the set F of all functions from ω_n to ω_m there is a subset S of power $\aleph_{n+1}$ such that *any* $f \in F$ is eventually majorized by some $g \in S$ (that is, for some ordinal $\alpha < \omega_n, g(\beta) > f(\beta)$ for all $\beta > \alpha$).[3] In the first two versions of his paper Gödel believed, erroneously, that the square axioms implied the rectangular ones — a mistake that he explicitly recognized in the third version; and, as Takeuti later showed, the rectangular axiom $A(\omega_1, \omega_0)$ *does* imply that $2^{\aleph_0} = \aleph_1$ (the conclusion of Gödel's second version). In the third version, however, Gödel admitted that he no longer found the rectangular axioms plausible, and Solovay and Ellentuck later independently proved that the square axioms imply no bound whatever on the size of $2^{\aleph_0}$.

On the other hand, the last two of Gödel's four original axioms together imply that $2^{\aleph_0} = 2^{\aleph_1}$, whence, by Cantor's theorem, $2^{\aleph_0} \geq \aleph_2$ — a result in agreement with Gödel's final statement to Morgenstern. But the consistency of axioms three and four relative to those of ZFC remains an open question, and, aside from Gödel, few if any logicians have found reason to believe in the intrinsic truth of those axioms.

$$\diamond \quad \diamond \quad \diamond$$

The mental crisis of 1970 was Gödel's worst since 1936. In many respects it was comparable to that earlier breakdown, but his recovery from it was much more rapid and dramatic, and, perhaps because of its more limited duration, the later episode caused much less interruption in his work.

In late August Morgenstern visited Gödel in his office, and in early October he spoke with him by telephone. On both occasions he found him "effervescent" and eager to talk about his latest endeavors. In addition to the continuum problem, Gödel continued to labor on revisions to his *Dialectica* translation (especially two lengthy footnotes thereto), and he announced that he was now fully satisfied with the ontological proof that he had obtained. He hesitated to publish it, however, for fear that a belief in God might be ascribed to him, whereas, he said, it was undertaken as a purely logical investigation, to demonstrate that such a proof could be carried out on the basis of accepted principles of formal logic [539]; and indeed, the manuscript found in his *Nachlaß*, published as item *1970* in volume III of his *Collected Works*, consists of two pages of formal symbols, with only a few explanatory words.

---

[3]Such a set S is sometimes called a "scale." Accordingly, Gödel's unpublished continuum manuscripts have often been referred to as his "scales of functions" papers.

◇ ◇ ◇

Gödel's health remained exceptionally good for the next three and a half years. Morgenstern, who continued to marvel at the change that had taken place, reported that Gödel "looked better than ever," was "lively," and appeared to be in "the best frame of mind." He was engaged with his research and eager to talk about politics and world events.

During that period the two had many long phone conversations. The discussions ranged widely over mathematics, philosophy, religion, and neurobiology, in addition to personal and institute affairs. In general, Morgenstern found the exchanges quite stimulating. On one occasion, however, he remarked — somewhat ambiguously — that whenever one spoke with Gödel one was thrust "immediately into another world." And on another he noted that Gödel imagined "too many plots." They were, he said, always logically grounded, but the premises underlying them were either wanting or false.

Notably absent from Morgenstern's record of those discussions is any mention of a significant mathematical breakthrough that occurred in 1970: the proof by the young Russian mathematician Yuri Matijasevich of the unsolvability of Hilbert's Tenth Problem.[4] One would expect that result to have been of great interest to Gödel because it concerned a decision problem for Diophantine equations that was related to unpublished results he had obtained years earlier as corollaries to his incompleteness theorems [540]. (Like other mathematicians before Matijasevich, Gödel had overestimated the problem's difficulty. In a letter to William Boone written some nine years before Matijasevich's work he had advised Boone not to tackle the problem unless he had "a genuine taste for number theory," as it was his own impression that "a great deal of number theory" would be required for its solution. In fact, however, Matijasevich's proof invoked properties of the well-known Fibonacci sequence, and was soon recast in terms of familiar facts about solutions of the Pell equation — both standard topics in courses on elementary number theory.)

In view of Gödel's health problems one could hardly expect him to have become aware of Matijasevich's work when it was first announced. What is surprising, though, is that he seems never to have mentioned it in any subsequent correspondence [541]. One might, in particular, have expected the subject to have come up in his correspondence with Bernays, since several

---

[4] The problem asked: Does there exist a general algorithm for determining whether an arbitrary polynomial (in several variables) with integer coefficients has a solution in integers?

of the letters the two exchanged between July 1970 and January 1975 were devoted to technical discussions of logical matters.

Most of that correspondence, however, was concerned with details of Gödel's English translation of his *Dialectica* paper, which Scott had forwarded to Bernays for publication. In July 1970 Gödel had written to Bernays to inquire when that translation was expected to appear, and Bernays had replied that it was scheduled for inclusion in the volume then in preparation. A short time later Gödel received two galley proofs of the article. But in the meantime he had once again become dissatisfied with one of the footnotes and had determined to revise it.

Twice thereafter Gödel promised to send the revisions right away. But in his diary entry for 11 April 1971 Morgenstern reported that the footnote, already "enormous," was still growing. When Gödel then told him that he had decided to introduce a new formal system, Morgenstern suggested instead that he shorten the note and rewrite the paper.

What happened after that is unclear. Morgenstern never mentioned the subject again. Bernays did, but not until March 1972, when he referred to Gödel's "new" English translation. Apparently he had not seen it but had heard from Kreisel that Gödel still had qualms about some of its details. Guessing that Gödel's reservations might concern the extent to which his proofs relied on impredicative principles, Bernays offered some observations in an attempt to resolve the matter. His remarks elicited one further exchange on the subject, which ended with Bernays' letter to Gödel of 21 February 1973. But no new manuscript was forthcoming, and Gödel never did return either of the galleys he had been sent. Extensively annotated, they were found in his *Nachlaß* amid a welter of fragmentary drafts and revisions, from which a reconstructed version was finally published in volume II of Gödel's *Collected Works* [542].

◇   ◇   ◇

While all that was going on, a new window on Gödel's thoughts was opened through his discussions with Professor Hao Wang of Rockefeller University. Wang had first met Gödel in 1949 and had corresponded with him at intervals thereafter. Over the years he had become very much interested in Gödel's philosophical views, and in the late 1960s he had begun writing a book (*From Mathematics to Philosophy*, published in 1974) in which he planned to discuss some of them.

The book presented an opportunity for Gödel to disseminate his thoughts indirectly — an idea that no doubt appealed to his innate cautiousness —

and that is probably the reason why, in July 1971, he expressed his willingness to talk with Wang. The two agreed to meet for conversation at Gödel's office every other Wednesday [543]. The sessions generally lasted about two hours; they began in October 1971 and continued, with occasional cancellations, until December 1972.

An overview of those discussions is given in Wang's 1987 book *Reflections on Kurt Gödel*. Detailed reconstructions of many of the conversations are to appear in his forthcoming *A Logical Journey: From Gödel to Philosophy*. The following remarks, taken from a preliminary draft of that work [544], offer tantalizing glimpses into Gödel's later views on a wide variety of mathematical and philosophical topics.

> I don't think the brain came in the Darwinian manner. . . . Simple mechanism can't yield the brain. . . . Life force is a primitive element of the universe and it obeys certain laws of action. These laws are not simple and not mechanical.   [13 October 1971]

> The intensional paradoxes (such as that of the concept of not applying to itself) remain a serious problem for logic, of which the theory of concepts is the major component.   [27 October 1971]

> Positivists . . . contradict themselves when it comes to introspection, which they do not recognize as experience. . . . The concept of set, for instance, is not obtained by abstraction from experience.   [27 October 1971]

> Some reductionism is correct, [but one should] reduce to (other) concepts and truths, not to sense perceptions. . . . Platonic ideas are what things are to be reduced to.   [24 November 1971]

> We do not have any primitive intuitions about language, [which] is nothing but a one-to-one correspondence between abstract objects and concrete objects. . . . Language is useful and even necessary to fix our ideas, but that is a purely practical affair. Our mind is more inclined to sensory objects, which help to fix our attention to abstract objects. That is the only importance of language.   [24 November 1971]

> Introspection calls for learning how to direct attention in an unnatural way.   [24 November 1971]

> What Husserl has done [is] . . . to teach . . . an attitude of mind which enables one to direct the attention rightly . . . .   [6 December 1971]

To develop the skill of introspection you have to know what to *disregard.*
[5 April 1972]

Husserl does what Kant did, only more systematically. . . . Kant recognized that all categories should be reduced to something more fundamental. Husserl tried to find that more fundamental idea which is behind all these categories. . . . [But] Husserl only showed the way; he never published what he arrived at, . . . only the method he used.[5]
[15 March 1972]

Mathematicians make mistakes.   [10 November 1971]

[But] every error is due to extraneous factors (such as emotion and education); reason itself doesn't err.   [29 November 1972]

Philosophy has as one of its functions to guide scientific research.[6]
[18 October 1972]

Wang quotes two other remarks that relate to items of correspondence of which he may have been unaware. In the first (18 October 1972) Gödel declared that "the 'new math' may be a good idea for students in the higher grades, but [it] is bad for those in the lower grades" — a remark perhaps stimulated by a query he had received in March 1971 from a special student in elementary education at Elmhurst College in Illinois. The student had sought Gödel's reaction to his own belief that "the worth and beauty of mathematics ... should be brought out to all beginning students in mathematics with the hope that it will stimulate and motivate them" [546].

Gödel found the issue interesting enough to draft a reply, which, however, he never sent off. "It seems to me," he began in a passage later crossed out, "that the worth and beauty of mathematics should not be pointed out to *beginning* students ... because one has first to know some mathematics before one can appreciate such a statement." Rather,

> What should be pointed out ... is the truly astonishing number of simple
> and non-trivial theorems and relations that prevail in mathematics. ...
> In my opinion, this property of mathematics somehow mirrors the order
> and regularity which prevails in the whole world, which turns out to be
> incomparably greater than would appear to the superficial observer.

---

[5] On this point Gödel gave further evidence of his paranoia: He suggested that Husserl was obliged to conceal his great discovery, because otherwise "the structure of the world might have killed him" [15 December 1972].

[6] At the same time, however, Gödel remarked to Morgenstern that "Philosophy today is, at best, at the point where Babylonian mathematics was" [545]!

"Abstract considerations," he believed, "are started too soon in today's schools (while formerly they were started too late or [were] omitted altogether)" [547].

The second remark Wang quotes (dated 19 January 1972) concerns the "psychological difficulties" that Gödel thought confronted anyone who attempted to pursue philosophy as a profession: Gödel suggested that many of those difficulties arose from "sociological factors" that might well disappear "in another historical period." On the surface the comment merely reiterates Gödel's long-held belief that certain intellectual prejudices had hindered the progress of philosophy. But there is also a distinctly Spenglerian air about it, an attitude that Gödel expressed more explicitly in a letter he wrote at about that same time to Hans Thirring, his former physics teacher at the University of Vienna.

Then eighty-four years old and crippled by a stroke, Thirring had written Gödel on 31 May 1972 with a query to which he urgently desired an answer [548]. He had a clear recollection, he said, of Gödel's having visited him in 1940, "just before your adventurous escape from Hitler across Russia and China." At that time he had directed Gödel's attention to an article in the journal *Die Naturwissenschaften* that he believed presaged Nazi Germany's imminent development of an atomic bomb. He had shared the same concern with Karl Przibram and Henry Hausner, two other émigrés who were escaping via different routes, and had asked all three to contact Einstein in America and urge him to warn President Roosevelt of the great danger should Hitler produce the bomb before the United States. He did not know, he said, whether Przibram or Hausner had ever spoken about it to Einstein, though of course he knew of Einstein's historic letter. His question was: Had Gödel carried out his request?

Gödel answered on 27 June, saying that he had seen Einstein very seldom during the years 1940–41 and merely recalled having conveyed Thirring's "greetings" to him. He pointed out that he had lost contact with physics and physicists some ten years before that and so had been unaware that work was under way toward achieving a chain reaction. Moreover, he remarked,

> When I later heard of such things, I was very skeptical, not on physical, but rather on sociological grounds, because I thought that that development would only ensue toward the end of our culture period, which presumably still lay in the distant future [549].

What Thirring thought of Gödel's response is unknown. In any case, the question he had posed was moot, since of Einstein's two letters to Roosevelt warning of the atomic peril (the earlier one drafted by Leo Szilard) the first was delivered to Roosevelt on 11 October 1939, some three months prior to

Gödel's emigration, while the second, urging greater haste with the weapons development program, was dated 7 March 1940, just three days after Gödel debarked in San Francisco.

◇ ◇ ◇

Throughout 1972 Morgenstern repeatedly mentioned Gödel's "high spirits," a state of mind no doubt reinforced by his receipt that year of three further academic honors: He was elected a corresponding member both of the Institut de France (Académie des sciences morales et politiques) and of the British Academy and, largely through Wang's efforts, was awarded an honorary doctorate by Rockefeller University (his fourth, following those from Yale, Harvard and Amherst).

The convocation at Rockefeller, which took place on 1 June 1972, was apparently the last time Gödel participated in a public event outside Princeton; but he was also present five days later at an IAS conference honoring the twenty-fifth anniversary of von Neumann's work on the development of electronic computers.

Among the invited speakers were two eminent logicians, Dana Scott and Michael Rabin. Morgenstern, however, thought that the most significant remarks at the conference were the questions that Gödel posed from the floor, of which Wang has given the following paraphrases [550]:

1. Is there enough specificity in genetic enzymatic processes to permit a mechanical interpretation of all the functions of life and the mind?

2. Is there anything paradoxical in the idea of a machine that knows its own program completely?

Gödel's outspokenness on that occasion — a marked departure from his usual public reticence — accords with Kreisel's report that Gödel became "more gregarious" in his later years [551]. Nevertheless, he was still considered unapproachable by most of his IAS colleagues, as well as by many logicians.[7]

---

[7] Gödel himself claimed that his colleagues sometimes discriminated against him. Given his paranoid outlook, however — his "all-pervading distrust," as Kreisel has called it [552] — it is hard to know what to make of his charges. He complained, for example, that other faculty were wont to hold meetings about him behind his back; and while the charge seems absurd on its face, it is true that he was never asked to serve on any institute committees, probably because it was feared that his indecision and legalistic ways would seriously impede the committees' work.

Morgenstern thought it a pity that so many had, out of fear, denied them-selves the opportunity of interacting with such a great intellect. Moreover, he believed that in so doing they had actively fostered Gödel's intellectual isolation. But others who had made an effort to get to know Gödel disagreed.

Deane Montgomery, for example, who knew Gödel well and regarded him as a good friend (albeit one burdened with grave psychological problems), testified that while Gödel liked to talk at length to a chosen few, he was hard to approach. And Atle Selberg, another of Gödel's longtime IAS colleagues, recalled that although Gödel enjoyed having lunch with Selberg, Montgomery, and Arne Beurling at the Nassau Tavern, Gödel's introverted nature made him ill at ease in larger social gatherings.

No doubt there is merit in all those views. In any case, however, Gödel shed his accustomed reserve on two further occasions the following year. In March of 1973 Abraham Robinson spoke at the institute on his work on nonstandard analysis (a latter-day vindication of the notion of infinitesimals), and at the end of the talk Gödel took the opportunity to express his opinion of the significance of Robinson's achievement. Nonstandard analysis was not, he said, "a fad of mathematical logicians" but was destined to become "the analysis of the future." Indeed, he predicted, "in coming centuries it will be considered a great oddity ... that the first exact theory of infinitesimals was developed 300 years after the invention of the differential calculus."[8]

Then, some seven months later, Gödel attended a garden party given by institute director Carl Kaysen. His appearance at such a social event was not, in itself, unusual, but his active participation was. According to Morgenstern, Gödel was in especially droll form that evening and ended up holding court in the midst of a group of younger logicians.

As it happened, Kaysen's party came just as his own relations with IAS faculty were approaching a crisis. The source of the conflict was Kaysen's effort to establish a School of Social Sciences, and especially his attempt to appoint Robert Bellah to its faculty. The appointment was opposed by most of the mathematics faculty (including Gödel), who viewed Bellah's work and much of sociology in general with disdain; and when Kaysen attempted to bypass their objections and present the appointment for approval to the institute's board of trustees, a bitter struggle ensued.

---

[8]Unfortunately, just a few months after his IAS presentation Robinson was discovered to have pancreatic cancer, from which he died the next April. His sudden demise recalls that of Clifford Spector, whose work Gödel had also praised unstintingly. Gödel's acquaintance with both men was all too brief, and their passings affected him deeply.

For Gödel the affair was upsetting in several respects: It shattered the peace and quiet of his intellectual haven; it attracted unfavorable public scrutiny and criticism [553]; and it forced him to take sides. He did not dislike Kaysen and was loath to become part of a "coup" against him;[9] yet he also thought Bellah's writings lacked merit. Ultimately, in line with his penchant for seeking "hidden causes" and his tendency to impute secret motives to others, he concluded that Bellah must possess some hidden worth that Kaysen saw but could not reveal to the faculty.

Besides the internal strife at the institute, the Watergate scandal and its aftermath also occupied Gödel's attention. While it was going on he frequently discussed politics and ethical concerns with Morgenstern, who found Gödel's views on world affairs "odd, but in part striking." Morgenstern reported, for example, that Gödel thought the dollar was actually worthless in Europe, its value only the result of a "conspiracy"; and he noted further that Gödel found it "strange" that in certain circumstances — those involving national security, for example — "the law could ... force one to lie" [555].

Such views reflect the conflicting mixture of naiveté, paranoia, and fundamental concern for truth that was central to Gödel's world view. While proclaiming the unerring power of human reason, he remained deeply suspicious of human motives. Like Euclid, he endeavored to "look upon beauty bare." Yet in the end he found truth only *behind* appearances. In all things he always looked *within himself* for explanations.

◊ ◊ ◊

Unfortunately, Gödel's distrust of others, coupled with his unshakable faith in the rectitude of his own judgment, posed a grave threat to his physical well-being. As a hypochondriac, he had for years been obsessively concerned with his body temperature and bowel habits and had indulged in excessive self-medication, especially with laxatives. He had also delved into the medical literature and, though distrusting their advice, had consulted numerous physicians. And yet, despite his intellect, he continued to hold bizarre ideas about nutrition and the etiology of disease. Because of his "little knowledge" he was reluctant to seek timely medical attention, and that in turn made treatment more difficult. Sometimes his recalcitrance even became life threatening.

---

[9]Writing to his mother at the time of Kaysen's appointment as director in 1966, Gödel said Kaysen had made "a very sympathetic impression" on him and he expected the transition to be "quite painless, unless in the end one of my dear colleagues finds a hair in the soup" [554].

Aside from the ulcer crisis of 1951, Gödel's health problems had stemmed primarily from psychiatric disturbance. In 1974, however, he was stricken by an acute physical emergency: His prostate became so enlarged that it blocked his urinary tract. Undoubtedly it was a condition that had been worsening for several years but for which he had never sought treatment. Instead, despite increasing pain and his wife's urgings that he see a doctor, he insisted that the problem could be controlled by milk of magnesia. Only when the pain became unendurable did he allow himself to be taken to Princeton Hospital. He was admitted on 4 April and underwent catheterization two days later.

The procedure provided immediate relief, but Gödel stubbornly refused to accept the diagnosis. He declined to discuss his illness, even with Morgenstern (whom he twice sent away), and he rejected his doctor's recommendation to undergo surgery. As the days passed he grew increasingly paranoid and uncooperative, until finally his urologist threatened to discontinue treating him. The conflict reached a climax a few days later, when Gödel defiantly removed the catheter. Without it he was once again unable to urinate, and it had to be reinserted against his will [556].

Had he possessed the strength to do so, Gödel would probably have checked himself out of the hospital (though his doctor had warned him that he would die in great pain if he did). Instead, through the efforts of IAS director Kaysen he was transferred to the University of Pennsylvania hospital in Philadelphia, where more specialized care was available and where his paranoia gradually subsided. Eventually he was able to come to terms with the nature of his affliction, but he remained steadfastly opposed to the idea of surgery. In the end, despite the discomfort and risk of infection, he chose to remain permanently catheterized.

◇ ◇ ◇

Gödel returned home toward the end of April. Morgenstern visited him there a few days later and found him looking "quite thin and miserable." Gödel was still unwilling to discuss his illness, and Morgenstern went away saddened and concerned about his friend's future [557].

The situation appeared especially bleak in view of Adele's condition. Already in September of 1970, just after his own recovery, Gödel had remarked that she found it necessary to lie down continually. Soon afterward she was confined to a wheelchair, and thus immobilized there was little she could do to counter the effects of her husband's meager diet.

Nevertheless, against all odds, Gödel rallied once more — for the last time, as it would turn out. By mid-May of 1974 Morgenstern reported that Gödel was doing "tolerably well" and that mentally he was "as good as ever." As spring passed into summer their long phone conversations resumed, and in late August Gödel visited Morgenstern at the latter's home, where the two conversed for an hour and a half.

Gödel was "lively and cheerful" that day, and Morgenstern declared that they had rarely had such a good discussion [558]. Still, there were some worrisome signs. The following month, for example, Morgenstern expressed amazement that Gödel was able to survive the "unbelievable amounts of strange medicines" he was then taking[10]; and when Gödel came over for tea a few days later he wore two sweaters and a vest, though the temperature was around 70 degrees. Still feeling cold, he then borrowed another sweater from Morgenstern's wife, as well as a cloak and two coats. But those too proved insufficient. Finally, Morgenstern made a fire in the fireplace [559].

It was much the same seven months later. In early April Gödel once again came for tea, wearing a vest, pullover, and winter coat. He was in a "splendid humor," yet despite the warmth of the Morgenstern's living room he kept his winter attire on throughout the visit.

There had apparently been little change in his physical condition over the intervening months. It was disturbing, though, that his contacts with outsiders had continued to decrease. Particularly regrettable was the lapse in his correspondence with Bernays, now in his mid-eighties: Since 1972 their exchanges had been limited to holiday greetings, and in January 1975 even those came to an end.

In May of that year Gödel's spirits were again briefly boosted by the news that at long last, in large part through the efforts of Professor Paul Benacerraf, Princeton University had voted to award him an honorary doctorate. As the day of the ceremony drew near, however, he began to hedge on grounds of health. He refused to make a definite commitment to attend and told Morgenstern privately that though "ten years ago such a thing would ... have been proper," he had in the meantime received honorary degrees from Harvard, Yale, Amherst, and Rockefeller and saw no reason why he should collect any more [560]. Benacerraf negotiated with him up to the last moment,

---

[10]Among them, in addition to the ever-present milk of magnesia and Metamucil, were various drugs for the control of kidney and bladder infections (Keflex, Mandelamine, Macrodantin, and Gantanol), the antibiotics Achromycin and Terramycin, the cardiac medications Lanoxin and Quinidine, and the laxatives Imbricol and Pericolase.

offering to chauffeur him to and from the ceremony and attend to various other concerns, but on the morning of commencement Gödel finally determined not to go. Consequently, though his name was listed in the program as the recipient of a Doctor of Science, the degree was never actually conferred.[11]

Three months later Gödel received a much greater honor: the National Medal of Science. Yet his response was much the same. He was notified of the award well beforehand and urged to come to Washington to receive it in person. Again an offer was made to chauffeur him and his wife to and from the event, and once again he demurred. On the day of the ceremony (18 September) Professor Saunders Mac Lane, then President of the American Mathematical Society, accepted the medals and certificate from President Ford on Gödel's behalf.

After the ceremony Mac Lane traveled to Princeton to deliver the items personally to Gödel, who reportedly "bubbled over with joy" when he received them [561]. Few others, though, seem to have taken much notice. Even the *New York Times* paid scant attention to Gödel's award. Inevitably, his name was overshadowed by those of some of the other recipients, especially Nobel laureate Linus Pauling; for the sad fact was that, quite apart from his absence at the ceremony, Gödel was still almost unknown outside the mathematical community.

◇ ◇ ◇

Less than two months later Adele became seriously ill, an event that, with hindsight, may be taken as the beginning of Gödel's own final decline. On 2 November Morgenstern visited the two of them at home and confronted what he described as "a real tragedy." Adele was bedridden, and Kurt was still suffering noticeably from his prostate condition; hired help was difficult to obtain, so responsibility for shopping and essential household tasks fell upon him; and though his mind remained "brilliant as ever," Gödel told Morgenstern that he could no longer muster the strength to write up his important new results on the continuum.

It soon became evident that Gödel could not continue to care for Adele without assistance, so a nurse who lived nearby was engaged to stay with her each weekday from 9 until 2. The nurse, Elizabeth Glinka, had become

---

[11] It is hard to say whether Gödel's health really was as precarious as he claimed (Morgenstern, for one, thought he looked bad a few days before the ceremony), or whether his initial pleasure at being offered the degree was tinged with lingering resentment over the tardiness of Princeton's recognition of him.

acquainted with Kurt and Adele some six years before through their gardener. A sympathetic person, she came to know the two of them on a personal level as well as anyone, and unlike most in Princeton she was able to look beyond the surface to recognize Adele's talents as well as her faults, to understand her need for companionship, and to appreciate her role as caretaker and stabilizing influence in Gödel's life.

Because she could not afford to devote the ever-increasing amounts of time to Adele that her condition came to require, Glinka's formal employment in the Gödel household lasted but a few months. But her friendship with the Gödels endured, and during the months she stayed with Adele her relation to her was as much that of friend and confidante as it was nurse. The physical assistance she provided was probably less important than the companionship she offered, and she also did what she could to overcome Gödel's growing reluctance to eat.

According to her firsthand testimony [562], Gödel usually ate only a single egg for breakfast, plus a teaspoon or two of tea and sometimes a little milk or orange juice. For lunch he generally had string beans (never meat); and when, on one occasion, the two women succeeded in getting him to eat carrots, Glinka regarded the event as "a real breakthrough." Countless times, she recalled, he would ask her to purchase oranges for him, only to reject those she bought as being "no good." Finally she refused to buy them for him any longer. But it made no difference: Even when he went to the store and selected the fruits himself, he usually discarded them once he had brought them home.

Toward the end of February 1976, not long after Glinka's arrival, Gödel's paranoia once again surfaced. He telephoned Morgenstern two to three times a day at all hours to express his fears of being committed, his worries about his catheter, and his suspicions about his doctors, with whom he asked Morgenstern to serve as intermediary. On 31 March, his weight down to eighty pounds, he was hospitalized, convinced that he would die within a few days. But then, barely a week later, he suddenly left the hospital and walked home, without checking out or obtaining his doctor's permission [563].

His return home left Adele in a state of despair — not least because he told Morgenstern and others that during his hospitalization she had given away all his money. Neither she nor anyone else could alleviate his delusions, which continued unabated through May and into June. At various times Gödel asked Morgenstern to become his legal guardian, said that the police were about to arrive, and asserted that the doctors were all conspiring against him. One day he said that he wanted to see his brother; the next, that he hated him [564].

During that same period Gödel was in frequent telephone contact with Hao Wang, to whom he spoke of various technical matters in set theory. He also told him about all the physical maladies from which he was then ostensibly suffering (among them a kidney infection, indigestion, and "not having enough red blood cells"), asserted that "antibiotics are bad for the heart," and confided that Adele had in fact had a light stroke the previous autumn [565].

Wang listened sympathetically, but he also took the opportunity to ask Gödel a number of historical questions concerning his earlier work. In view of Gödel's mental state, there could hardly have been a more inauspicious time to pursue such inquiries; but Wang persisted, drawing up a list of written questions that he sent to Gödel in early June. Later that same month he wrote up his interpretation of Gödel's replies (subsequently published as "Some Facts About Kurt Gödel") and sent the manuscript to Gödel for his correction and approval [566].

It is doubtful that Gödel ever checked Wang's article in detail, for had he done so he surely would have corrected a number of factual errors in it. But he was too preoccupied with concerns about his own health and that of Adele. Around the middle of June she was admitted to the hospital, purportedly (so Gödel told Wang) for "tests." Morgenstern, however, reported that she was being fed intravenously and was delirious. She remained hospitalized for some time and upon her release was transferred to a nursing facility for further convalescence. She did not return home until around the beginning of August [567].

How Gödel managed in her absence is hard to imagine. Wang, who spoke with him during that time, had the impression that he cooked for himself only once every few days. It seems likely that he spent much of his time with Adele, for the Morgensterns tried repeatedly to reach him by telephone, without success. They were much relieved when he finally called them and "sounded quite normal."

The improvement, however, was only temporary. On 28 September Gödel telephoned from New York to say that he expected to be "thrown out" of the IAS. (He had retired as Professor Emeritus on 1 July, having reached the mandatory retirement age of seventy the previous April.) A few days later he entered a hospital in Philadelphia, and on 3 October he called Morgenstern to say that someone wanted to kill him, that the streptomycin he was being given was too strong, and that the doctors understood nothing about his case [568].

Morgenstern did what he could to help. He consulted with Adele and, at Kurt's own urging, contacted Gödel's urologist. But it was difficult for him to do much more, for by then it was clear to everyone but Gödel that Morgenstern was himself succumbing to the ravages of a metastasizing cancer, against which he had been struggling vainly for several years.

After a short time Gödel returned home, but his condition did not improve. By May of 1977 several of his friends, including Wang, Deane Montgomery, and Carl Kaysen, had become concerned enough about his physical and mental deterioration to try to persuade him to reenter the hospital at the University of Pennsylvania. Wang broached the idea several times, but Gödel repeatedly refused.

Matters reached a climax in July, when Adele was rushed to the hospital to undergo an emergency colostomy. For several weeks after the surgery she was in intensive care, and more than once she came close to dying [569]. Gödel remained devoted to his wife throughout her ordeal, but its effect on him was devastating. Altogether her convalescence lasted some five months, first in the hospital and then in a nursing home. She did not return home until just before Christmas, and during her long absence Gödel's paranoia and anorexia continued to grow without restraint.

His lamentable state, even at the beginning of that period, can be gauged from a poignantly despairing memorandum that Morgenstern wrote on 10 July, just sixteen days before his own death:

> Today . . . Kurt Gödel called me again . . . and spoke to me for about 15 minutes . . . . After briefly asking how I was and asserting that . . . my cancer would not only be stopped, but recede, and that furthermore the paralysis in [my] legs would completely redress itself . . . he went over to his own problem[s].
> He asserted that the doctors are not telling him the truth, that they do not want to deal with him, that he is in an emergency (exactly what he told me with the same words a few weeks ago, a few months ago, two years ago), and that I should help him get into the Princeton Hospital. . . .
> [He] also assured me that . . . perhaps two years ago, two . . . men appeared who pretended to be doctors. . . . They were swindlers [who] were trying to get him in the hospital . . . and he . . . had great difficulty unmasking them. . . .
> It is hard to describe what such a conversation . . . means for me: here is one of the most brilliant men of our century, greatly attached to me, . . . [who] is clearly mentally disturbed, suffering from some kind of paranoia, expecting help from me, . . . and I [am] unable to extend it to him. Even while I was mobile and tried to help him . . . I was unable to accomplish anything. . . . [Now,] by clinging to me — and he has nobody else, that is quite clear — he adds to the burden I am carrying [570].

In turn, Morgenstern's death added to *Gödel's* burdens. Gödel learned of the event just hours after it happened, when he called Morgenstern's home expecting to speak with him. When informed that his friend had died he hung up without saying another word [571] — a highly uncharacteristic action that revealed his profound shock in the face of an outcome that, however obvious it may have been to others, he had simply never foreseen.

In part, perhaps, his failure to recognize that Morgenstern was terminally ill reflected his distrust of doctors: He could not accept that their prognosis had proved correct. In addition there was probably an element of personal denial since Morgenstern's cancer had begun as prostate trouble.

He was, of course, preoccupied with his own troubles. Nevertheless, he was still capable of expressing concern for others. Indeed, just two months before his own death he called Deane Montgomery to inquire about Montgomery's son, who was then undergoing cancer chemotherapy [572].

Morgenstern's death, coupled with Adele's absence and her critical illness and incapacitation, must be reckoned as a major factor in the swift and steady decline in Gödel's condition that followed over the next few months.[12]

There were few witnesses to that final tragedy, though Elizabeth Glinka visited Adele regularly during and after her hospitalization, and Gödel's next-door neighbor, Mrs. Adeline Federici, looked in on him and brought him groceries. She recalled that he asked only for navel oranges, white bond bread, and soup, and quit buying the latter when the price went up by two cents [573]. Those of Gödel's colleagues at the IAS who remained aware of his situation grew increasingly alarmed, but there was little they could do for him without his acquiescence.

Outside of Princeton, Hao Wang was almost the only one who attempted to keep in touch with Gödel. During 1977, however, he too was busy with other activities and concerned with a variety of problems of his own, so his contacts with Gödel after Adele was stricken were limited.

The two seem not to have talked at all during the months of July and August, and from mid-September until mid-November Wang was out of the country [574]; but he did call Gödel shortly before his departure and after his return, and on one of his trips to Princeton he brought Gödel a chicken that his wife had prepared. He had called beforehand to advise Gödel of his coming, yet when he arrived Gödel eyed him suspiciously and refused to open the door. Finally Wang left the offering on the doorstep.

---

[12] On 18 September, less than two months after Morgenstern's demise, Bernays also died, but it seems unlikely that Gödel ever learned of the event.

Wang did manage to visit Gödel at his home on 17 December, a few days before Adele's return. To Wang's surprise Gödel did not then appear very sick, though he must surely have been severely emaciated, for at his death less than a month later he weighed only sixty-five pounds. His mind, too, appeared to Wang unaffected. Gödel told him, however, that he had "lost the power for making positive decisions" and could make only negative ones [575].

Among the latter was his continuing refusal to enter the hospital. He was finally persuaded to do so by Adele; but by then there was little his doctors could do to reverse the effects of his prolonged self-starvation.

Gödel was admitted to Princeton Hospital on 29 December, and Wang spoke with him there by telephone on 11 January: "He was polite but sounded remote" [576]. Three days later he died of "malnutrition and inanition" resulting from "personality disturbance" [577].

XIII
Aftermath
(1978-1981)

GÖDEL'S DEMISE WAS fraught with Pyrrhic irony: Unable to escape from the inner logic of his paranoia — to adopt, as it were, a "metatheoretical" perspective — he succumbed to starvation in the grip of his obsessive fear of being poisoned. Like the actions of an inhabitant of one of the timeless universes he had envisioned, his death equally was, and was not, of his own volition.

He was buried in Princeton Cemetery on 19 January 1978, following a private funeral at the nearby Kimble mortuary. A public memorial service, chaired by IAS director Harry Woolf, was held at the institute on 3 March, and there Hao Wang, Simon Kochen, André Weil, and Hassler Whitney paid tribute to Gödel and his work. The event was little publicized, however, and attracted only a small audience. In death, as in life, Gödel remained for most an enigmatic, impersonal figure. Apart from local obituary notices and articles in some of the major international newspapers [578], few outside the mathematical community took note of his passing. To the great majority, his achievements — to the extent they were known at all — remained arcane intellectual mysteries [579].

Among those present at the funeral was Louise Morse, whose husband Marston had been one of Gödel's illustrious colleagues. She recalled how utterly bereft Adele appeared on that occasion: The death of her beloved "Kurtele" was for her not only a personal calamity but a tragic defeat. She had served as his protectress for nearly fifty years, since early in their courtship; she had tasted his food, fended off the attack by young Nazi rowdies, and helped him to cope with his manifold fears. She had nursed him, coaxed him to eat, and, prior to her crippling strokes, taken charge of the management and upkeep of their home, living all the while within his shadow.

255

FIGURE 15. Adele Gödel in later years.

In return for her devotion she had received his love and affection, his financial and emotional support, and, not least, the opportunity to bask in the reflected glory of his achievements — in which, however little she understood them, she took particular pride. And rightly so: For without her to minister to his psychic and physical needs, he might never have accomplished what he did. Without him, she was now left an invalid, with few friends and little ability to manage his estate.

From the beginning he had taken responsibility for all their financial affairs. In later years they had maintained joint checking and savings accounts, but even so, Adele reportedly learned to write checks only after his death. It is thus rather surprising that in his will of 20 November 1963 Gödel named his wife not only as sole beneficiary but as executrix as well. Later, however, he had second thoughts; a codicil of 2 November 1976 left settlement of the estate to a qualified attorney.

The will made special mention of two items of property in which he claimed a one-fourth interest: the home in Brno, long since confiscated by the Czechoslovakian state, for which he apparently still held out some hope of indemnification; and a stamp collection, compiled by his father, in the possession of his brother, Rudolf. (In the codicil Gödel noted that much of that collection had since been sold. He directed that the remainder be appraised by Rudolf, whose valuation was to be accepted by all parties concerned.) But he made no special provision for the disposition of his papers.[1] Like everything else, they were left to Adele to do with as she saw fit.

Happily, her concern for posterity was greater than his. Within a few months after her husband's death Adele enlisted the help of Elizabeth Glinka in sorting through the mountain of documents Gödel had stored in the basement of the house. Inevitably, some items of biographical interest were discarded at that time (notably, all financial records after 1940). Adele also deliberately destroyed all the letters from Gödel's mother, despite Rudolf Gödel's urgent plea for them. No doubt she wanted no one else to read her mother-in-law's opinions of her; and, as she told Mrs. Glinka, she suspected that Rudolf's interest in them was monetary rather than sentimental.[2] But she left intact Gödel's scientific *Nachlaß*, amounting to some sixty boxes of books and papers. After the sorting had been completed she invited the IAS to take possession of all that material, and in the will she had drawn up in June 1979 she bequeathed the literary rights to all of Gödel's papers to the institute in his memory.

As the sole beneficiary of Gödel's estate, Adele should have been left with a comfortable inheritance. Retired IAS faculty members received substantial pensions, amounting, in 1970, to some $24,000 per year [580], and Gödel had also accumulated funds in a TIAA annuity. But on the other hand, after coming to America she had never earned any income of her own, and prior to Gödel's promotion to professor the couple's financial resources appear to

---

[1] The Library of Congress had solicited their donation some years before, but Gödel had failed to respond.

[2] Some years later Rudolf did in fact sell the letters Gödel had written to his mother. They were purchased by Dr. Werner DePauli-Schimanovich, who donated them to the Wiener Stadt- und Landesbibliothek to ensure their preservation.

have been quite limited. Furthermore — at least in the eyes of Gödel's mother — Adele had at times been something of a spendthrift. After the war she had made frequent trips to Europe; she had insisted on purchasing the home on Linden Lane at a substantially inflated price; and in furnishing it she had indulged her fondness for chandeliers and oriental rugs. But she had also made substantial improvements to the house, and, apart from their occasional trips by chauffered limousine to attend opera performances in New York, the Gödels had otherwise lived quite unostentatiously and without the expenses involved in childrearing. By 1976 their savings thus reportedly amounted to some $90,000 [581].

The primary threat to Adele's economic security was the cost of her own continuing medical care. For some years she had been confined to a wheelchair, and after her return home following her colostomy she required round-the-clock nursing assistance. A neighbor recalled that after Kurt's death Adele spent most of her waking hours lying on a sofa in the living room, dependent on house servants toward whom she was often temperamental and demanding.

A few years before, she had capriciously fired two reliable employees, and thereafter she found it difficult to obtain trustworthy household help. Consequently, she became an easy target for thieves. In one instance she discovered a maid handing some of her jewelry through an open window to an accomplice outside — an incident that was no doubt among the factors that prompted her to move to a nursing facility.

In December of 1978 Adele sold the adjacent vacant lot and shortly thereafter placed her home on the market as well. It was sold in April of 1979, some months after she had moved to Pine Run in nearby Doylestown, Pennsylvania, an establishment that promised its residents lifetime medical care in exchange for their life's savings.

At Pine Run she reportedly received excellent care. To her delight, she also discovered that her doctor there had once studied mathematics and was acquainted with her husband's work. But she soon became dissatisfied. Against the advice of all concerned she determined to move out, though doing so meant the forfeiture of most of the proceeds she had received from the sale of the house. Perhaps with that eventuality in mind, while at Pine Run she had had Rudolf send her the income from the sale of her share of the aforementioned stamp collection.

Most unwisely, she moved not to another care facility, but to Rossmoor, a retirement complex in Hightstown, New Jersey, where, as her doctor had predicted, she did not fare well. She died there on 4 February 1981.

FIGURE 16. Gravesite of Kurt and Adele Gödel, Princeton Cemetery, Princeton, New Jersey.

◇ ◇ ◇

Adele was laid to rest beside her husband and mother (Figure 16). What remained of her estate passed to Elizabeth Glinka, who donated a few more of Gödel's books and personal effects to Princeton University and the IAS.

At that time the final disposition of Gödel's *Nachlaß* had still not been determined. Ever since their donation the papers had lain undisturbed and uncatalogued in a cage in the basement of the institute's Historical Studies Library. The institute had no proper facilities for their preservation and little desire to establish an archive, which, inevitably, would pose problems of administration and would attract outside scholars in numbers that would be likely to disturb the serenity of the members. But the boxes into which Gödel's papers had been stuffed had begun to burst under the pressure of their contents and the weight of those stacked above them, and the longer the papers remained unorganized the more impatient scholars became to gain access to them. It was necessary, then, to confront the issue of their retention.

In the end, just at the time that I began my own inquiries about access to the *Nachlaß*, a committee of three IAS faculty, Professors Armand Borel,

Enrico Bombieri, and John Milnor, was formed to address the problem. In due course, on the strength of an annotated bibliography I had prepared of Gödel's publications, I was asked if I would be willing to undertake the cataloguing. I commenced work in June of 1982 and completed the task in July of 1984. Some months later, in accordance with my recommendation, the papers were placed on indefinite loan to the manuscript division of the Firestone Library at Princeton University. There, on 1 April 1985, they were opened to scholars.

XIV

Reflections on Gödel's Life and Legacy

IN A PERCEPTIVE essay on Gödel's life and work Solomon Feferman has rightly cautioned that "with any extraordinary thinker, the questions we would [most] like to see answered ... are the ones which prove to be the most elusive. What we arrive at instead is a mosaic of particularities from which some patterns clearly emerge" [582]. By identifying such patterns we may hope in the end to arrive at a better understanding of the thinker's being: in Gödel's case, of the essential tension between his scientific rationality and personal instability, of the interconnections between his life and achievements, and of the difficulties his discoveries posed for his own world view and for the subsequent development of mathematical philosophy. The aim of this final chapter is thus to look at Gödel's life and work as a whole in the hope of discerning certain common threads that run through it.

◇ ◇ ◇

Central to Gödel's life and thought were four deeply held convictions: that the universe is rationally organized and comprehensible to the human mind; that it is causally deterministic; that there is a conceptual and mental realm apart from the physical world; and that conceptual understanding is to be sought through introspection. Taken together, those beliefs served both to inspire and to circumscribe his accomplishments, and I believe they were also the source of much of the angst he experienced throughout his life.

At the beginning of chapter I, I stressed the role that rationalism played in the development of Gödel's personality. It is important to note, however, that

261

rationalism by itself need not lead one to seek answers to all questions, nor to search for hidden meanings and causes. One may, for example, believe that the world is an orderly creation amenable to rational examination but fully comprehensible only to the mind of God. Or one may, as in quantum mechanics, interpret order and causation in terms of statistical regularities, without presuming strict determinism.[1] But the rationalist who believes, as Gödel did, that the powers of human reason are potentially unbounded, and who rejects the notion of chance[2] as anything but an admission of our present state of ignorance, is faced with the daunting and ultimately frustrating task of looking within himself to discover the truth about all things. It is that quest, if pursued too far, that can lead to paranoia, which the psychiatrist Emil Kraepelin described as "a permanent unshakable delusional system" that is logically coherent but is based on premises that, viewed from the *outside*, are seen to be false.

Many of Gödel's interpretations — of historical events, films, literature, political and economic affairs, and seemingly mundane happenings — struck his contemporaries as far-fetched or even bizarre. In his mathematical endeavors, however, his willingness to countenance possibilities that others were wont to dismiss or overlook served him very well. Unlike Russell, for example, he took seriously Hilbert's idea of using mathematical methods to investigate metamathematical questions; but he also criticized Hilbert for equating the existence of mathematical entities with the consistency of formal theories describing them because he recognized, as Hilbert had not, that Peano arithmetic might fail to *decide* all propositions formalizable within it and so might admit consistent but mutually incompatible interpretations. Gödel's cosmological

---

[1] Gödel explicitly rejected such a point of view. In a footnote to the German translation of his contribution to Schilpp's Einstein volume he suggested that the uncertainty principle amounted to mistaking "a practical difficulty" for "an impossibility in principle"; and in his manuscript "The Modern Development of Mathematics in the Light of Philosophy" he declared that "[I]n physics, ... to a large extent, the possibility of knowledge of ... objectivizable states of affairs is denied, and it is asserted that we must be content to predict results of observations. This is really the end of all theoretical science in the usual sense." One wonders what he would have thought of chaos theory, which has shown how, in practice, unpredictability can arise from the iteration of perfectly well-defined functions because of sensitive dependence on initial conditions. No doubt he would have appreciated the prominent role that self-similarity plays within that theory.

[2] Gödel did sometimes make ad hoc estimates of the likelihood of events. In his letter to his mother of 21 September 1953, for example, he noted that Stalin and Taft ("Eisenhower's two principal opponents"), as well as the president of the Senate ("one of Truman's creatures") had all died within six months of one another — a circumstance whose probability he claimed was 1 in 2,000. Presumably, though, he regarded such ratios just as empirical statements of the frequency of occurrence of *determinate* happenings.

models, too, challenged some of the most deep-seated physical assumptions. Yet his aims were certainly not iconoclastic. He shared Einstein's belief that we live in an ordered universe, created by a God who "doesn't play dice"; he embraced Leibniz's visions of a *characteristica universalis* and a *calculus ratiocinator*; and, his incompleteness theorems notwithstanding, he clung throughout his life to the presumption that Hilbert had expressed in his 1900 address:

> Is the axiom of the solvability of every problem a peculiar characteristic of mathematical thought alone, or is it possibly a general law inherent in the nature of the mind, that all questions which it asks must be answerable? [583]

In line with that presumption, Gödel was not content merely to demonstrate the *possibility* of alternative models — to establish, that is, the consistency of the underlying theories; he sought to determine which model was the *correct* one. His consummate faith (so often vindicated) in the power of his own mathematical intuition led him to search for axioms from which the Continuum Hypothesis might be decided, to find a consistency proof for arithmetic based on constructively evident, though abstract, principles, and to expect that astronomical observations would eventually confirm that the universe in which we live is a rotating one of the sort he had envisioned. What is most striking about Gödel's philosophical stance is his resolute optimism in confronting the implications of his own results. Having demonstrated that the axiomatic method is fundamentally inadequate for number theory insofar as the truths of arithmetic cannot all be obtained as theorems within any fixed recursively axiomatized system, he saw no reason for despair. In contrast to Post and Turing [584], he construed the incompleteness theorems *not* as establishing limitations on the power of human reasoning, but rather as showing "that the kind of reasoning necessary in mathematics cannot be completely mechanized" [585] and thus as *affirming* the role of the human intellect in mathematical research. Because of his belief that our minds are "not static, but constantly developing," he was confident that new mathematical insights — principles that "force themselves upon us as being true" [586] — would continue to emerge.

Gödel's belief in the power of introspection was balanced, though, by his recognition of the necessity of stepping outside a system in order fully to comprehend it. As I have stressed in earlier chapters, distinctions between internal and external points of view are crucial both to the incompleteness theorems, in which the truth of the undecidable statements can be seen only metamathematically, and to Gödel's consistency proofs in set theory, which rest on the invariance of certain notions under internal relativization. Long before anyone else, Gödel saw and clearly understood the need for making such distinctions. But his way of resolving the difficulties that undecidability appeared to pose for continued progress in mathematics was to look inward — a move in character with his fundamentally introspective nature.

In his essay *1984* Solomon Feferman has identified conviction and caution as two traits that are keys to Gödel's personality and to the form in which he presented his scientific discoveries. Both, in my view, sprang from his underlying self-absorption, which was integral to his very being.

Gödel's hypochondria was one obvious manifestation of his obsessive self-concern, and I have already suggested that his paranoia may have arisen from overly intense introspection. As he withdrew more and more into his own mental world, his fears intensified — and so, too, did his Platonistic convictions and his faith in man's (and especially his own) ability to perceive conceptual truths. The skepticism he expressed in his lecture to the Mathematical Association of America (*1933a*), for example, contrasts sharply with his later advocacy of mathematical Platonism (in *1947* and, more pronouncedly, in the 1964 supplement thereto).

Gödel himself never acknowledged that such a shift had occurred, and nothing so far has come to light to explain it. I conjecture, however, that it came about in response to the results of his own investigations, particularly his unexpected discovery of the second incompleteness theorem and of cosmological models possessing closed time-lines. On the surface at least, those discoveries, as well as the consistency and independence results in set theory, appeared to rebut the first two of Gödel's four fundamental tenets. Platonism, however, provided a way to integrate his results with his belief in rational comprehensibility and deterministic causality.

In his Gibbs Lecture and his remarks on Turing's "error" Gödel addressed the philosophical ramifications of his incompleteness theorems [587]. In his essay on Cantor's continuum problem he endeavored (not entirely convincingly) to justify his belief in the falsity of the Continuum Hypothesis despite its formal undecidability. And in his contribution to Schilpp's Einstein volume he recognized the problem that temporal circularity posed for the notion of causality — a problem that he chose to evade[3] through appeal to considerations of practical feasibility.

---

[3] That it was a conscious choice is indicated by the more considered response quoted in chapter IX, which for unknown reasons Gödel deleted from the manuscript prior to publication. What is most interesting, though, is that he did *not* choose the "easy" way out — that of denying the reality of the models he had found.

Reading those papers one is impressed by Gödel's nimbleness in adducing arguments to reconcile his philosophical views with seemingly incompatible facts. His writings are characterized by clarity and precision, coupled with unexpected and subtly intricate twists of reasoning that give them an elusive profundity. Because of his skill in anticipating objections the arguments carry considerable persuasive force; and if they sometimes seem tendentious or slippery, they are always thought provoking. Nevertheless, the fact that he kept returning to the same issues, revising and refining his analyses, suggests that despite the strength of his convictions Gödel was plagued by gnawing dissatisfaction with the cogency of some of his efforts.

To what extent, then, did he ultimately succeed in upholding his system of beliefs against the countercurrents that he himself had helped to set in motion? The answer, I think, is that he succeeded very well within the sphere of mathematical logic. Outside that sphere, however, he did so only abberationally, by developing a paranoid belief structure that was logically consistent (and therefore impregnable to logical assault) but outwardly absurd.

Paranoia is a disease of judgment, to which Gödel was susceptible precisely *because* he was so meticulously logical. His affliction was a quintessential example of the so-called paradox of paranoia, in which "on the one hand we recognize the formal defects in the general structure of the intellectual apparatus of the paranoic, [while] on the other we consider [that] mental apparatus superior" [588]. As the psychiatrist Robert J. Lifton has observed, "In paranoia, ideas ... tend to be logically systematized, and therefore convincing to the afflicted individual" [589]. Moreover, because the paranoic's "fundamental assumptions ... [are at times] no worse than ... those which are accepted by his society," they can serve as "integrative principles which make his image of the world an integrated one, better than the average" [590]. I believe they did so for Gödel, whose paranoia may be seen as the culmination of a lifelong quest for a consistent world view.[4]

Hypochondria, too, "is capable of safeguarding a person from psychotic disintegration" [591] and may have exerted a stabilizing effect on Gödel much as Darwin acknowledged that his illnesses, real or imagined, had upon him. ("I know well," Darwin declared, "that my head would have failed years ago had not my stomach saved me from a minute's over-work" [592].) Indeed, there are marked parallels between the two: Through the force of their intel-

---

[4] In perceiving his assumptions — especially that of the mind's boundlessness — to be consistent, Gödel exemplified the situation he had described in his Gibbs Lecture: He was either transcending the powers of a finite machine or else deluding himself.

lects both men made profound discoveries that ran counter to their own prior expectations and that, as each was acutely aware, placed them in conflict with the philosophic spirit of their times; both persevered in their quests despite chronic (psycho)somatic disturbances; both lived reclusive lives, cared for by devoted wives and insulated from the buffetings of the outside world by the protective efforts of a small circle of professional colleagues; and both published with great caution and deliberation, determined to make their arguments as detailed and compelling as possible.

There was, however, an important difference: Because of its evident conflict with theological doctrines, Darwin's theory of evolution immediately generated widespread controversy; it was clear that natural selection vitiated the argument from design, and the idea that man was descended from apes blurred the distinction between human and animal natures and seemed to demean human intelligence. The broader implications of Gödel's incompleteness theorems, however, were neither so readily nor so widely apprehended. Only later were those theorems brought to the fore, in the debate over human versus *machine* intelligence.

The contrasting responses of Gödel and Darwin to what they themselves had wrought is striking. The force of Darwin's observations and deductions caused him to renounce his religious faith and become an agnostic. Gödel, however, remained steadfast in his beliefs. He was not only a theist, but one committed to a rigidly deterministic view of the world; and whereas Darwin discredited the argument from design, Gödel attempted to give a formal justification for the ontological argument.

◇   ◇   ◇

In his remarks at the IAS memorial service for Gödel, Simon Kochen recalled a question that Professor Kleene had put to him during his oral examination for the doctorate: "Name five theorems of Gödel." The point of the question was that "Each of th[os]e theorems ... [was] the beginning of a whole branch of modern mathematical logic" [593]: The completeness and compactness theorems are cornerstones of model theory; the incompleteness theorems showed what obstacles proof theory had to confront, and the techniques used in their proofs became central to recursion theory; the constructible sets have remained an important focus of research in set theory[5]; and Gödel's results

---

[5]Two examples of particular importance are Ronald B. Jensen's investigations of the fine structure of the constructible hierarchy and studies by various logicians of relative notions of constructibility, especially the class of sets constructible from the real numbers.

in intuitionistic logic, beginning in 1933 and culminating in the consistency proof sketched in the *Dialectica* paper, stimulated important investigations in constructive mathematics.

That logic is now firmly established as a specialty within mathematics is due in no small part to the precise mathematical methods that Gödel and a few other pioneers (Church, Tarski, Post, Kleene, and Turing, among others) brought to bear on the subject. Gödel taught few courses, had no disciples, and published relatively little; but his works exemplify Gauss's motto ("few but ripe") and have had a pervasive influence.

Among mathematicians who are not logicians there are still many, including some of the highest stature, who have little awareness or understanding of what Gödel accomplished. One reason is that at neither the undergraduate nor the graduate levels do most mathematics departments require their students to take a course in formal logic — a reflection of a lingering notion that the subject is largely irrelevant to the concerns of "working" mathematicians.

That attitude has been reinforced by the very extent to which logic has come to be regarded as a part *of* mathematics — not as a basis for all the rest, but as one among many self-centered technical specialties. Recently, though, a number of factors have begun to make the disregard of formal logic less tenable. One is the increasingly important role that computers have come to play in mathematical research — a development that Gödel had already foreseen and welcomed as early as the 1940s [594]. Another has been the protean character of the incompleteness theorems themselves, which over the years have lent themselves to ongoing reinterpretation.[6]

In particular, several examples are now known of "strictly mathematical" statements (ones not involving the numerical coding of logical notions) that are undecidable within the system of Peano arithmetic. The first such example was discovered and published in 1977 [595], when Gödel was too physically debilitated and mentally disturbed to appreciate it — if, indeed, he ever became aware of it at all. Known as the Paris-Harrington Theorem, the statement is a strengthening of the Finite Ramsey Theorem, a principle provable in Peano arithmetic, which states that given any three natural numbers n, m, and s, there is a natural number k with the property that if the unordered n-tuples of natural numbers less than k be partitioned in any way whatever into m nonoverlapping classes, there will always be a set H of s or

---

[6]A recent example is the demonstration, via a diagonalization argument, that no program designed to detect computer viruses can be both safe (guaranteed not to alter the code of the operating system of the machine on which it is run) and 100% effective (guaranteed to detect all programs that do alter the operating system code). For details see *Dowling 1989.*

more natural numbers less than k all of whose unordered n-tuples belong to the same partition class. (Such a set H is said to be *homogeneous* for the given partition.) The Paris-Harrington Theorem adds that among the H satisfying the conclusion of the Finite Ramsey Theorem there must always exist at least one of size greater than or equal to its own least element. That evidently self-referential (and somewhat ad hoc) condition makes the statement undecidable within arithmetic.

What is most notable about the Paris-Harrington Theorem is that although its statement refers only to natural numbers, its proof requires the use of infinite *sets* of natural numbers. As such it calls to mind Gödel's reproof in the Gibbs Lecture that mathematics had up to that time "not yet learned to make use of the set-theoretical axioms for the solution of number-theoretical problems," though it followed from his own work "that axioms for sets of high levels ... have consequences even for ... the theory of integers."

Likewise, Gödel repeatedly expressed the hope that suitable axioms concerning large transfinite cardinal numbers might ultimately settle the continuum problem. Since his death a great many such large cardinal axioms have been studied, some of which — notably those asserting the existence of "supercompact" and "Woodin" cardinals — have been justified by their proponents on the very grounds Gödel envisioned in his paper *1947*: Despite having "no intrinsic necessity" they have proved "fruitful" in their consequences, have "shed much light" on the discipline of descriptive set theory, and have furnished "powerful methods for solving given problems." The continuum problem itself, however, has so far proved unyielding. Of the large cardinal axioms studied hitherto, most are consistent with both the Generalized Continuum Hypothesis and its negation. The rest have not been shown to imply either.

Were he alive today, Gödel would likely be ambivalent concerning the directions that logic and philosophy have taken in recent years. On the one hand, he would surely applaud the great progress that has been achieved within the former field. On the other, he would undoubtedly rue the loss of foundational focus that has occurred as logic has become more mathematical and less philosophical in character.

Whether the latter trend will continue remains to be seen. Platonism has remained alive and well as the uncritical working philosophy of many practicing mathematicians, but among philosophers it has until recently won few adherents. Of late, however, several prominent philosophers of mathematics have endeavored to defend certain "compromise" versions of Platonism that are less radical than that Gödel espoused [596].

The appearance of the third volume of Gödel's *Collected Works*, devoted to previously unpublished manuscripts and lecture texts, may perhaps renew the debate among philosophers about Gödel's realistic philosophy of mathematics, his modal proof of the existence of God, and the ramifications of his cosmological work for the philosophy of time. To date, only a single volume (*Yourgrau 1991*) has examined the latter in any detail.

Outside the scientific community, Gödel has shared the fate of most mathematicians: His name has until recently remained largely unknown. But that has slowly begun to change, due in no small part to Douglas Hofstadter's Pulitzer prize-winning book *Gödel, Escher, Bach: An Eternal Golden Braid.*

Though dispraised by some specialists, Hofstadter's book attracted a great deal of attention in the years immediately following its publication. It sold well and has remained in print ever since. One suspects that few have read it in detail all the way through. Nonetheless it deserves credit for having brought Gödel's work to the attention of a much wider audience.

Since then Gödel's incompleteness theorems have been popularized in a number of works by Raymond Smullyan (*The Lady or the Tiger?* and *Forever Undecided: A Puzzle Guide to Gödel,* to name two), and in Roger Penrose's controversial books *The Emperor's New Mind* and *Shadows of the Mind.* They are also discussed in Hugh Whitemore's play *Breaking the Code,* based on Andrew Hodges' biography of Turing [597]. Remarkably, Whitemore not only gives a correct statement of the incompleteness theorem, but sketches the idea behind its proof and places it in the context of Hilbert's program, all within the space of a two-page monologue.[7]

Perhaps one day Gödel's life, too, will be dramatized, with all its elements of triumph, tragedy, and eccentricity. It will require a playwright of exceptional sensitivity to capture the intellectual excitement, the inner turmoil, and, not least, the *humanity* that lay behind the impassive face and recondite thought of the genius who was Kurt Gödel.

---

[7] For his efforts in increasing public awareness of mathematics Whitemore was honored by the American Mathematical Society.

Notes

For ease of reference, the following abbreviations have been employed throughout these notes: FC for the family correspondence between Gödel, his mother, and brother, preserved at the *Wiener Stadt- und Landesbibliothek*; GN for Gödel's *Nachlaß*, held by the Institute for Advanced Study, Princeton, and available to scholars at the Firestone Library of Princeton University; JD for the author; KG for Kurt Gödel; OMD for the diaries of Oskar Morgenstern, part of the Morgenstern papers at the Perkins Memorial Library of Duke University; and RG for Rudolf Gödel.

Chapter I

[1] "Meine philosophische Ansicht" (My philosophical outlook), GN 060168, folder 06/15.

[2] KG, unsent reply to a letter from Ralph Hwastecki of 17 March 1971 (GN 010897 and 010898, folder 01/69).

[3] Otto Neugebauer, conversation with JD, 1983; Deane Montgomery, interview with JD, 17 August 1981.

[4] Information in this paragraph was provided by the late Dr. Rudolf Gödel in 1982 (undated letter to JD).

[5] According to Kurt Gödel's *Taufschein*, preserved in GN, and the videotape *Schimanovich and Weibel 1986*, based on recollections of Dr. Rudolf Gödel. In his "History of the Gödel Family" (*R. Gödel 1987*, p. 19), however, Dr. Gödel states that his father was born in Vienna. The latter source (a translation of an unpublished German original entitled "Biographie meiner Mutter") provides much valuable information about the Gödel family, but it must be used with caution, since it is inaccurate with regard to some details (especially dates). In particular, appended to the end of the German text is a list of data not included in the English translation, and there we read — contrary to the earlier statement — "Unsere Eltern waren beide in Brünn geboren."

[6] *R. Gödel 1987*, p. 24 (both quotations).

[7] Another of the tenants there was a boy named Leo Slezak, who would become one of the great Lieder singers. (Ibid., p. 17.)

[8] Ibid., p. 14. Interior views of the building may be seen in the videotape *Schimanovich and Weibel 1986*.

[9] Ibid., p. 15.

[10] Ibid.

[11] GN 010729 (hereinafter referred to as the Grandjean questionnaire). A cover letter, intended to accompany the reply, is dated 19 August 1975. Both documents, together with some ancillary material (all from GN folder 01/55), have been published in *Wang 1987*, pp. 16–21.

[12] "Ich glaube in der Religion, wenn auch nicht in den Kirchen, liegt viel mehr Vernunft als man gewöhnlich glaubt, aber wir werden von frühester Jugend an zum Vorurteil dagegen erzogen, durch die Schule, den schlechten Religionsunterricht, durch Bücher und Erlebnisse" (FC 175, 14 August 1961). In a similar vein, in his private notes (GN 060168, folder 06/15) Gödel remarked that "Die Religionen sind zum größten Teil schlecht, aber nicht die Religion." (Religions, for the most part, are bad, but not religion itself.)

[13] Published in *Wang 1987*, p. 12.

[14] RG to JD, 10 January 1984.

[15] *R. Gödel 1987*, p. 25.

[16] RG to JD, 10 January 1984.

[17] "Ja an den Namen 'Achensee' erinnere ich mich sehr gut aus meiner Kindheit, noch besser allerdings an den Sandhaufen, an dem ich in Mayrhofen immer gespielt habe." (FC 172, 28 May 1961)

[18] *R. Gödel 1987*, p. 16, and FC 179 (18 December 1961).

[19] Information in this paragraph is taken from the transcript of an interview with Dr. Rudolf Gödel, conducted in the summer of 1986 by Werner Schimanovich and Peter Weibel (hereinafter referred to as the SW interview).

[20] The SW interview. Cf. also segment II of the videotape *Schimanovich and Weibel 1986*.

[21] The history of the Austrian school system sketched here is based on the account in *Gulick 1948*, pp. 545–549. The quotation is taken from pp. 548–549.

[22] Ibid., p. 551.

[23] Ibid., pp. 551–552.

[24] Ibid., p. 552.

[25] *R. Gödel 1987*, p. 16.

[26] The Grandjean questionnaire.

[27] Harry Klepetař to JD, 30 December 1983.

[28] *Johnston 1972*, p. 335.

[29] *Janik and Toulmin 1973*, p. 13.

[30] Data in this paragraph are taken from the fifty-first and fifty-second *Jahresberichte des deutschen Staats-Realgymnasium* (GN 070019.5/.6, folder 07/00).

[31] *R. Gödel 1987*, p. 25.

[32] Klepetař to JD, 30 December 1983.

[33] *R. Gödel 1987*, p. 25.

[34] *Givant 1991*, p. 26.

[35] Klepetař to JD, 30 December 1983.

[36] Data and quotations in this paragraph are taken from *Komjathy and Stockwell 1980*, p. 17. According to the the 1921 Austrian census, cited there, the Germans in Czechoslovakia were "the largest minority group," amounting to "23.4 percent of the whole population."

[37] FC 105 and 106.

[38] The latter essay, in which Gödel heroically contrasted "the austere life led by Teutonic warriors" with "the decadent habits of civilized Rome," has been adduced by Kreisel (*1980*, p. 152) as further evidence of the nationalism cultivated within the Gödel family. The case seems overstated, however, since the poem itself manifestly expresses such sentiments and the topic for the essay may very well have been chosen not by Gödel but by his teacher, whose duties undoubtedly included inculcating patriotism in the students.

[39] *Zweig 1964*, p. 32.

[40] Klepetař to JD, 30 December 1983; RG to JD, 10 January 1984 ("Wir hatten gerade in Mathematik einen Professor der kaum geeignet war für sein Fach zu interessieren").

[41] "Von Realgymnasium steht dafür kein Wort. Wahrscheinlich ist seine Ver-
gangenheit wenig ruhmreich oder sogar unrühmlich, was mich in Anbetracht
der Verhältnisse zur Zeit, als ich es besuchte, gar nicht wundern würde" (FC
166).

[42] FC 16 (2[?] August 1946 — the second digit is illegible) and FC 145 (7 June
1958). In the Grandjean questionnaire Gödel reiterated that he was fourteen
when his interest in mathematics began, and he noted that the following
year (1922) he began to read Kant. According to his brother (SW interview,
p. 13), the Gödel parents were quite fond of Marienbad and its wonderful
surroundings: They went there "six or seven times," and on one or two of
those occasions the boys joined them.

Chapter II

[43] R. Gödel 1987, p. 19.

[44] The Grandjean questionnaire.

[45] Weyl 1953, pp. 543 and 561. See also Fermi 1971, p. 38.

[46] Weyl 1953, p. 546.

[47] Einhorn 1985, p. 107. The popularity of Furtwängler's lectures is all the more
remarkable given that he was paralyzed from the neck down and depended
on an assistant to write the proofs on the blackboard for him.

[48] Feigl 1969, p. 634, n. I.

[49] Weyl 1953, p. 547.

[50] Richard Nollan, interview with Carl G. Hempel, 17 March 1982. See also
Fermi 1971, p. 37.

[51] Weyl 1953, p. 546.

[52] R. Gödel 1987, p. 18. See also Wang 1987, p. 12.

[53] Einhorn 1985, p. 107. I presume that course was the same one Taussky-Todd
recalled as having been devoted to class field theory. If so, however, it must
have been an "introduction" at a very high level.

[54] Wang (1987, p. xx) and Moore (1990, p. 349) have reported that Gödel enrolled for Hans Thirring's course on relativity theory, but when he did so
remains uncertain. Of the physics notebooks in GN other than that for Kottler's course, those that are dated are all from 1935.

[55] According to the Grandjean questionnaire, Gödel had already begun to read Kant in 1922. He later expressed disagreement with many of Kant's views, but he nonetheless deemed Kant to have been important for the development of his own interests and general philosophical outlook.

[56] *Taussky-Todd 1987*, pp. 31 and 35.

[57] In the Grandjean questionnaire Gödel stated that he did not study *Principia Mathematica* until 1929. But bookseller's receipts in GN show that he ordered and received a personal copy of the book in July of 1928, and in a letter to Herbert Feigl that he wrote the following September (published in *Wiedemann 1989*, pp. 432–439) Gödel remarked that he had spent the summer in Brünn, where, among other things, he "had read a part of *Principia Mathematica*." Despite its reputation, however, he was "less enchanted" with the work than he had expected to be.

[58] For further details on Hahn's work, including a bibliography of all his publications, see the obituary memoir *Mayrhofer 1934*. Less detailed but more personal appreciations may be found in *Menger 1934* and in Karl Menger's introduction to *Hahn 1980*.

[59] Collected in *Hahn 1980*.

[60] Menger, introduction to *Hahn 1980*, p. x.

[61] *Feigl 1969*, p. 637. The date of Gödel's first attendance is given in the Grandjean questionnaire.

[62] "Ich habe oft darueber nachgedacht, warum wohl Einstein an den Gespraechen mit mir Gefallen fand, und glaube eine der Ursachen darin gefunden zu haben, dass ich haeufig der entgegengesetzten Ansicht war und kein Hehl daraus machte." (KG to Carl Seelig, 7 September 1955). The statement ostensibly expresses Gödel's view of Einstein's reaction, but it very likely expresses Gödel's own attitude as well.

[63] *Feigl 1969*, p. 635.

[64] See *Wang 1987*, p. 22. Gödel may also have attended Carnap's course on "Axiomatics" the previous summer.

[65] *Einhorn 1985*, pp. 179–183.

[66] Information in this paragraph is taken largely from *Menger 1981*, a memoir that contains much information about Gödel unavailable elsewhere.

[67] *Feigl 1969*, p. 640. Gödel's night-owl habits are also attested by Taussky-Todd, who mentions his "having to sleep long in the morning" (*Taussky-Todd 1987*, p.32).

[68] *Feigl 1969*, p. 633.

[69] *Hempel 1979*, p. 22, and *Hempel 1981*, p. 208.

[70] Menger, introduction to *Hahn 1980*, p. xi.

[71] *Feigl 1969*, p. 637.

[72] "Carnap [war] introvertiert, cerebrotonisch und überaus systematisch"; "Neurath war ... extravertiert [und] endlos energisch .... Er war ein lebhafter, humorvoller Mann" (*Hempel 1979*, pp. 22 and 21, respectively; see also his *1981*).

[73] Menger, introduction to *Hahn 1980*, pp. xii–xiii.

[74] Ibid., p. xviii, n. 10.

[75] *Carnap 1963*, p. 23.

[76] Menger, introduction to *Hahn 1980*, p. xv.

[77] Ibid., pp. xv–xvii.

[78] Emphasis added. "Deine Abneigung gegen das Gebiet der okkulten Erscheinungen ist ja insofern sehr berechtigt, als er sich dabei um ein schwer entwirrbares Gemisch von Betrug, Leichtgläubigkeit u[nd] Dummheit mit echten Erscheinungen handelt. Aber das Resultat (u[nd] der Sinn) des Betruges ist meiner Meinung nach nicht, echte Erscheinungen vorzutäuschen, sondern die echten Erscheinungen zu verdecken" (FC 60, 3 April 1950).

[79] GN 060774, folder 06/52.

[80] "Das gehört auch in das Kapitel der okkulten Erscheinungen, die man in einer hiesigen Universität mit grösster wissenschaftliches Strenge untersucht hat, mit dem Resultat, dass jeder Mensch diese Fähigkeiten besitzt, die meisten aber nur in ganz geringem Grade" (FC 85, 20 September 1952).

[81] Oskar Morgenstern papers, Special Collections, Duke University Libraries, folder "Gödel, Kurt, 1974–1977," memorandum of 17 September 1974. In his obituary memoir of Gödel, Kreisel has asserted that demonology was also a subject that "occupied a great deal of [Gödel's] attention" (*1980*, p. 218). Evidence to support that claim, however, seems to be wanting. Gödel did have a strong interest in comparative religion, as attested by the many books in his library devoted to different, and sometimes strange, religious sects. But that interest was of a piece with the rest of his exceptionally wide range of intellectual concerns.

[82] Supplement to the second edition of *Gödel 1964*, pp. 271–272. (See *Gödel 1986–*, vol. II, p. 268.)

[83] *Zweig 1964*, p. 39.

[84] According to Franz Alt, one of Gödel's colleagues in Menger's colloquium. (Interview with JD, 14 June 1983)

[85] According to Rudolf Gödel's recollections in the SW interview.

[86] *Kreisel 1980*, p. 153. In the SW interview Rudolf Gödel offers a similar account. He places the event during Kurt's Gymnasium years and recalls an even greater age disparity.

[87] FC 135, 9 August 1957.

[88] *Taussky-Todd 1987*, p. 32. Taussky-Todd even relates an instance of Gödel's "showing off" a particularly attractive "catch."

[89] The SW interview.

[90] *Gulick 1948*, pp. 735–746.

[91] *Schimanovich and Weibel 1986*, segment III. In addition to the information about the café, the building itself is shown on that videotape.

[92] The Vienna Police Registry shows July 4 as the date of the move, but we know from Gödel's letter to Feigl that he himself spent that summer in Brno, returning to Vienna only at the resumption of the university term in October. Presumably, his brother remained in Vienna to pursue his medical studies.

[93] *R. Gödel 1987*, p. 19.

[94] All deriving, apparently, from the statement of Professor Edmund Hlawka of the University of Vienna (see *Einhorn 1985*). The story is repeated in *Kreisel 1980*, *Schimanovich and Weibel 1986*, and *Wang 1987*.

[95] *Zweig 1964*, p. 84.

[96] *Johnston 1981*, p. 102.

Chapter III

[97] *Kneale and Kneale 1962*, p. 44. Their text is the most comprehensive general reference on the history of logic.

[98] Ibid., p. 40. The use of circles to represent class inclusion and overlap was popularized by Leonhard Euler in the eighteenth century.

[99] Cf. *van Heijenoort 1985*, p. 22. Van Heijenoort notes further that although "Aristotle's syllogistic does not rest on the subject-predicate" distinction, in the *Categories* (another part of the *Organon*) Aristotle did distinguish "between the individual referred to and what is said of that individual."

[100] *Putnam 1982*, p. 298.

[101] *Goldfarb 1979*, p. 351.

[102] *van Heijenoort 1967*, p. 1.

[103] For more detailed accounts of the genesis of Cantor's ideas see *Moore 1989* or *Dauben 1979*, ch. 5.

[104] *Dauben 1979*, p. 50.

[105] See *Ferreirós 1993*.

[106] The summary here is based on the account in *Moore 1989*, pp. 83–85.

[107] Gösta Mittag-Leffler to Georg Cantor, 9 March 1885, quoted in *Moore 1989*, p. 96.

[108] In *Peano 1888*.

[109] See *Quine 1987* for further details.

[110] *van Heijenoort 1967*, p. 84.

[111] For further discussion of Peano's contributions see *Kennedy 1980*, pp. 25–27, or *Torretti 1978*, pp. 218–223; for an overview of Hilbert's work on the foundations of geometry see *Bernays 1967*.

[112] See *Heck 1993*, on which the statements in this paragraph are based.

[113] An expanded version of Hilbert's address, containing an additional thirteen problems, was published as *Hilbert 1900*. The passages quoted here are from the English translation by Mary Winston Newsom, published originally in the *Bulletin of the American Mathematical Society* (vol. 8, 1902, pp. 437–479) and reprinted in *Browder 1976*, part 1, pp. 1–34.

[114] The historian Gregory H. Moore has argued that it was *not* the discovery of the paradoxes, nor Russell's proposals (in his *1906*) of three ways to avoid them, that impelled Zermelo to axiomatize set theory, but rather his determination to secure the acceptance of his well-ordering theorem. In support of that contention he points out that Zermelo had independently discovered "Russell's" paradox himself but had not found it troubling enough to publish, and he remarks that in his paper *1908a* Zermelo employed the paradoxes "merely as a club with which to bludgeon [his] critics" (*Moore 1982*, pp. 158–159).

[115] Quotations are from the translation of Hilbert's paper in *van Heijenoort 1967*, pp. 369–392.

[116] *Simpson 1988*, p. 351.

[117] *Putnam 1982*, p. 296.

[118] *Goldfarb 1979*, p. 357.

[119] *Löwenheim 1915*.

[120] Post's results were published in his *1921*. The assessment of them given here is based on the remarks by Burton Dreben and Jean van Heijenoort in *Gödel 1986–*, vol. I, pp. 46–47. But see also the earlier remarks by van Heijenoort (*1967*, p. 264.)

[121] Bernays presented his results in his *Habilitationsschrift* of 1918, but they first appeared in print in his *1926*.

[122] Information in this paragraph is paraphrased from the account in *Moore 1988a*, p. 116.

[123] The address was published the following year both in the Congress *Proceedings* (*Hilbert 1929a*) and, in somewhat revised form, in *Mathematische Annalen* (*Hilbert 1929b*).

[124] According to Feigl (*1969*, p. 639), Brouwer's lectures also galvanized Wittgenstein to resume his work in philosophy. For the texts of both lectures see *Brouwer 1975*, pp. 417–428 and 429–440, respectively.

Chapter IV

[125] *Skolem 1923b*. In connection with Gödel's dissertation it is important to note that that paper deals *not* with notions of satisfiability, but with the axiomatic grounding of set theory. Gödel's work is also closely related to another paper of Skolem (*1923a*), which, however, does not appear among Gödel's library request slips until 1930. In the dissertation itself (*Gödel 1929*), as opposed to its published revision (*Gödel 1930a*), Gödel cited no paper at all by Skolem.

[126] The latter, but not the former, *is* cited in *Gödel 1930a*, p. 349, n. 2. To judge from the surviving library request slips, Gödel does not seem to have requested the volume of the *American Journal of Mathematics* in which Post's paper appeared until 1931.

[127] *Wang 1981*, p. 654, n. 2.

[128] Quotations from *Gödel 1929* and *1930a* are from the English translations in volume I of *Gödel 1986–*.

[129] Quotations in this and the following paragraph are from *Gödel 1986–*, vol. I, pp. 61 and 63.

[130] *Feferman 1984.*

[131] Carnap Papers, University of Pittsburgh, doc. 102-43-22 (quoted by permission of the University of Pittsburgh; all rights reserved).

[132] KG to George W. Corner, 19 January 1967 (GN 021257, folder 01/35). The letter is in English, so there is no question of its mistranslation.

[133] For a succinct analysis of the change in perspective that was needed before metamathematical issues could be considered, see the essay "Logic as calculus and logic as language," in *van Heijenoort 1985*, pp. 11–16.

[134] *Gödel 1986–*, vol. I, pp. 51 and 55. Herbrand's work was nearly contemporary with Gödel's own, while Skolem's work preceded the appearance of *Hilbert and Ackermann 1928.*

[135] *Wang 1974*, pp. 8–9.

[136] The circumstances of that neglect are explored in *Dawson 1993*. As several commentators have noted, it is of particular interest that in his review (*1934b*) of one of Skolem's papers (in which Skolem established the existence of non-standard models of arithmetic), Gödel himself failed to see that Skolem's main result could be obtained by a compactness argument.

[137] KG to Yossef Balas, undated (GN 010015.37, folder 01/20). Balas's letter of inquiry is dated 27 May 1970.

[138] *Tarski 1933* (published in abstract as *Tarski 1932* and better known in the German translation *Tarski 1935*), wherein Tarski both demonstrated the impossibility of *formally* defining the notion of truth in arithmetic and showed, informally, how to do so (by induction on the complexity of formation of the formulas).

[139] *Carnap 1963*, pp. 61–62.

[140] *Gulick 1948*, vol. I, pp. 724–725.

[141] Carnap Papers, doc. 025-73-04 (quoted by permission of the University of Pittsburgh; all rights reserved). The conversation recorded there took place during a stroll together on 10 September 1931.

[142] "Das [Vermögen] haben wir damals verbraucht um gut zu leben zu können" (SW interview, p. 7).

[143] Mentioned in *Ergebnisse eines mathematischen Kolloquiums*, vol. 2, p. 17. This seems to have been the only occasion during that academic year when Gödel actually made a presentation to the colloquium, though he had begun attending its sessions on 24 October. He did, however, contribute another short article (*1932b*) to the colloquium's *Gesammelte Mitteilungen* for that year. Concerned with a special case of the decision problem, it, too, drew on techniques and ideas employed in the dissertation.

[144] See note 137.

[145] *Menger 1981*, p. 2.

[146] *Carnap 1963*, p. 30.

[147] *Menger 1981*, p. 2. Nearly a year later, on 20 January 1931, Gödel wrote to Tarski to tell him about the incompleteness theorems (which, he said, would appear "in a few weeks"). He enclosed two offprints of its abstract (*1930b*) and five of his revised dissertation (*1930a*), as well as a spare copy of the dissertation itself.

[148] This tabular way of illustrating Gödel's argument is adapted from the discussion in *van Heijenoort 1967*, p. 439.

[149] Carnap Papers, doc. 023-73-04 (quoted by permission of the University of Pittsburgh; all rights reserved).

[150] Especially *Wang 1981*.

[151] Information supplied by Eckehart Köhler (letter to JD, 24 October 1983).

[152] Texts of the addresses were published in vol. 2 (1931) of the journal *Erkenntnis*, on pp. 122–134, 156–171 and 172–182, respectively. English translations are available in *Benacerraf and Putnam 1964*, pp. 31–54.

[153] An edited transcript of the roundtable discussion, including Gödel's remarks, was published on pp. 147–151, vol. 2 (1931) of *Erkenntnis*. An English translation of that transcript appears in *Dawson 1984a*.

[154] Vol. 18 (1930), pp. 1093–1094. An abstract of Gödel's twenty-minute talk on his dissertation results appeared on page 1068 of that same volume. In it there is no mention of his incompleteness discovery. Yet, in the last paragraph of what is believed to be the text of that talk (item **1930c* in vol. III of *Gödel 1986-*) , Gödel did forthrightly announce the existence of undecidable statements in number theory. It is not known whether he also made that announcement in his oral presentation.

[155] *Wang 1981*, pp. 654–655.

[156] Interviewed by Richard Nollan, 24 March 1982.

[157] Reported in *Crossley 1975*, p. 2, and *Kleene 1987a*, p. 52. According to the latter account, at the time Church, too, had been unaware of Gödel's work; and even after the colloquium, "Church's course [unlike von Neumann's] continued uninterruptedly . . .; but on the side we all read Gödel's paper."

[158] *Goldstine 1972*, p. 174.

[159] "Für den Mathematiker gibt es kein Ignoramibus, und meiner Meinung nach auch für die Naturwissenschaft überhaupt nicht. . . . Der wahre Grund, warum es [niemand] nicht gelang, ein unlösbares Problem zu finden, besteht meiner Meinung nach darin, daß es ein unlösbares Problem überhaupt nicht gibt. Statt des törichten Ignoramibus heiße im Gegenteil unsere Losung: Wir müssen wissen, Wir werden wissen." (The full text of the lecture was published in *Die Naturwissenschaften*, 28 November 1930, pp. 959–963.)

[160] KG to Constance Reid, 22 March 1966.

[161] "Von Prof. Courant und Prof. Schur hörte ich, dass Sie neuerdings zu bedeutsamen und überraschenden Ergebnissen im Gebiete der Grundlagen-Probleme gelangt sind und dass Sie diese demnächst publizieren wollen. Würden Sie die Liebenswürdigkeit haben, mir, wenn es Ihnen möglich ist, von den Korrekturbogen ein Exemplar zu schicken" (Paul Bernays to KG, 24 December 1930).

[162] With his usual caution, Gödel did not express that opinion in his published review (*Gödel 1931b*) of *Hilbert 1931a*. He did to Carnap, however, who recorded it in his diary (Carnap papers, doc. 025-73-04, entry for 21 May 1931. Cited by permission of the University of Pittsburgh; all rights reserved).

[163] "Das ist wirklich ein erheblicher Schritt vorwärts in der Erforschung der Grundlagenprobleme" (Bernays to KG, 18 January 1931).

[164] "Über seine Arbeit, ich sage, daß sie doch schwer verständlich ist" (Carnap papers, doc. 025-73-04. Quoted by permission of the University of Pittsburgh; all rights reserved).

[165] Ibid., doc. 081-07-07.

[166] Ibid., doc. 028-06-19.

[167] A summary of his remarks there was later published as *Gödel 1932c*.

[168] *Menger 1981*, p. 3.

[169] His undated letter, but not the accompanying query, is preserved in GN 011491, folder 01/105.

[170] Introductory note to *Gödel 1932d*, in *Gödel 1986–*, vol. I, p. 239.

[171] *Taussky-Todd 1987*, p. 38.

[172] Abstracted in *Zermelo 1932*.

[173] *Taussky-Todd 1987*, p. 38.

[174] For the full German text of Zermelo's first letter, together with my English translation of it, see *Dawson 1985b*.

[175] For the texts of Gödel's reply and Zermelo's response, see *Grattan-Guinness 1979*.

[176] *Zermelo 1932*.

[177] Carnap papers, doc. 102-43-13 (quoted by permission of the University of Pittsburgh; all rights reserved).

[178] For a detailed discussion of contemporary reactions to the incompleteness theorems, see *Dawson 1985a*.

[179] But see also *Floyd 1995*.

[180] Russell to Leon Henkin, 1 April 1963. Quoted with the copyright at the Russell Archives, McMaster University, Hamilton, Ontario.

[181] In the videotape *Schimanovich and Weibel 1986* and in the SW interview, p. 9.

[182] *Wang 1987*, pp. xxi and 91, and *Moore 1990*, p. 350.

[183] As recorded in Carnap's diary entries for 7 February, 14 and 17 March, 21 April and 21 May 1931 (cited by permission of the University of Pittsburgh; all rights reserved).

[184] "Sagen wir wegen seiner schwachen Nerven war er also zweimal in Sanatorien. . . . Also Purkersdorf und Rekawinkel" (SW interview, p. 8).

Chapter V

[185] Published in full in *Christian 1980*, p. 261.

[186] For further discussion, see the introductory note by A. S. Troelstra in *Gödel 1986–*, vol. I, pp. 282–287; the introductory note by Hao Wang to the English translation of Kolmogorov's paper in *van Heijenoort 1967*, pp. 414–416; and *Bernays 1935*, p. 212, n. 2.

[187] *Menger 1981*, p. 10.

[188] For further details see *Boolos 1979*.

[189] The correspondence on which this account is based is preserved in the *Nach-lässe* of Gödel and Heyting, the latter housed at the Heyting Archief in Amsterdam.

[190] Preserved in GN folder 04/10.

[191] Gödel's letter is preserved among the Carnap papers. Carnap's letters are items 010280.87/.88 in GN folder 01/22 (quoted by permission of the University of Pittsburgh; all rights reserved).

[192] In oversize folder 7/04.

[193] *Weyl 1953*, pp. 550–551.

[194] Ibid., p. 549.

[195] *Einhorn 1985*, p. 248.

[196] Extracted in *Christian 1980*, p. 263.

[197] Werner DePauli-Schimanovich to JD, 17 May 1984.

[198] *Taussky-Todd 1987*, p. 35.

[199] "Dabei legen Sie jedoch einen engeren und deshalb schärferen Formalismus zugrunde, während ich, um den Beweis kürzer führen und nur das Wesentliche hervorheben zu können, einen allgemeinen Formalismus angenommen habe. Es ist natürlich von Wert, den Gedanken auch in einem speziellen Formalismus wirklich durchzuführen, doch hatte ich diese Mühe gescheut, da mir das Ergebnis doch schon festzustehen schien und ich deshalb für die Formalismen selbst nicht genügend Interesse aufbringen konnte" (Paul Finsler to KG, 11 March 1933, GN 010632, folder 01/53).

[200] "Das System ... mit dem Sie operieren ist überhaupt nicht definiert, denn Sie verwenden zu seiner Definition den Begriff des 'logischen einwandfreien Beweises' der ohne nähere Präzisierung der Willkür den weitesten Spielraum läßt. ... Die von Ihnen p. 681 oben definierte Antidiagonalfolge und daher auch der unentscheidbare Satz ist ... *niemals* in demselben formalen System P darstellbar, von dem man ausgeht" (KG to Finsler, 25 March 1933, GN 010632.5, folder 01/53).

[201] KG, unsent reply to Yossef Balas, 27 May 1970 (GN 010015.37, folder 01/20).

[202] See especially *Finsler 1944*.

[203] "Wenn man über ein System γ Aussagen machen will, so ist es durchaus nicht notwendig, dass dieses System scharf definiert vorgelegt ist; es genügt, wenn man es als gegeben annehmen kann und nur einige Eigenschaften desselben kennt, aus denen sich die gewünschten Folgerungen ziehen lassen" (Finsler to KG, 19 June 1933, GN 010633, folder 01/53).

[204] *Gulick 1948*, p. 6.

[205] The summary of events given in this paragraph is based on the account in *Pauley 1981*, pp. 104ff. For greater detail see *Gulick 1948*, ch. 23.

[206] *Menger 1981*, p. 12.

[207] Ibid., p. 21.

[208] Gustav Bergmann to JD, 4 March 1983.

[209] Oswald Veblen to Menger, 11 November 1932 (Veblen papers, Library of Congress).

[210] Veblen to Menger (telegram), 7 January 1933; Menger to KG, n.d.; Veblen to KG, 21 January; KG to Veblen, 25 January; Veblen to KG, 7 February (GN 013024.5, 011495.5, 013024.6, 013025, and 013025.5, folders 01/105 and 01/197).

[211] John von Neumann to KG, 14 February 1933; KG to von Neumann, 14 March 1933 (GN 013031 and 013034, folder 01/198).

[212] *Pauley 1981*, p. 105.

[213] Translated text excerpted from J. Noakes and G. Pridham, eds., *Nazism 1919–1945: A History in Documents and Eyewitness Accounts* (Exeter: University of Exeter Press, 1983), vol. I, pp. 223–224.

[214] Statistics quoted from Arno J. Mayer, *Why Did the Heavens Not Darken?* (New York: Pantheon Books, 1988), p. 136.

[215] Ibid., p. 135.

[216] *Menger 1981*, p. 8.

[217] Cf. *Gulick 1948*, vol. II, p. 1068. To say the least, Gödel was not an enthusiastic supporter of Dollfuss's cause: His total monetary contribution to the Front appears to have been 230 Groschen, equivalent to about $5.60 today.

[218] Recalled in his letter to her of 30 April 1957 (FC 132).

[219] Quoted from Warren Goldfarb's introductory note to Gödel's contributions to the decision problem (*Gödel 1986–*, vol. I, p. 226). The discussion there provides an admirably concise, detailed survey of all the surrounding issues and developments.

[220] Full details of the proof were given in *Goldfarb 1984*.

[221] This speculation is based on Beatrice Stern's unpublished manuscript "A History of the Institute for Advanced Study, 1930–1950" (*Stern 1964*, p. 157). Though commissioned by the IAS and based on its own internal documents, Stern's 764-page typescript failed to win the institute's imprimatur, probably because it was deemed to be too muckraking. Nevertheless, it is the most detailed history of the IAS ever undertaken and as such is a valuable source of information. A microfilm copy of the manuscript is available to scholars as part of the Oppenheimer papers at the Library of Congress.

[222] Harry Woolf, foreword to *Mitchell 1980*, p. ix.

[223] Ibid., p. x.

[224] *Borel 1989*, p. 122.

[225] *Stern 1964*, pp. 155–156; see also *Borel 1989*, pp. 124–126. The issue concerning the hiring of Princeton faculty was to arise again and arouse great bitterness in 1962, when the School of Mathematics proposed to offer an IAS professorship to John Milnor; see chapter XI.

[226] The roster of members published on the occasion of the institute's fiftieth anniversary (*Mitchell 1980*) lists only nineteen others besides Gödel: A. Adrian Albert, Mabel Schmeiser Barnes, Willard E. Bleick, Leonard M. Blumenthal, Robert L. Echols, Gustav A. Hedlund, Anna Stafford Henriques, Ralph Hull, Nathan Jacobson, Börge C. Jessen, Egbertos R. van Kampen, Derrick H. Lehmer, Arnold N. Lowan, Thurman S. Peterson, Harold S. Ruse, Isaac J. Schoenberg, Tracy Y. Thomas, John Arthur Todd, and Raymond L. Wilder. But the roster published twenty-five years earlier (*Sachs 1955*) lists four more: Robert H. Cameron, Meyer Salkover, Charles C. Torrance, and Leo Zippin.

[227] *Stern 1964*, p. 195, n. 111.

[228] IAS archives, file "Gödel, pre–1935 memberships."

[229] *Taussky-Todd 1987*, p. 32.

[230] IAS archives, file "Gödel, pre–1935 memberships."

[231] RG to JD, 5 January 1983.

[232] Virginia Curry, interview with JD, spring 1981.

[233] Abraham Flexner to KG, 18 November 1935 (GN 010648.658, folder 01/48).

[234] See especially *Kleene 1981* and *1987a*, and Kleene's remarks in *Crossley 1975*, pp. 3–6.

[235] KG to Alonzo Church, 17 June 1932 (GN 010329.09, folder 01/26).

[236] Church to KG, 27 July 1932 (GN 010329.1, folder 01/26).

[237] Church to JD, 25 July 1983.

[238] *Kleene 1981*, p. 57.

[239] Ibid., p. 59.

[240] Contained in his letter to S. C. Kleene of 29 November 1935.

[241] Ibid.

[242] See, in particular, *Wang 1974*, pp. 8–11, and the statement in Gödel's reply to the Grandjean questionnaire, reproduced in *Wang 1987*, p. 18.

[243] For further discussion of these and other salient points in the 1933 lecture see the introductory note by Solomon Feferman in *Gödel 1986–*, vol. III, pp. 36–44.

[244] *Kleene 1987a*, p. 53.

[245] The original papers are *Ackermann 1928* (reprinted in *van Heijenoort 1967*, pp. 493–507) and *Sudan 1927*. On the relations between them, including the question of priority, see *Calude, Marcus, and Tevy 1979*.

[246] KG to Martin Davis, 15 February 1965. Quoted in *Gödel 1986–*, vol. I, p. 341.

[247] *Davis 1965*, pp. 71–73; reprinted in *Gödel 1986–*, vol. I, pp. 369–371.

[248] See especially *Davis 1982*; *Gandy 1980* and *1988*; *Sieg 1994*.

[249] For the technical details of Gödel's modifications see *Gödel 1986–*, vol. I, pp. 368–369, as well as S.C. Kleene's introductory note to the 1934 lectures (ibid., pp. 338–345).

[250] *Sieg 1994*, p. 82.

[251] Abraham Flexner to KG, 7 March 1934 (IAS archives, file "Gödel, pre-1935 memberships"). The offer was repeated in Flexner's letter to KG of 11 May (GN 010648.654, folder 01/48).

[252] *Gulick 1948*, vol. II, p. 1556.

[253] *Menger 1981*, p. 12.

[254] *Gulick 1948*, vol. II, p. 1555.

[255] Ibid., p. 1556.

[256] KG to Veblen, 1 January 1935 (Veblen papers, Library of Congress). An undated draft, differing in minor details, is preserved in GN (010648.655, oversize portfolio 1).

[257] FC 195, 20 October 1963.

[258] KG to Veblen, 1 January 1935 (Veblen papers, Library of Congress).

[259] Historical and architectural descriptions, as well as Figures 11 and 12, are taken from *Hoffman n.d.*

[260] RG to JD, 25 August 1982.

[261] Veblen to KG, 25 January 1935 (GN 013027.12, folder 01/197).

[262] *Menger 1981*, p. 11.

[263] Ibid., p. 12.

[264] The notebooks and notes in question are in GN folders 03/74–03/79.

[265] See *Wang 1987*, p. 97.

[266] Several sources, including Menger's own memoir (*1981*), have mistakenly dated Gödel's length-of-proof presentation as 1934, perpetuating a typographical error that appeared in the published paper (*1936b*). The error can be recognized as such through comparison with the dates of other sessions of the colloquium.

[267] For further details, including mention of some open problems, see the introductory note to *Gödel 1936b* by Rohit Parikh, in *Gödel 1986–*, vol. I, pp. 394–397.

[268] KG to Flexner, 1 August 1935 (IAS archives, file "Gödel, pre-1935 memberships").

[269] Flexner to KG, 22 August 1935 (GN 010648.660, folder 01/48).

[270] Wolfgang Pauli to KG, "on board *Georgic*," 23 September 1935 (GN 011709.3, folder 01/126).

[271] Georg Kreisel, interview with JD, 16 March 1983.

[272] Preserved in GN folders 03/52 and 03/53.

[273] Flexner to KG, 21 November 1935 (GN 010648.659, folder 01/48); Veblen to KG, 27 November 1935 (GN 013027.13, folder 01/197).

[274] Veblen to KG, 3 December 1935 (GN 013027.14, folder 01/197).

[275] Veblen to Paul Heegard, 10 December 1935 (Veblen papers, Library of Congress).

[276] This account is based on the recollections of Rudolf Gödel, interviewed by JD at Baden bei Wien on 21 July 1983. It agrees with the accounts in *Kreisel 1980* and in *Wang 1987*. However, in an interview with Eckehart Köhler in 1986 Dr. Gödel claimed that he *had* gone to Paris and brought his brother back with him.

[277] Menger to Veblen, undated, but before December 1935 (Veblen papers, Library of Congress).

[278] Menger to Veblen, 17 December 1935 (Veblen papers, Library of Congress).

[279] KG to Veblen, draft of letter of 27 November 1939 (GN 013027.29, folder 01/197).

[280] According to the receipt from the hotel Aflenzer Hof (GN folder 09/25).

[281] To her neighbor Dorothy Brown (now Mrs. Dorothy Paris), whose husband, George, took the notes for Gödel's 1938 lectures at the IAS (Paris to JD, 4 August 1983).

[282] *Wang 1987*, p. 98.

Chapter VI

[283] *Moore 1982*; see especially sec. 1.5, 3.2, and 4.9 therein.

[284] Ibid., p. 151.

[285] Quotations from Zermelo's paper *1908b*, here and below, are from the English translation in *van Heijenoort 1967*, pp. 200–202.

[286] *Moore 1982*, p. 167.

[287] Ibid., p. 261. Recently, logics more general than first-order logic have once again begun to be taken seriously. According to Lindström's Characterization Theorem (1969), however, first-order logic is the strongest logic for which *both* the compactness and Löwenheim-Skolem theorems hold. With that result in mind, the late Hao Wang echoed Zermelo's sentiments: "When we are interested in set theory or classical analysis, the Löwenheim theorem is usually taken as a sort of defect ... [of] first-order logic. [From that point of view,] what [Lindström] established is not that first-order logic is the only possible logic, but rather that it is [such] only ... when we in a sense deny reality to the concept of uncountability ... [while yet requiring] that logical proofs be formally checkable" (*Wang 1974*, p. 154).

[288] *van Heijenoort 1967*, p. 368. Gödel explained the analogy more fully in his Göttingen lecture of December 1939 (item *1939b* in vol. III of *Gödel 1986–*), and later, in a lecture at Brown University in November 1940 (item *1940a* in that same volume), recast his definition of the constructible sets in a form closer to Hilbert's conception. The essential differences between Hilbert's and Gödel's approaches are explained by Robert M. Solovay in his introductory note to those papers.

[289] In particular, his library request slips and his correspondence with Bernays.

[290] In his posthumously published *1940a*, cited in note 288.

[291] Mostowski's recollections appear on pp. 41–42 of *Crossley 1975*. One other known enrollee was Shiann-Jiun Wang of Nanking, China, one of the teachers of Hao Wang (no relation).

[292] *Menger 1981*, p. 29.

[293] The shorthand reads "Kont. Hyp. im wesentlichen gefunden in der Nacht zum 14 und 15 Juni 1937." It is one indication among many of Gödel's night-time working habits.

[294] According to KG's letter to Menger of 15 December 1937, quoted in translation on p. 15 of *Menger 1981*.

[295] Von Neumann's letters are items 013038 and 013039 in GN folder 01/198.

[296] GN 011497, folder 01/105.

[297] John F. O'Hara to KG, 3 August 1937 (Notre Dame President 1933–40: O'Hara (UPOH) General Correspondence 1937–38, "Gh–Gy" folder, University of Notre Dame Archives).

[298] GN 011498, folder 01/105.

[299] GN 013027.17, folder 01/197.

[300] Quoted in translation in *Menger 1981*, p. 14.

[301] "Zusammenkunft bei Zilsel," part of the "Prot[okoll]" notebook (GN 030114, folder 03/81).

[302] *R. Gödel 1987*, pp. 20–21.

[303] Filed in GN folders 04/125–04/127.

[304] Quoted in translation in *Menger 1981*, p. 15.

[305] Von Neumann to KG, letter of 13 January 1938 and telegram of 22 January (GN 013041/2, folder 01/198)

[306] KG to von Neumann, 12 September 1938 (von Neumann papers, Library of Congress, container 8). Gödel stated those results in his abstract *1938*, but not in his paper *1939b*, where the proofs of his other consistency results were sketched. He did discuss the details in his IAS lectures, but the "cryptic proof of a few lines [that he gave in his *1940* was] comprehensible only to cognoscenti" (R. M. Solovay, introductory note to *Gödel 1940*, pp. 13–14 in vol. II of *Gödel 1986–*).

[307] Menger to KG, 20 May 1938 (GN 011503, folder 01/105).

[308] KG to Menger, 25 June 1938 (quoted in translation in *Menger 1981*, pp. 15–16).

[309] The letter is preserved among the Veblen papers at the Library of Congress.

[310] KG to Esther Bailey, 3 September 1938 (IAS archives).

[311] KG to Flexner, 3 September 1938 (IAS archives).

[312] Gödel file, IAS archives.

[313] GN 010648.662, folder 01/48.

[314] The two *Vollmachten* are preserved in GN folder 08/13. As it happened, on that same day Gödel was also granted Viennese citizenship (as distinguished from the Austrian *Landschaft* he had obtained nine years before).

[315] *Kreisel 1980*, p. 154.

[316] The SW interview, pp. 6 and 8.

[317] Menger to Veblen, n.d. (Veblen papers, Library of Congress). In a letter of congratulations to Gödel dated 12 October, Flexner mentioned having received the news "a few days ago in the form of a card," adding that "I assume that Mrs. Gödel is coming to America with you" (GN 010648.663, folder 01/48).

[318] It was communicated by Veblen, who stressed the importance of Gödel's note and urged its immediate publication (Veblen to E. B. Wilson, 8 November 1938; Veblen papers, Library of Congress, box 16).

[319] Emil L. Post to KG, postcard, 29 October 1938 (GN 011717.3, folder 01/120).

[320] Post to KG, 30 October 1938 (GN 011717.4, folder 01/120).

[321] Post to KG, 12 March 1939 (GN 011717.5, folder 01/120). It was only after receipt of that third communication that Gödel finally replied to Post. As one of the reasons for his delay he cited his desire "first to read in detail your letters to Church, which you mentioned to me." He went on to say that "Your method of treating formal systems . . . is certainly very interesting and worth while to follow up in its consequences," and he assured Post that it had been a pleasure to meet him and that he had "noticed nothing of what you call egotistical outbursts" (KG to Post, 20 March 1939; Emil Post papers, American Philosophical Society, Philadelphia).

[322] In GN they occupy five notebooks (folders 04/39–04/43) plus two sets of loose sheets (folders 04/44–04/45). By contrast, the much more fragmentary drafts of his 1934 lectures occupy but a single folder (04/28). From the labeling of the notebooks it appears that Gödel devoted a total of seven lectures to the presentation of his consistency results.

[323] George W. Brown to JD, 31 May 1983.

[324] Ibid.

[325] For example, in his paper *1940a* (unpublished in his lifetime, but included in vol. III of *Gödel 1986–*), he commented that his "former proofs . . . contain[ed] the heuristic viewpoints."

[326] *Menger 1981*, p. 14.

[327] Ibid., p. 20.

[328] *Stritch 1981*, p. 26.

[329] Menger to Veblen, n.d. (Veblen papers, Library of Congress).

[330] *Menger 1981*, p. 20.

[331] *Stritch 1981*, p. 26.

[332] *Menger 1981*, p. 12. On a few of his visits to Notre Dame Emil Artin also dropped in on Gödel's lectures.

[333] Ibid., p. 19. For more on the subject see chapters VIII and X.

[334] The ICM invitation was tendered 14 January by William C. Graustein, with follow-up correspondence from Haskell Curry (GN folder 02/51). The invitation from the IAS has not been found, but from Gödel's reply of 18 February (preserved in the Notre Dame archives), we know that it came from Flexner.

[335] KG to Flexner, 18 February 1939 (Notre Dame archives).

[336] KG to Flexner, 24 April 1939 (Notre Dame archives).

[337] "Ich möchte Ihnen vor Allem meine Bewunderung ausdrücken: Sie haben diese enorme Problem mit einer wirklich meisterhaften Einfachheit erledigt. Und die unvermeidlichen technischen Komplikationen der Beweisdetails haben Sie . . . aufs Minimum reduci[e]rt. Die Lektüre Ihrer Untersuchung war wirklich ein ästhetischer Genuss erster Klasse" (von Neumann to KG, 2 April 1939, GN 013044, folder 01/198).

[338] Flexner to KG, 7 June 1939 (GN 010648.666, folder 01/48).

Chapter VII

[339] Von Neumann to Flexner, 16 October 1939 (IAS archives, file "Gödel—visa, immigration").

[340] "Kommen Sie nicht bald wieder einmal nach Princeton?" (KG to Bernays, 19 June 1939; Bernays papers 975:1692, ETH). Bernays did in fact reply just two days later, welcoming the resumption of their correspondence but advising regretfully that he saw no prospect of accepting an invitation to return to Princeton in the near future. (Having been dismissed from Göttingen in 1933, he had relocated to Zürich, where up to that time he had eked out an existence through a series of temporary appointments at the ETH).

[341] "Bei mir gibt es nicht viel Neues; ich hatte in letzter Zeit eine Menge mit Behördern zu tun. Ende September hoffe ich wieder in Princeton zu sein." (GN 013047; von Neumann's letter, dated 19 July, is item 013046. Both letters are filed in GN folder 01/198.)

[342] "Ich bin seit Ende Juni wieder hier in Wien u. hatte in den letzten Wochen eine Menge Laufereien, so dass es mir bisher leider nicht möglich war, etwas für das Kolloquium zusammenzuschreiben. Wie sind die Prüfungen über meine Logikvorlesung ausgefallen?" (KG to Menger, 30 August 1939, GN 011510, folder 01/105)

[343] KG to Devisenstelle Wien, 29 July 1939 (GN 090303, folder 09/15).

[344] Quoted from a letter of 6 February 1939 from F. Demuth, chairman of the Notgemeinschaft deutscher Wissenschaftler in Ausland, to Betty Drury, chairman of the Emergency Committee in Aid of Displaced Foreign Scholars; retained in the records of the Committee (box 153, folder "Situation in Austria"), Rare Books and Manuscripts Division, The New York Public Library, Astor, Lenox, and Tilden Foundations.

[345] "Ich ersuche . . . mir mitzuteilen, ob sich Dr. Gödel aus privaten Grunden in USA aufhält oder ob es sich hiebei um eine ausgesprochene Lehrtätigkeit handelt." (Signed "Dorowin," the letter is among the items in the administrative file on Gödel at the University of Vienna.)

[346] The letter is preserved in Gödel's personal file in the *Dekanats-Bestand* of the philosophical faculty at the University of Vienna. It bears only the initial "F" as signature, below the typed title "Der Dekan."

[347] KG to Veblen, draft of 27 November 1939 (GN 013027.29, folder 01/197). Since no copy of the letter has been found among Veblen's papers at the Library of Congress, it is not certain that it was ever sent.

[348] A. Marchet to the dean of the philosophical faculty, 30 September 1939: "Der bisherige Dozent Dr. Kurt Gödel ist wissenschaftlich gut beschrieben. Seine Habilitierung wurde von dem jüdischen Professor Hahn durchgeführt. Es wird ihm vorgeworfen immer in liberal-jüdischen Kreisen verkehrt zu haben. Es muß hier allerdings erwähnt werden, daß in der Systemzeit die Mathematik stark verjudet war. Direkte Äusserungen oder eine Betätigung gegen den Nationalsocialismus sind mir nicht bekannt worden. Seine Fachkollegen haben ihn nicht näher kennengelernt, sodaß weitere Auskünfte über ihn nicht zu erhalten sind. Es ist daher auch nicht möglich seine Ernennung zum Dozenten neuer Ordnung ausdrücklich zu befürworten, ebensowenig habe ich aber die Grundlagen mich dagegen auszusprechen" (retained in Gödel's personal file in the *Dekanats-Bestand* of the philosophical faculty at the University of Vienna).

[349] Von Neumann to Flexner, 27 September 1939 (von Neumann papers, Library of Congress, container 4).

[350] *Fermi 1971*, pp. 25–26.

[351] Figures quoted from p. 66 of Roger Daniels, "American Refugee Policy in Historical Perspective," in *Jackman and Borden 1983*.

[352] 43 Stat. 153.

[353] In *Jackman and Borden 1983*, pp. 61–77.

[354] Flexner to Avra M. Warren, 4 October 1939; Warren to Flexner, 10 October 1939 (IAS administrative archives, file "Gödel—visa, immigration").

[355] Von Neumann to Flexner, 16 October 1939 (IAS archives, file "Gödel—visa, immigration")

[356] *Borel 1989*, p. 129. See also *Stern 1964*, chs. 7 and 8.

[357] Frank Aydelotte to Warren, 10 November 1939 (IAS archives, file "Gödel—visa, immigration").

[358] Warren to Aydelotte, 24 November 1939 (IAS archives, file "Gödel—visa, immigration").

[359] "Gödel ... besitzt kaum ein inneres Verhältnis zum Nationalsozialismus. Er macht den Eindruck eines durchaus unpolitischen Menschen. Es wird daher auch aller Voraussicht nach schwierigeren Lagen, wie sie sich für einen Vertreter des neuen Deutschland in USA sicherlich ergeben werden, kaum gewachsen sein.

"Als Charakter macht Gödel einen guten Eindruck .... Er hat gute Umgangsformen und wird gesellschaftlich gewiss keine Fehler begehen, die das Ansehen seiner Heimat im Auslande herabsetzen können.

"Falls Gödel aus politischen Grunden die Ausreise nach Amerika versagt werden sollte, erhebt sich allerdings die Frage des Lebensunterhaltes für ihn. Gödel verfügt hier über keinerlei Einkommen und will die Einladung nach USA nur annehmen, um seinen Unterhalt bestreiten zu können. Die ganze Frage der Ausreise wäre hinfällig, wenn es gelänge, Gödel innerhalb des Reiches eine entsprechend bezahlte Stellung zu bieten" (Dean of the philosophical faculty to the rector, 27 November 1939; part of Gödel's personal file in the *Dekanats-Bestand* of the philosophical faculty at the University of Vienna).

[360] GN 09006, folder 09/02. The statement is dated 11 December 1939.

[361] These estimates are based on the exchange rates in effect when last quoted by the *New York Times*. For the Czech crown that was 15 March 1939, shortly after Hitler's takeover of Czechoslovakia; for the Reichsmark, 31 August 1939, just prior to the Nazi's invasion of Poland. Those rates were, respectively, $.0343 and $.395. The rate for the Swiss franc remained essentially unchanged at $.226. Present-day equivalents have been computed using the price deflators for the U.S. gross domestic product tabulated by the U.S. Department of Commerce.

[362] *Clare 1980*, p. 159.

[363] "Ich habe in nächsten Zeit in Berlin zu tun und beabsichtige mich auf der Rückreise ... in Göttingen aufzuhalten. Ich könnte diese Gelegenheit benützen, in einem Vortrag über meinen Beweis für die Widerspruchsfreiheit der Cantorschen Kontinuum-hypothese zu referieren, wenn dafür Interesse besteht" (KG to Helmut Hasse, 5 December 1939, GN 010807.4, folder 01/68).

[364] Gödel's lecture was published posthumously in volume III of his *Collected Works* (pp. 126–155). For more detailed analysis of its contents, see the introductory note therein by Robert M. Solovay (pp. 114–116 and 120–127).

[365] Aydelotte to the German chargé d'affaires, Washington, 1 December 1939 (IAS archives, file "Gödel—visa, immigration").

[366] KG to Aydelotte, 5 January 1940 (IAS archives, file "Gödel—visa, immigration").

[367] FC 212–213, 29 November and 16 December 1964.

[368] The manifest appears on microfilm roll 360, collection M1410 in the National Archives.

[369] *Fermi 1971*, p. 26.

[370] FC 63, 30 July 1950.

Chapter VIII

[371] "Gödel ist aus Wien gekommen. . . . Über Wien befragt: 'Der Kaffee ist erbärmlich'(!) Er ist sehr spassig, in seiner Mischung von Tiefe und Welt-fremdheit" (OMD, box 13, entry for 11 March 1940).

[372] OMD, 4 July 1940: "Sie ist ein W[iene]r Wäschermädeltyp. Wortreich, unge-bildet, resolut und hat ihm wahrscheinlich das Leben gerettet. Das er sich für Gespenster interessiert, war neu! Er war so gut aufgelegt, wie ich ihn noch nie gesehen habe."

Ibid., 12 July 1940: "Ein Rätsel. Wie soll das bloß in der Princet. Gesellschaft werden. . . . Man kann mit ihm kaum sprechen, wenn sie dabei ist."

Ibid., 7 October 1941: "Gestern Abend bei Gödels zum Nachtmal. Das er ein bedeutender Mann ist zeigt sich immer wieder; aber er ist leicht verrückt."

[373] KG to Veblen, 24 July 1940 (Veblen papers, Library of Congress). Adele apparently knew no English at all when she first arrived, and her German accent raised suspicions. Because of it she, like many others (including Ein-stein's secretary, Helen Dukas), was challenged more than once by passersby on the streets.

[374] The quotation is taken from the minutes of the meeting of 6 December 1939, and the figure for 1941–42 from those of the meeting of 19 October 1940, both of which are preserved among the von Neumann papers in the Library of Congress (container 11). The figure for 1940–41 is given in a letter to KG from director Aydelotte, dated 14 March 1940 (IAS archives, file "Gödel—visa, immigration").

[375] Information in this paragraph is based on records of the Emergency Com-mittee in Aid of Displaced Foreign Scholars, boxes 11 ("Scholars receiving aid") and 13 ("Grant work sheets"), Rare Books and Manuscripts Division, The New York Public Library, Astor, Lenox, and Tilden Foundations.

[376] IAS archives, file "Gödel—visa, immigration."

[377] KG to Veblen, 24 July 1940.

[378] So prestigious, in fact, that authors received no royalties from them. In that regard it is of interest to note that, contrary to the statement printed in *1940*, Princeton University Press did not hold the rights to the volume during its initial copyright term. Gödel retained them until 1968, when the press offered to pay him a $100 honorarium if, to save trouble for himself(!), he would sign the rights over to them for the renewal term. Understandably suspicious of such a "generous" offer, Gödel waited until the last minute to reply. In the end, however, he did as they requested; the honorarium was the only recompense he ever received for a volume the press acknowledged to be one

of their best sellers. (Subsequently, the press insisted on sharing in the royalties of volume II of Gödel's *Collected Works* before granting permission to reprint *1940* there.)

[379] KG to Frederick W. Sawyer III, unsent draft reply to Sawyer's letter of 1 February 1974 (GN 012108.98, folder 01/166).

[380] KG to Veblen, 4 July 1941; FC 16, twenty-[?] August 1946.

[381] A snapshot of the two with their hosts, in which Kurt's stiff posture suggests that of a toy soldier, is reproduced on p. 59 of *Kleene 1981*.

[382] "Die Wohnung habe ich dreimal gewechselt, weil ich die schlechte Luft von der Zentralheizung nicht vertrug" (FC 2, 7 September 1945). Another reason for the moves, cited by Adele and discussed further below, was Gödel's fear of "foreign agents" (mentioned by Klepetar in his letter to JD of 30 December 1983).

[383] OMD, 7 October 1941. The bed incident is confirmed in FC 30, 8 June 1947.

[384] Aydelotte to Dr. Max Gruenthal, 2 December 1941 (IAS archives, file "Gödel—visa, immigration").

[385] Gruenthal to Aydelotte, 4 December 1941; Aydelotte to Gruenthal, 5 December 1941 (IAS archives, file "Gödel—visa, immigration").

[386] The correspondence is preserved in boxes 11 and 115 of the records of the Emergency Committee in Aid of Displaced Foreign Scholars, Rare Books and Manuscripts Division, The New York Public Library, Astor, Lenox, and Tilden Foundations.

[387] Veblen's letter is filed in container 11 of the von Neumann papers. A copy of the IAS minutes is preserved in box 51 of the Aydelotte papers at Swarthmore College.

[388] "Er sagt, daß er mit dem Unabhängigkeitsbew. des Kontin.problems gute Fortschritte macht and viell. in einigen Monaten fertig sein wird!" (OMD, 7 October 1941)

[389] The heading "Blue Hill House" together with the date 10 July 1942 appears at the top of page 26 of *Arbeitsheft* 15.

[390] In an interview in July 1989 with Professor Peter Suber of Earlham College, published in the Ellsworth (Maine) *American*, 27 August 1992, sec. 1, p. 2.

[391] "Ich [war] bloss in Besitze gewissen Teilresultate . . ., nämlich von Beweisen für die Unabhängigkeit der Konstruktibilitäts- und Auswahlaxioms in der Typentheorie. Auf Grund meiner höchst unvollständigen Aufzeichnungen von

damals (d.h. 1942) könnte ich ohne Schwierigkeiten nur den ersten dieser beiden Beweise rekonstruieren" (KG to Wolfgang Rautenberg, 30 June 1967, GN 011834, folder 01/141; published in *Mathematik in der Schule*, vol. 6, p. 20).

[392] Brown to JD, 31 May 1983.

[393] Interviewed by JD at Stanford, 31 July 1986.

[394] Paris to JD, 4 August 1983.

[395] Brown to JD, 22 June 1983. The kitten incident may explain the appearance of the words "bobtail cat" at the top of p. 31 of Gödel's *Arbeitsheft* 15.

[396] RG to JD, 15 February 1982.

[397] Elizabeth Glinka, interview with JD, 16 May 1984. Gödel himself referred to their support of the foster child in his letter to his mother of 22 August 1948 (FC 43).

[398] The Glinka interview, confirmed in part by Gödel's letter to his mother of twenty-[?] August 1946 (FC 16).

[399] KG to Paul A. Schilpp, GN 012132, folder 01/145 (undated by Gödel, but cited as 13 September 1943 by Schilpp in his reply). Gödel also mentioned Adele's surgery in a letter he wrote to Tarski at about the same time. In neither letter, however, did he specify the nature of her operation.

[400] GN 010120.3, folder 01/21.

[401] *The Autobiography of Bertrand Russell: The Middle Years, 1914–1944*, pp. 326–327.

[402] *Gödel 1944*, pp. 127 and 137.

[403] From a paper read by Russell 9 March 1907, published in his *Essays in Analysis*.

[404] *Chihara 1973*. For a more recent analysis of Gödel's Platonism see *Parsons 1995*.

[405] The passages quoted are taken from *Chihara 1973*, pp. 62–63, 76, and 78.

[406] OMD, entries for 17 September and 18 November 1944 and 23 July and 30 October 1945.

[407] *Menger 1981*, p. 20.

[408] OMD, 27 June 1945; *Menger 1981*, p. 23. Gödel's voluminous bibliographic notes on Leibniz (GN folders 05/27 through 05/38) presumably date from this period.

[409] FC 21 (22 November 1946) and 23 (5 January 1947).

[410] "Meine gegenwärtigen Magenbeschwerden sind etwa die folgenden: Schmerzen (niemals sehr starke) rechts rückwärts in Magenhöhe, wenn ich etwas mehr als gewöhnlich esse (u. Du weisst ja dass das 'Gewöhnliche' bei mir nicht viel ist). . . . Ausserdem habe ich meine gewöhnliche hartnäckige Verstopfung, so dass ich immerfort *Abführmittel* nehmen muss. . . . Das ganze führt natürlich zu einem ständige Untergewicht: ich bin in den letzten Jahren *niemals* über *54kg* hinausgekommen" (FC 14, 8 August 1946).

[411] FC 7 (28 April 1946), 12 (21 July 1946), and 15 (15 August 1946).

[412] FC 21 (22 November 1946) and 23 (5 January 1947).

[413] For a recent historical appraisal of the session on logic see *Moschovakis 1989*.

[414] *Princeton University 1947*, p. 11.

[415] In *Myhill and Scott 1971* and *McAloon 1971*.

Chapter IX

[416] T.S. Eliot, *The Complete Poems and Plays, 1909–1950* (New York: Harcourt, Brace & World, 1971).

[417] Lester R. Ford to KG, 30 November 1945 and 21 February 1946.

[418] *Gödel 1947*, p. 521.

[419] Hempel discussed Oppenheim's role as an intellectual go-between in an interview with Richard Nollan 17 March 1982.

[420] The quotations are taken from Straus's "Reminiscences" (in *Holton and Elkana 1982*, p. 422) and from remarks he contributed to a panel discussion entitled "Working with Einstein" (published in *Woolf 1980*, p. 485).

[421] KG to Carl Seelig, 7 September 1955.

[422] "Ich habe oft darueber nachgedacht, warum wohl Einstein an den Gespraechen mit mir Gefallen fand, und glaube eine der Ursachen darin gefunden zu haben, dass ich haeufig der entgegengesetzten Ansicht war und kein Hehl daraus machte." Ibid.

[423] *Woolf 1980*, p. 485.

[424] Straus, "Reminiscences," p. 422.

[425] Ibid.

[426] The quotation is taken from the text of Gödel's lecture at the IAS on 7 May 1949.

[427] FC 12, 21 September 1946.

[428] FC 35, 7 November 1947.

[429] OMD, 23 September 1947.

[430] Howard Stein, Introductory Note to *1946/9, in *Gödel 1986–*, vol. III, pp. 203–206. In the opening paragraph of his technical paper *1949a* Gödel explicitly noted the equivalence of those two properties.

[431] The story as recounted here is based on Morgenstern's diary entry for 7 December 1947 and on an interview with his widow Dorothy 17 October 1983. Morgenstern claimed to have written up a separate account of the incident suitable for publication, but I have not found it among his papers.

[432] "Der Beamte . . . war ein äusserst sympathischer Mensch, ein Richter und persönlicher Freund Einsteins. Er hielt nachher eine lange Rede von ca. 1 Stunde, die gerade durch ihre Einfachheit u. Natürlichkeit ihre Wirkung nicht ganz verfehlte. Er erzählte von den gegenwärtigen u. vergangenen Verhältnissen hierzulande u. man ging mit dem Eindruck nach Hause, dass die amerikanische Staatsbürgerschaft im Gegensatz zu den meisten andern wirklich etwas bedeutet" (FC 40, 10 May 1948).

[433] FC 24, 19 January 1947.

[434] OMD, 13 January 1948.

[435] Freeman Dyson, interview with JD, 2 May 1983.

[436] I am indebted to Howard Stein's "Introductory Note to *Gödel *1946/9*" (in *Gödel 1986–*, vol. III, pp. 202–229), for clarifying my own understanding of some of the points discussed here. For further detailed exploration of the philosophical ramifications of Gödel's discoveries, see *Yourgrau 1991*.

[437] *Schilpp 1949*, pp. 687–688.

[438] OMD, 7 and 12 May 1949.

[439] *Chandrasekhar and Wright 1961*.

[440] *Stein 1970*, p. 589.

[441] Ibid.

[442] FC 51–58, 12 July 1949–18 January 1950.

[443] OMD, 8 December 1948.

[444] FC 29, 26 May 1947.

[445] FC 155, 7 June 1959.

[446] *Kritischer Katalog der Leibniz-Handschriften zur Vorbereitung der interakademischen Leibniz-ausgabe.*

[447] Information obtained from those agencies through the Freedom of Information Act.

[448] Gödel's original German reads: "Wovon Einstein warnte, ist, dass man den Frieden durch Aufrüstung u. Einschüchterung der 'Gegners' zu erreichen suchte. Er sagte, dass dieses Verfahren notwendig zum Krieg (u. nicht zum Frieden) führt, womit er ja recht hatte. Und es ist ja bekannt, dass das andere Verfahren (auf gütlichem Weg eine Einigung zu erzielen) von Amerika gar nicht versucht, sondern von vorneherein abgelehnt wurde. Wer angefangen hat, ist nicht die einzige Frage u. meistens auch schwer festzustellen. Aber sicher ist jedenfalls, dass Amerika unter dem Sch[l]agwort der 'Demokratie' einer Krieg für ein vollkommen unpopuläres Regime führt u. unter dem Namen einer 'Polizeiaktion' für die V.N. Dinge tut, mit denen selbst die V.N. nicht einverstanden sind."

[449] The translation is again that of the censors. The German original here reads: "Die politische Lage hat sich hier während der Feiertage wunderbar weiterentwickelt u. man hört überhaupt nichts mehr als: Vaterlandsvertei[d]igung, Wehrpflicht, Steuererhöhung, Preissteigerung etc. Ich glaube selbst im schwärzesten (oder braunsten) Hitler-deutschland war das nicht so arg. Die Leute, die bei Euch wieder so blöd reden wie zu Hitler's [*sic*] Zeiten, sind doch wahrscheinlich in der Minorität u. ich hoffe die Deutschen werden nicht so dumm sein, sich als Kanonenfutter gegen die Russen verwenden zu lassen. Ich habe den Eindruck, dass Amerika mit seinen Irrsinn bald isoliert dastehen wird."

[450] FC 65, 29 September 1950.

Chapter X

[451] OMD, 10 February 1951. See also *Borel 1989*, p. 130.

[452] Robert Oppenheimer to Lewis L. Strauss, 25 January 1951 (von Neumann papers, Library of Congress, container 22).

[453] The circumstances that prompted the change in the committee's recommendation are known only from Morgenstern's diary entry of 24 February 1951.

[454] OMD, 14 March 1951. A typescript of von Neumann's tribute is preserved among his papers at the Library of Congress (container 22).

[455] FC 71–72, 13 May and 28–30 June 1951.

[456] FC 75, 27 September 1951.

[457] He confirmed his intention in a letter of 21 May 1953 to one Rita Dickstein
       and another of 7 January 1954 to Yehoshua Bar-Hillel. According to an entry
       in Morgenstern's diary, he was still working on revisions to the lecture in
       October of 1953.

[458] OMD, 7 February 1954.

[459] Two of the six versions, both entitled "Is Mathematics Syntax of Language?,"
       were published posthumously in volume III of Gödel's *Collected Works*.

[460] OMD, Thanksgiving [25 November] 1954 and FC 109, 10 December 1954.

[461] KG to Paul Arthur Schilpp, 14 November 1955 (Open Court Archives, collec-
       tion 20/21/4, Special Collections, Morris Library, Southern Illinois University
       at Carbondale).

[462] *Ulam 1976*, p. 80. It is worth noting that Gödel himself seems never to have
       complained about his status, either publicly or in private remarks or corre-
       spondence.

[463] KG to C. A. Baylis, 14 December 1946 (GN 010015.46, folder 01/20).

[464] FC 126, 30 September 1956.

[465] OMD, 5 October 1953, based on von Neumann's report.

[466] *Borel 1989*, p. 130.

[467] Information in this paragraph is based on the account in *Pais 1982*, pp. 476–
       477.

[468] "Ist er nicht merkwürdig, dass Einsteins Tod kaum 14 Tage nach dem 25-
       jährigen Gründungsjubiläum des Instituts erfolgte?" (FC 114, 25 April 1955)

[469] *Pais 1982*, p. 497.

[470] Herman H. Goldstine has given a detailed account of the IAS computer
       project in his book *1972*. In assessing the significance of von Neumann's con-
       tributions to computer science, Goldstine states (pp. 191–192) "Von Neumann
       was the first person, as far as I am concerned, who understood explicitly that
       a computer essentially performed logical functions. . . . Today this sounds so
       trite as to be almost unworthy of mention. Yet in 1944 it was a major advance
       in thinking." It is now clear that Turing had come to the same realization
       independently and at about the same time. Just how revolutionary an insight
       it was may be judged from a statement made by another computer pioneer,
       Howard Aiken, in 1956: "If it should turn out that the basic logics of a ma-
       chine designed for the numerical solution of differential equations coincide

with the logics of a machine intended to make bills for a department store, I would regard this as the most amazing coincidence I have ever encountered" (*Ceruzzi 1983*, p. 43; quoted in *Davis 1987*, p. 140).

[471] *Borel 1989*, p. 131.

[472] KG to von Neumann, 20 March 1956 (von Neumann papers, Library of Congress, container 5). The original German text is to appear in a forthcoming volume of Gödel's *Collected Works*.

[473] For further discussion see *Hartmanis 1989*.

[474] KG to Bernays, 6 February 1957 (Bernays papers, ETH, Hs. 975:1698).

[475] KG to George E. Hay, 23 February 1961 (GN 012425, folder 01/164).

[476] "Provably Recursive Functionals of Analysis: A Consistency Proof of Analysis by an Extension of Principles Formulated in Current Intuitionistic Mathematics," pp. 1–27 in J. C. E. Dekker, ed., *Recursive Function Theory*, Proceedings of Symposia in Pure Mathematics, vol. 5, American Mathematical Society, Providence, 1962.

[477] "Wir leben eben leider in einen Welt in der 99% von allem Schönen schon in Entstehen (oder noch vorher) zerstört wird" (FC 88, 14 January 1953). "Es müssen also irgendwelche Kräfte sein, die das Gute direkt unterdrücken. Man kann sich auch leicht ausmalen woher die stammen" (FC 89, 20 February 1953).

[478] "Gerade hier (im Gegensatz zu Europa) hat man das Gefühl, von lauter guten u. hilfsbereiten Menschen umgeben zu sein. Insbesondere gilt das auch von allem, was mit Staat u. Ämtern zu tun hat. Während man in Europa den Eindruck hat, dass Ämter überhaupt nur dazu da sind, um den Menschen das Leben sauer zu machen, ist es hier umgekehrt" (FC 89, 20 February 1953).

[479] "Unter den 14 [waren] auch der gegenwärtige Verteidigungsminister u. der Urheber des Friedensvertrags mit Japan. Ich bin also da ganz unverschuldet in eine höchst kriegerische Gesellschaft geraten" (FC 83, 22 July 1952).

[480] *Woolf 1980*, p. 485.

[481] FC 100, 6 January 1954.

[482] "Dass er die Rosenbergs nicht begnadigte, hat mich zwar etwas enttäuscht . . . . Andrerseits aber muss man sagen, dass in erster Linie die Verteidigung ihre Hinrichtung verschuldet hat, inden sie ganz unglaubliche Fehler machte, offenbar absichtlich, denn die Kommunisten brauchten doch einen neuen Beweis für die amerikanische 'Barbarei'" (FC 94, 26 July 1953).

[483] FC 105, 27 June 1954; FC 128, 12 December 1956. Nine years later he remarked that it had been twenty years since the last war and so was "slowly becoming time" for another (FC 236, 7 December 1965).

[484] FC 223, 21 April 1965.

[485] FC 61 and 62, 11 May and 25 June 1950.

[486] "Es wäre auch ganz unberechtigt zu sagen, dass man gerade in diesem Gebiet mit dem Verstande nichts ausrichten kann. . . . Man ist natürlich heute weit davon entfernt, das theologische Weltbild wissenschaftlich begründen zu können, aber ich glaube . . . dürfte es möglich sein rein verstandesmässig . . . einzusehen, dass die theologische Weltanschauung mit allen bekannten Tatsachen . . . durchaus vereinbar ist" (FC 177, 6 October 1961).

[487] FC 174, 23 July 1961.

[488] FC 176, 12 September 1961.

[489] FC 175, 14 August 1961.

[490] FC 90–92, 25 March, 14 April, and 10 May 1953.

Chapter XI

[491] The review, by Stefan Bauer-Mengelberg, appeared in vol. 30 (1965), pp. 359–362.

[492] Gödel's correspondence with Oliver and Boyd and Basic Books occupies GN folders 02/26–02/28. In addition, Gödel spelled out his criticisms of the venture in a letter he wrote to Church on 2 March 1965 (GN 010332, folder 01/26): He was particularly upset that the introduction "entirely disregard[ed] . . . subsequent developments" concerning "the concept of computable function" and "thereby . . . withh[eld] from the philosophically interested reader" what was of greatest interest, namely, that the incompleteness theorems could "now be proved rigorously for *all* formal systems." He worried that the reader might get "the wrong impression that my theorems apply only to 'arithmetical systems' . . . and that . . . [the system] P is only the 'arithmetical part' of PM," whereas "in fact it contains all the mathematics of today, except for a few papers dealing with large cardinals."

[493] KG to Davis, 15 February 1965 (GN 010479, folder 01/38).

[494] The extraordinary saga of van Heijenoort's life is vividly recounted in Anita Burdman Feferman's biography of him, *Politics, Logic, and Love* (Wellesley, Mass.: A.K. Peters, 1993).

[495] Ibid., pp. 261–262 and 274.

[496] George W. Corner to KG, 13 December 1961 (GN folder 02/37). Gödel had been elected to membership the previous April and had attended the meeting in November at which he was formally inducted. Afterward, unusually, he took time to write a letter of appreciation, in which he told Corner that he had "enjoyed the meeting . . . very much and . . . [found] some of the lectures . . . very interesting."

[497] FC 165, 6 July 1960.

[498] FC 181, 14 February 1962.

[499] *Borel 1989*, p. 137.

[500] Atle Selberg, interview with JD at Stanford, 25 August 1986. According to Selberg, Gödel sided with the other mathematicians on the issue but felt that one must always respect authority.

[501] *Borel 1989*, p. 138.

[502] Ibid., pp. 138–139, supplemented by information supplied by Selberg.

[503] *Shepherdson 1953*.

[504] In a letter of 29 July 1963 to David Gilbarg, then chairman of the Stanford mathematics department, Gödel stated that he had not met Cohen until the previous May. The earliest correspondence between the two is a letter from Cohen dated 24 April.

[505] Details of Cohen's path to discovery are taken from *Moore 1988b*.

[506] *Cohen 1966*, pp. 108–109.

[507] Cohen to KG, 6 May 1963 (GN folder 01/27).

[508] The quotes are taken from Gödel's letters to Cohen of 5 June and 20 June 1963 (GN folder 01/27).

[509] All the items quoted or paraphrased are in GN. The letter to Church (010334.36) and its draft (010334.34) are in folder 01/26; that to Rautenberg, subsequently excerpted in *Mathematik in der Schule*, vol. 6, p. 20, is item 011834 in folder 01/141. See also chapter VIII, note 391.

[510] See "Hilbert's first problem: the Continuum Hypothesis," by Donald A. Martin, in *Browder 1976*, vol. I, pp. 81–92.

[511] *Gödel 1964*, p. 271. Gödel reiterated the same point in a letter of 10 December 1969 to Professor George A. Brutian (GN 010280.5, folder 01/19).

[512] KG to Cohen, 22 January 1964 (GN 010390.5, folder 01/30).

[513] The translation never did appear in *Dialectica*. It was published only posthumously, in volume II of Gödel's *Collected Works*.

[514] In his letter to his mother of 13 May 1966 (FC 242), Gödel told her that he had raised that objection with the Austrian ambassador in connection with the award of the medal. He also invoked it explicitly in the letter of 6 June 1966 (reproduced in *Christian 1980*, p. 266) in which he declined Academy membership. The posthumous honorary doctorate is mentioned both in Christian's memoir (ibid.) and in *Kreisel 1980* (p. 155).

[515] *R. Gödel 1987*, p. 23.

[516] KG to RG, 18 August 1966 (GN 010711.2, folder 01/54.6).

Chapter XII

[517] *Kreisel 1980*, p. 160.

[518] FC 231, 232, and 235 (19 August, 12 September, and 6 October 1965); OMD, 28 April 1966.

[519] GN folder 01/54.6.

[520] OMD, 18 May 1968.

[521] KG to Bernays, 17 December 1968 (Bernays papers, ETH, Hs. 975:1742).

[522] OMD, 11 January 1969.

[523] In his probing, insightful paper *1994*, Wilfried Sieg has argued on the contrary that the principal result of the analysis that Turing gave in his 1936 paper was "the axiomatic formulation of limitations" on the behavior of *human* calculators. He also notes that Post, who that same year proposed a similar model for how a "worker" carries out computations (*Post 1936*), viewed Gödel's incompleteness theorems as "a fundamental discovery in the limitations of the mathematizing power of Homo Sapiens."

[524] OMD, 9 December 1969.

[525] Gödel's criticism of Turing is quoted from *Wang 1974*, p. 325, where it first appeared in print. Another version of those remarks, found appended to Gödel's English revision of the *Dialectica* paper, was published posthumously in *Gödel 1986–*, p. 306.

[526] OMD, 9 and 25 December 1969.

[527] OMD, 26 January 1970.

[528] "Gödel ist ein grosses Problem (ich glaube auch Theater) . . . . Dagegen ist nicht aufzukommen. Aber . . . er wird schon wieder herauskommen" (OMD, 6 February 1970).

[529] Perhaps the first two of the remarks published in *Gödel 1986–*, pp. 305–306.

[530] The last four items are all included in *Gödel 1986–*, vol. III.

[531] OMD, 10 February 1970.

[532] OMD, 12 March 1970.

[533] OMD, 13 March 1970.

[534] OMD, 4 and 29 August 1970. In the draft of a letter to his brother dated 17 September Gödel said that he weighed 105 pounds, compared with 80 the previous May (GN 010711.43, folder 01/54.5).

[535] GN 012783, folder 04/151. The document in question is an undated draft and bears the shorthand annotation "nicht abgeschickt" (not sent). In it Gödel stated that he "had been sleeping very purely [*sic*] and had been taking drugs impairing the mental functions."

[536] Ibid.

[537] OMD, 17 September 1975.

[538] For a survey of these later developments see the discussion by Gregory H. Moore in *Gödel 1986–*, vol. II, pp. 173–175. Further commentary is contained in the introductory note (by Robert M. Solovay) to Gödel's three manuscripts, published in volume III of those same *Collected Works*.

[539] OMD, 29 August 1970: "über seinen ontologischen Beweis—er hatte das Resultat vor einigen Jahren, ist jetzt zufrieden damit aber zögert mit der Publikation. Es würde ihm zugeschickt werden das er wirkl. an Gott glaubt, wo er doch nur eine logische Untersuchung mache (d.h. zeigt, daß ein solcher Beweis mit klassischen Annahmen (Vollkommenheit usw.), entsprechend axiomatisiert, möglich sei)."

[540] Those results are now available in volume III of *Gödel 1986–* (item *193?*, pp. 164–175).

[541] So far as I have been able to determine, Matijasevich's result was not brought to Gödel's attention until 1973, at which time, apparently in response to a request from Gödel, Abraham Robinson listed it as one of the three most outstanding achievements in logic during the preceding decade (Robinson to KG, 6 April 1973; GN 011966, folder 01/136).

[542] *Gödel 1986–*, vol. II, pp. 271–280.

[543] *Wang 1987*, p. 131.

[544] Wang, "Conversations with Gödel" (unpublished typescript).

[545] OMD, 20 November 1971.

[546] Ralph Hwastecki to KG, 17 March 1971 (GN 010897, folder 01/69).

[547] KG, unsent reply to Hwastecki (GN 010898, folder 01/69).

[548] GN 012867, folder 01/188.

[549] "Als ich später von diesen Dingen hörte, war ich sehr skeptisch, nicht aus physikalischen, sondern aus soziologischen Gründen, weil ich glaubte, dass diese Entwicklung erst gegen das Ende unserer Kulturperiode erfolgen wird, die vermutlich noch in ferner Zukunft liegt" (KG to Hans Thirring, 27 June 1972). The original of this letter is among Thirring's papers at the Zentralbibliothek für Physik in Wien. I am grateful to Dr. Wolfgang Kerber, director of that library, for kindly providing me with a copy while the Thirring papers were still in the process of being catalogued.

[550] Wang, "Conversations," reported in the entry for 14 June 1972.

[551] *Kreisel 1980*, p. 160.

[552] Ibid., p. 158.

[553] A particularly snide and anti-intellectual account of the confrontation appeared in the *Atlantic Monthly*. Entitled "Bad Days on Mount Olympus," it featured a photograph of Gödel and Kaysen together, taken at the von Neumann conference.

[554] FC 239, 24 February 1966.

[555] OMD, 27 December 1973.

[556] OMD, entries of 9, 10, and 13 April 1974.

[557] OMD, 30 April 1974.

[558] OMD, 24 August 1974.

[559] OMD, 14 and 17 September 1974.

[560] OMD, 4 June 1975.

[561] OMD, 20 September 1975.

[562] Glinka, interview with JD in her home, 16 May 1984.

[563] OMD, entries of 23 and 28 February and 2 and 4 April 1976.

[564] OMD, 8 April, 9 May, and 11 June 1976.

[565] Wang, "Conversations," items dated 19 April and 1 June 1976.

[566] Wang, "Conversations," 1 June 1976.

[567] OMD, 19 June and 24 July 1976.

[568] OMD, entries of 24 July, 6 September and 1 and 3 October 1976.

[569] Glinka, interview with JD, 16 May 1984.

[570] Morgenstern papers, Perkins Memorial Library, Duke University, folder "Gödel, Kurt, 1974-1977".

[571] Dorothy Morgenstern, interview with JD, 17 October 1983.

[572] Deane Montgomery, interview with JD, 17 August 1981.

[573] Telephone interview with JD, 2 March 1988.

[574] Wang, "Conversations," 1977.

[575] *Wang 1987*, p. 133.

[576] Ibid.

[577] The terminology is taken from Gödel's death certificate, signed by Dr. Harvey Rothberg.

Chapter XIII

[578] The *New York Times* published a standard-format obituary, with accompanying photograph, on page 28 of the issue for 15 January. There was also a perfunctory editorial about Gödel's work on page 18, section 4, of the next Sunday edition (22 January). Neither article mentioned any of his contributions except the incompleteness theorems, and the dates given for some events in his life were erroneous. Worst of all, the editorial perpetuated the idea that the incompleteness theorems had created a *dilemma* for mathematics: "there are no guarantees, it turns out, that our cherished edifices of logic and mathematics *are* free of contradictions, and our daily assumptions to that effect are mere acts of faith." It was against just that view that Gödel himself had argued long and hard.

[579] The account in this chapter is based largely on an interview with Mrs. Elizabeth Glinka (16 May 1984), supplemented by information obtained from legal records and from conversations with Mrs. Louise Morse and Mrs. Adeline Federici.

[580] OMD, 14 April 1970. Morgenstern noted that that amount exceeded his own current salary as a professor at Princeton.

[581] OMD, 8 April 1976.

Chapter XIV

[582] *Feferman 1986*, p. 2.

[583] *Browder 1976*, part 1, p. 7. The translation of Hilbert's lecture from which this quotation is taken is by Mary Winston Newson.

[584] Cf. [523].

[585] KG to David F. Plummer, 31 July 1967 (GN 011714, folder 01/126).

[586] *Gödel 1964*, p. 271.

[587] For a detailed analysis of Gödel's criticism of Turing see the commentary by Judson Webb in *Gödel 1986–*, pp. 292–304.

[588] *Fried and Agassi 1976*, pp. 5–6.

[589] *Lifton 1986*, p. 440.

[590] *Fried and Agassi 1976*, pp. 4–6.

[591] *Baur 1988*, p. 74.

[592] Ibid., p. 183; quoted from Ralph Colp, Jr., *To Be an Invalid: The Illness of Charles Darwin* (Chicago: University of Chicago Press, 1977), p. 70.

[593] Kochen's remarks are quoted from an audiotape of the memorial service preserved by the IAS.

[594] In his diary entry for 17 July 1965 Morgenstern recalled an occasion twenty years before, when he had taken Gödel to a lecture by von Neumann and Gödel had remarked, "Now you will see real progress in logic." He asked Gödel whether he thought his prediction had in the interim been fulfilled, to which Gödel replied: "Leider nicht, da die Logiker die Comp. nicht so benützen & letztere viell. noch nicht imstande seien die enorme langen Ketten von Schlüssen wirkl. durchzuführen. Auch hätten sie sich mehr mit der Logik des design von C. beschäftigen sollen." (Unfortunately not, since logicians

use computers much and the latter are perhaps still not capable of carrying out the enormously long chains of inferences. Also, they should have paid more attention to the logic of the design of computers.) Morgenstern went on to say that Gödel had "a thoroughly positive attitude" toward such new developments, in sharp contrast to the haughty "pure" mathematicians at Princeton.

[595] *Paris and Harrington 1977.*

[596] *Kitcher 1983* and *Maddy 1990* have attracted particular notice.

[597] *Alan Turing: The Enigma* (New York: Simon and Schuster, 1983)—now regrettably out of print.

Appendix A
Chronology

1906   Gödel is born in Brünn (Brno), Moravia, 28 April

1912   Enters Lutheran primary school in Brünn

1916   Enters German-language Staats-Realgymnasium in Brünn

1924   Enters the University of Vienna, intending to study physics

1926   Switches to mathematics, and begins attending meetings of the Vienna Circle

1929   Father dies unexpectedly on 23 February, aged fifty-four

        Becomes citizen of Austria, 6 June

        Doctoral dissertation, establishing the completeness of first-order logic, is approved 6 July and submitted for publication 22 October

        Participation in Karl Menger's mathematics colloquium begins; Gödel contributes thirteen short articles to its published proceedings, which he helps edit

1930   Granted degree of Phil.D., 6 February

        Presents dissertation results at conference in Königsberg (6 September); announces the existence of formally undecidable propositions during a discussion session the following day

1931   Publishes his incompleteness theorems in *Monatshefte für Mathematik und Physik*

1932   Submits the incompleteness paper to the University of Vienna as his *Habilitationsschrift*

1933   Becomes *Privatdozent* at the University of Vienna and teaches a summer course there on foundations of arithmetic

            In the fall, travels to Princeton, where he spends the academic year 1933–34 at the newly established Institute for Advanced Study

1934   Delivers lecture course at the IAS on the incompleteness results

            After return to Austria, enters sanatorium for treatment of depression

1935   Offers a second summer course at Vienna, on selected topics in mathematical logic

            Succeeds in proving the consistency of the Axiom of Choice with the other axioms of set theory

            Briefly returns to the IAS that fall, but resigns after suffering a relapse of depression; remains incapacitated until spring 1937

1937   Teaches last course at the University of Vienna, on axiomatic set theory

            Establishes the consistency of the Generalized Continuum Hypothesis with the axioms of set theory

1938   Marries Adele Nimbursky (née Porkert), 20 September

            Returns to the IAS for the fall semester, where he lectures on his results in set theory. The lectures are published as a monograph two years later.

1939   Spends the spring semester at Notre Dame, at Menger's invitation

            After return to Austria, is declared fit for German military service; begins quest for U.S. non-quota immigrant visa

1940   Emigrates to America via trans-Siberian railway and ship from Yokohama to San Francisco (January-March)

            Takes up temporary membership at the IAS, renewed annually until 1946

1941   Lectures at Yale and the IAS on a new type of consistency proof for arithmetic

1942   Attempts to prove the independence of the Axiom of Choice and the Continuum Hypothesis relative to the other axioms of set theory; obtains some partial results, but soon abandons set theory and turns to work in philosophy

1944   Publishes essay, "Russell's Mathematical Logic"

1946   Appointed permanent member at the IAS

1948   Becomes U.S. citizen

1949   Publishes results in general relativity theory, demonstrating the existence of universes in which "time travel" into the past is possible

1950   Delivers invited address on cosmological results to International Congress of Mathematicians, 31 August

1951   Nearly dies of bleeding ulcer

          Shares first Einstein Award with Julian Schwinger, and receives honorary Litt.D. from Yale

          Delivers Gibbs Lecture to the American Mathematical Society, 26 December

1952   Awarded honorary Sc.D. by Harvard

1953   Elected to the National Academy of Sciences

          Promoted to professor at the IAS

1958   Consistency proof for arithmetic published in the journal *Dialectica* (last paper published during his lifetime)

1961   Elected to the American Philosophical Society

1966   Mother dies, 23 July, aged eighty-six

1967   Elected honorary member of the London Mathematical Society; awarded honorary Sc.D. by Amherst College

1972   Awarded honorary Sc.D. by Rockefeller University

1975   Awarded National Medal of Science

1976   Retires from IAS, 1 July

1978   Dies of self-starvation, 14 January

Appendix B
Family Genealogies

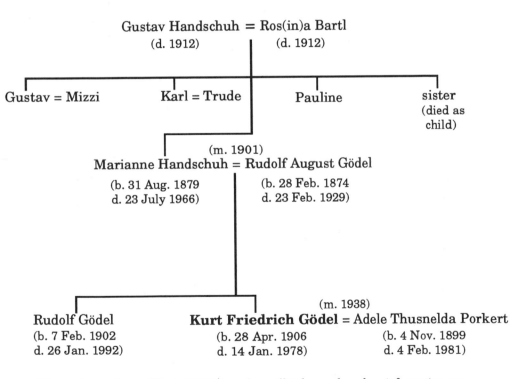

Gustav Handschuh = Ros(in)a Bartl
(d. 1912)   (d. 1912)

Gustav = Mizzi   Karl = Trude   Pauline   sister
                                                (died as
                                                 child)

(m. 1901)
Marianne Handschuh = Rudolf August Gödel

(b. 31 Aug. 1879   (b. 28 Feb. 1874
d. 23 July 1966)   d. 23 Feb. 1929)

(m. 1938)
Rudolf Gödel   **Kurt Friedrich Gödel** = Adele Thusnelda Porkert
(b. 7 Feb. 1902   (b. 28 Apr. 1906   (b. 4 Nov. 1899
d. 26 Jan. 1992)   d. 14 Jan. 1978)   d. 4 Feb. 1981)

Maternal genealogy of Kurt Gödel (based on official records and on information supplied by Dr. Rudolf Gödel).

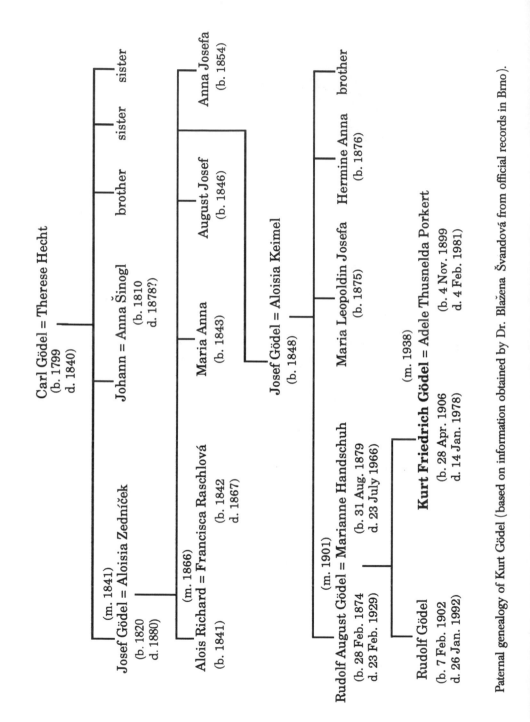

Carl Gödel = Therese Hecht
(b. 1799
d. 1840)

brother   sister   sister

(m. 1841)
Josef Gödel = Aloisia Zedníček
(b. 1820
d. 1880)

Johann = Anna Šinogl
(b. 1810
d. 1878?)

Anna Josefa
(b. 1854)

(m. 1866)
Alois Richard = Francisca Raschlová
(b. 1841)
(b. 1842
d. 1867)

Maria Anna
(b. 1843)

August Josef
(b. 1846)

Josef Gödel = Aloisia Keimel
(b. 1848)

Maria Leopoldin Josefa
(b. 1875)

Hermine Anna   brother
(b. 1876)

(m. 1938)

(m. 1901)
Rudolf August Gödel = Marianne Handschuh
(b. 28 Feb. 1874
d. 23 Feb. 1929)
(b. 31 Aug. 1879
d. 23 July 1966)

Kurt Friedrich Gödel = Adele Thusnelda Porkert
(b. 28 Apr. 1906
d. 14 Jan. 1978)
(b. 4 Nov. 1899
d. 4 Feb. 1981)

Rudolf Gödel
(b. 7 Feb. 1902
d. 26 Jan. 1992)

Paternal genealogy of Kurt Gödel (based on information obtained by Dr. Blažena Švandová from official records in Brno).

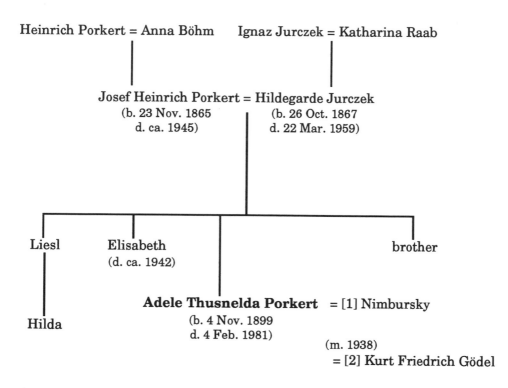

Genealogy of Adele Thusnelda Gödel (based on official records and on information supplied by Dr. Rudolf Gödel).

Appendix C
Biographical Vignettes

Paul Bernays (1888–1977) wrote his dissertation in analytic number theory at Göttingen under Edmund Landau. Subsequently he submitted two separate *Habilitationsschriften*: the first, also in number theory, at Zürich under Ernst Zermelo; the second, in logic, at Göttingen, to which Hilbert had invited him to return to assist in Hilbert's research on foundations. Made an untenured professor at Göttingen in 1922, Bernays was purged as a non-Aryan eleven years later and never again held a full-time professorial position, though he taught at the ETH in Zürich from 1939–59. A founder of the journal *Dialectica*, Bernays was principal co-author of the book *Grundlagen der Mathematik*, the second volume of which contained the first complete proof of Gödel's second incompleteness theorem. Bernays's system of axioms for set theory, based on an earlier formulation by von Neumann, was employed by Gödel with slight modifications for his work on the Axiom of Choice and the Continuum Hypothesis.

◇ ◇ ◇

Luitzen Egbertus Jan Brouwer (1881–1966) received his doctorate in 1907, for a dissertation in which he criticized alike the foundational views of Russell, Hilbert, and Poincaré. Between 1907 and 1912 he obtained a series of important results in topology, most notably his celebrated Fixed-Point Theorem and a proof of the invariance of dimension under topological mappings. As early as 1908 he campaigned against the use of the Law of Excluded Middle in proofs, and after 1912, the year in which he was appointed a professor at the University of Amsterdam, he abjured classical mathematics to develop the constructive philosophy of mathematics he called intuitionism. For many years his radical views met with little support outside Holland, and the polemical tone of his writings, as well as his combative personality, led him into conflict with many of his peers, including Hilbert and Karl Menger.

◇ ◇ ◇

The creator of transfinite set theory, **Georg Cantor** (1845–1918) studied at Berlin under Weierstrass, Kronecker, and Kummer. As outlined in chapter III, his set-theoretic conceptions grew out of his work in harmonic analysis, to which he made a number of important contributions. Thwarted by the opposition of Kronecker, Cantor failed ever to obtain an appointment at the University of Berlin. His career was spent instead at the University of Halle, where he became full professor in 1879. Cantor served as first president of the Deutsche Mathematiker-Vereinigung, which he helped to found, and was one of the organizers of the first International Congress of Mathematicians, held in Zürich in 1897. In later years he suffered episodes of depression, for which, at the time of his death, he was undergoing clinical treatment.

◇ ◇ ◇

One of the most prominent of the logical positivists, **Rudolf Carnap** (1891–1970) studied under Frege at the University of Jena, from which he received his doctorate in 1921. In 1926 he became *Privatdozent* at the University of Vienna, and in 1931 he accepted the chair in natural philosophy at the German University in Prague. He emigrated to America in 1935, where he taught at the University of Chicago (1936–52) and UCLA (1954–61). With Hans Reichenbach, Carnap founded the journal *Erkenntnis*. His principal works were the books *Der logische Aufbau der Welt* (The Logical Construction of the World, 1928), *Logische Syntax der Sprache* (The Logical Syntax of Language, 1934), and *Logical Foundations of Probability* (1950).

◇ ◇ ◇

Alonzo Church (1903–1995) is best known for having introduced the λ-calculus, a formalism for defining what are now called recursive functions, and for his Thesis (1934) that all effectively computable functions are λ-definable. In 1936, drawing on Gödel's work, Church showed that the valid sentences of Peano arithmetic do not form a recursive set. One of the founders of the Association for Symbolic Logic, he served as editor of its *Journal* from 1936 to 1979 and compiled a comprehensive bibliography of symbolic logic (*1936*) that occupied much of the *Journal*'s first issue. Church was a member of the faculty of Princeton University from 1929 to 1967 and at UCLA from then until his retirement in 1991.

◇ ◇ ◇

Though his works had little initial impact, the German mathematician and philosopher **Gottlob Frege** (1848–1925) is recognized today as the foremost figure in the development of modern logic. In his *Begriffsschrift* (Ideography, 1879) Frege enunciated the modern conception of the universal quantifier, introduced the fundamental syntactical notions of function and argument, and gave a formalization of second-order quantification theory; in *Die Grundlagen der Arithmetik* (The Foundations of Arithmetic, 1884), he proclaimed the logicist thesis that arithmetic is reducible to pure logic; and in his two-volume work *Die Grundgesetze der Arithmetik* (The Basic Laws of Arithmetic, 1893 and 1903) he drew the distinction between the sense (*Sinn*) and reference (*Bedeutung*) of a linguistic expression and presented a formal system, based on naive set theory, for carrying out the program outlined in the *Grundlagen*. The attempt failed because of the paradox of self-membership discovered by Bertrand Russell.

◇ ◇ ◇

Philipp Furtwängler (1869–1940) studied mathematics, physics, and chemistry at Göttingen, intending a career as a high school teacher. Stimulated by the lectures of Felix Klein, he went on to write a dissertation in number theory. He did not, however, submit a *Habilitationsschrift*, normally a prerequisite for a university career. Instead, after working for six years for the Prussian Geodetic Institute he was invited to teach surveying, first at an academy in Bonn and then at the Technische Hochschule in Aachen. While thus employed he continued his research in number theory, and in 1907 his proof of the reciprocity law for algebraic number fields was awarded a prize by the Göttingen Academy of Sciences. In 1910 he returned to Bonn to teach applied mathematics at the university there, but left two years later to take up a professorship in number theory at the University of Vienna. Shortly afterward he became paralyzed from the neck down, but despite his disability became renowned as a lecturer as well as a researcher. His most famous result, obtained at the age of sixty, was a proof of a conjecture of Hilbert, the so-called *Hauptidealsatz* (Prime Ideal Theorem) for class fields.

◇ ◇ ◇

Hans Hahn (1879–1934), Gödel's dissertation adviser, received his doctorate from the University of Vienna in 1905, where he served as *Privatdozent* from 1906 to 1909. Severely wounded in World War I, Hahn went to Bonn after his release from military service and within two years was appointed full professor there. He returned to Vienna in 1921, where he remained until his sudden death following cancer surgery. He made important contributions to set theory, the calculus of variations, and the theory of functions, and served for many years as an editor of the journal *Monatshefte für Mathematik und Physik*.

◇ ◇ ◇

In his short life, **Jacques Herbrand (1908–1931)** published important papers in both logic and number theory. His doctoral dissertation at the Sorbonne (1929) contained the statement (though with a faulty proof) of his Fundamental Theorem, according to which the validity of a prenex formula of quantification theory is finitarily reducible to the *propositional* validity of a disjunction of instances of its quantifier-free matrix.

Herbrand was one of the first to consider the class of what are now called general recursive functions; he did so in a paper published shortly after his death, in which he demonstrated, by finitary means, that a certain fragment of Peano arithmetic is formally consistent.

◇ ◇ ◇

Arend Heyting (1898–1980) received his doctorate in 1925 under the direction of L. E. J. Brouwer and spent most of his subsequent career at the University of Amsterdam. Though he published textbooks in projective geometry as well as logic, Heyting devoted most of his efforts to developing, formalizing, and popularizing the intuitionistic conception of mathematics. His explication of the intuitionistic interpretation of the logical operators, together with his axiomatization of various parts of intuitionistic mathematics, helped to make Brouwer's ideas intelligible to classical mathematicians and to ensure the survival of intuitionism as an active area of mathematical research.

◇ ◇ ◇

David Hilbert (1862–1943) was one of the two most preeminent mathematicians of his age. His breadth of mathematical achievements, spanning the fields of invariant theory, algebraic number theory, integral equations, axiomatics of geometry and logic, and mathematical physics, was rivaled only by that of Henri Poincaré, and his books *Theorie der algebraischen Zahlkörper* (1896), *Grundlagen der Geometrie* (1899), *Grundzüge der theoretischen Logik* (1928, co-authored with Wilhelm Ackermann), and *Grundlagen der Mathematik* (1934, co-authored with Paul Bernays) became mathematical classics. Born in Königsberg, where he also received his doctorate and began his professorial career, Hilbert accepted a call to Göttingen in 1895 and there carried on the mathematical tradition established a century before by Carl Friedrich Gauss. In 1900, in his address to the International Congress of Mathematicians, he posed a famous list of problems as challenges for the century ahead, and in 1917 he initiated the study of proof theory, whose development he hoped would secure the foundations of mathematics.

◇ ◇ ◇

Stephen C. Kleene (1909–1994), a student of Alonzo Church at Princeton, was one of the principal figures in the development of recursion theory. Among his many fundamental contributions to the field were the notion of partial recursive function, the definition of the arithmetical and analytical hierarchies, and the proofs of the Recursion, Normal Form, and Hierarchy Theorems. He contributed as well to interpretations of intuitionism and was author of the textbook *Introduction to Metamathematics* (1952), one of the basic references on modern logic.

Karl Menger (1902–1985), son of the distinguished Austrian economist Carl, is best known for his creation of dimension theory (discovered independently and almost simultaneously by the Russian Pavel Uryssohn). After receiving his doctorate from the University of Vienna in 1924, Menger served as a *Dozent* in Amsterdam for two years. In 1927 he returned to the University of Vienna as professor of geometry, and ten years later he emigrated to America, where he accepted a position at Notre Dame. One of Gödel's principal mentors, Menger was the founder of the mathematical colloquium at the University of Vienna and a member of the Schlick Circle. His work focused on curve theory, the algebra of functions, and mathematical pedagogy, especially notational reform. From 1948 until his retirement he was on the faculty of the Illinois Institute of Technology.

◇ ◇ ◇

A universal mathematician on a par with Hilbert, **(Jules) Henri Poincaré (1854–1912)** contributed profoundly to analytic function theory, algebraic geometry, number theory, differential equations and celestial mechanics. In addition, he defined the basic notions of simplicial complex, barycentric subdivision, homology group, and fundamental group, and thereby almost singlehandedly created the discipline of algebraic topology. His impact on logic was indirect, the result of his advocacy of mathematical constructivism and of his criticisms of the logicist and formalist programs.

◇ ◇ ◇

At the age of seven **Emil L. Post (1897–1954)** emigrated with his parents from Poland to New York City. After receiving his B.S. in 1917 from City College, to which he returned in 1935 to spend the rest of his career, he went on to Columbia for doctoral study; his dissertation (1920) introduced the method of truth tables as a decision procedure for propositional logic. Though his employment and research productivity were severely hampered by his affliction with manic-depression, he nonetheless initiated the study of degrees of recursive unsolvability, introduced notions central to the theory of automata and formal languages, and, independently of Gödel, both recognized the incompleteness of *Principia Mathematica* and proposed the notion of ordinal definability. In addition to logic he also published papers in analysis and algebra; the notion of polyadic group is due to him, as is that of what are now called Post algebras.

◇ ◇ ◇

Pacifist, moral iconoclast, and prolific author, **Bertrand Russell (1872–1970)** became in the eyes of the public the most famous philosopher of the twentieth century. His contributions to philosophy ranged widely over epistemology, metaphysics, ethics, and the foundations of mathematics, and his writings on political freedom earned him the Nobel Prize for Literature in 1950. Within mathematics Russell is known for the discovery of the paradox that bears his name, for his creation of the theory of types as a way of circumventing that paradox, and for his authorship, jointly with Albert North Whitehead, of the monumental treatise *Principia Mathematica*, in which the theory of types was employed as a means of resurrecting Frege's logicist program for the foundation of arithmetic.

◇ ◇ ◇

The Norwegian **Thoralf Skolem (1887–1963)** contributed to several areas of mathematics, most notably logic and number theory. Most of his career was spent at the University of Oslo, where he did his undergraduate work and to which he returned as *Dozent* in 1918 after two years study in Göttingen. He received his doctorate there in 1926 and became a full professor in 1938, following an eight-year stint at a private research institute in Bergen. Skolem's name is attached to a number of important theorems and concepts in logic, including, besides the Löwenheim-Skolem Theorem, the Skolem Paradox (that set theory, though it refers to uncountable collections, admits countable models); Skolem functions (function symbols added to a formal language as a means of eliminating existential quantifiers); and the Skolem normal form for satisfiability (a prenex formula whose satisfiability is equivalent to that of a given formula). Skolem was also the first to establish the existence of nonstandard models of arithmetic (structures that satisfy the same first-order sentences as the integers but are not isomorphic to them).

◇ ◇ ◇

The only twentieth-century logician whose stature rivaled that of Gödel, **Alfred Tarski (1901–1983)** began his career in Poland but came to the United States in 1939, where he eventually secured an appointment at the University of California at Berkeley. His landmark paper "The Concept of Truth in Formalized Languages" (*1956*), originally published in Polish in 1933, introduced the now-standard inductive definition of the satisfaction relation between sentences of and structures for a formal language, and pointed out that the related notion of truth was formally undefinable within the language itself. In contrast to Gödel, Tarski was a prolific author, was active in professional organizations, had many students, and had a strong interest in the interplay between logic and other fields of mathematics, especially algebra. Among the works for which he is best known are studies of equivalents and consequences of the Axiom of Choice (especially the Banach-Tarski paradox); the demonstration that first-order Euclidean geometry is decidable, whereas various other first-order theories (of groups, lattices, and so on) are not; and numerous results in cardinal arithmetic. Under his influence and direction Berkeley became a world center for research in logic.

◇ ◇ ◇

Alan Turing (1912–1954) was a seminal figure in the development of recursion theory and computer science, as well as a brilliant cryptanalyst who led the successful British assault on the German "Enigma" cipher during World War II. His abstract model of a universal computing machine and his demonstration of the unsolvability of the halting problem provided a new and remarkably perspicuous way of interpreting the undecidability results of Gödel and Church, and led to the general acceptance of Church's Thesis. Turing also contributed to the practical design and construction of two of the earliest general purpose digital computers.

◇ ◇ ◇

Oswald Veblen (1880–1960) received his Ph.D. from the University of Chicago in 1903 for a dissertation in which he presented an axiomatization of Euclidean geometry different from that given by Hilbert. His subsequent research focused on projective geometry, differential geometry, and topology, about each of which he wrote important textbooks. Among his many notable accomplishments was the first fully rigorous proof that a simple closed curve divides the plane into two disjoint, arcwise connected regions (the Jordan curve theorem). He taught at Princeton from 1905 to 1932, and at the IAS, where he was a driving force, from 1932 to 1950.

◇ ◇ ◇

The brilliant Hungarian mathematical prodigy **John von Neumann (1903–1957)** contributed profoundly to many areas of both pure and applied mathematics, including operator theory, measure theory, quantum mechanics, numerical analysis, and game theory. Early in his career he worked in set theory, where his definition of ordinal number and class formulation of the set-theoretic axioms are well known. In 1933 he was appointed as one of the original faculty members of the IAS, in whose employ he remained until his death. In the years during and after World War II he became heavily involved in various defense-related activities, including the Manhattan Project and the development of the EDVAC and IAS computers.

◇ ◇ ◇

A student of Hilbert, **Hermann Weyl (1885–1955)** was professor at the University of Zürich from 1913 until 1930. Following Hilbert's retirement he briefly returned to Göttingen, where he had begun his career as a *Privatdozent*; but when the Nazis came to power a few years later he emigrated to America to become one of the original five members of the mathematics faculty at the IAS. His research ranged widely over harmonic analysis, Lie groups, analytic number theory, general relativity theory, geometry, and topology, and he wrote a number of influential texts, including *Die Idee der Riemannschen Fläche* (The Concept of Riemann Surfaces, 1913); *Das Kontinuum* (1918), in which he espoused a constructivist philosophy of mathematics akin to Brouwer's intuitionism; *Raum, Zeit und Materie* (Space, Time and Matter, 1918); and *Gruppentheorie und Quantenmechanik* (Group Theory and Quantum Mechanics, 1928).

◇ ◇ ◇

Best known for having recognized the role of the Axiom of Choice in mathematical arguments and for having used it to prove that every set can be well ordered, **Ernst Zermelo (1871–1953)** was awarded a doctorate by the University of Berlin in 1894 for a dissertation on the calculus of variations. Ill health forced him to resign from a professorship at the University of Zürich, but in 1926 he was appointed honorary professor at the university in Freiburg im Breisgau. He renounced that position in 1935 in protest of Hitler's policies, but was reinstated in 1946. Zermelo's greatest achievement was his axiomatization of set theory, which, after modification by Abraham Fraenkel, became accepted as the standard formalization of Cantor's ideas.

Bibliography

Ackermann, Wilhelm
    1928   Zum Hilbertschen Aufbau der reellen Zahlen. *Mathematische Annalen* 99:118–133. English translation in *van Heijenoort 1967*, 493–507.

Baur, Susan
    1988   *Hypochondria: Woeful Imaginings.* Berkeley: University of California Press.

Benacerraf, Paul, and Hilary Putnam
    1964   *Philosophy of Mathematics: Selected Readings.* Englewood Cliffs, N.J.: Prentice-Hall.

Bernays, Paul
    1926   Axiomatische Untersuchung des Aussagen-Kalkuls der *Principia Mathematica. Mathematische Zeitschrift* 25:305–320.
    1935   Hilberts Untersuchungen über die Grundlagen der Arithmetik. In *Hilbert 1935*, 196–216.
    1967   Hilbert, David. In Paul Edwards, ed., *The Encyclopedia of Philosophy* 3:496–504. New York: Macmillan and the Free Press.

Borel, Armand
    1989   The School of Mathematics at the Institute for Advanced Study. In Peter Duren, ed., *A Century of Mathematics in America, Part III*, 119–147. Providence: American Mathematical Society.

Boolos, George
    1979   *The Unprovability of Consistency.* Cambridge: Cambridge University Press.

Brouwer, Luitzen E. J.
    1975   *Collected Works*, ed. Arend Heyting. Amsterdam: North-Holland Publishing Co.

329

Browder, Felix, ed.
1976   *Mathematical Developments Arising from the Hilbert Problems.* Proceedings
        of Symposia in Pure Mathematics XXVIII, pts. 1 and 2. Providence:
        American Mathematical Society.

Calude, Cristian, Solomon Marcus, and Ionel Tevy
1979   The First Example of a Recursive Function Which Is Not Primitive
        Recursive. *Historia Mathematica* 6:380–384.

Cantor, Georg
1870   Beweis, daß eine für jede reellen Wert von x durch eine Reihe gegebe-
        nen Funktion f(x) sich nur auf eine einzige Weise in dieser Form darstellen
        läßt. *Journal für die reine und angewandte Mathematik* 72:139–142.
1872   Über die Ausdehnung eines Satzes aus der Theorie der trigonometri-
        schen Reihen. *Mathematische Annalen* 5:123–132.
1874   Über eine Eigenschaft des Inbegriffes aller reelen algebraischen Zahlen.
        Journal für die reine und angewandte Mathematik 77:258–262.
1878   Ein Beitrag zur Mannigfaltigkeitslehre. *Journal für die reine und angewandte
        Mathematik* 84:242–258.
1891   Über eine elementare Frage der Mannigfaltigkeitslehre. *Jahresbericht der
        Deutschen Mathematiker-Vereinigung* I:75–78.

Carnap, Rudolf
1963   Intellectual Autobiography. In Paul A. Schilpp, ed., *The Philosophy of
        Rudolf Carnap*, 3–84. La Salle, Ill.: Open Court Publishing Co.

Ceruzzi, Paul E.
1983   *Reckoners, the Prehistory of the Digital Computer, from Relays to the Stored
        Program Concept, 1933–1945.* Westport, Conn.: Greenwood Press.

Chandrasekhar, Subrahmanyan, and James P. Wright
1961   The Geodesics in Gödel's Universe. *Proceedings of the National Academy
        of Sciences, U.S.A.* 47:341–347.

Chihara, Charles
1973   *Ontology and the Vicious-circle Principle.* Ithaca, N.Y.: Cornell University
        Press.

Christian, Curt
1980   Leben und Wirken Kurt Gödels. *Monatshefte für Mathematik* 89:261–273.

Church, Alonzo
1932   A Set of Postulates for the Foundation of Logic. *Annals of Mathematics*,
        2d ser., 33:346–366.

1933   A Set of Postulates for the Foundation of Logic (Second Paper). *Annals of Mathematics*, 2d ser., 34:839–864.

1936   A Bibliography of Symbolic Logic. *The Journal of Symbolic Logic* 1:121–218.

Clare, George
1980   *Last Waltz in Vienna: The Rise and Destruction of a Family, 1842–1942.* New York: Avon Books.

Cohen, Paul J.
1966   *Set Theory and the Continuum Hypothesis.* New York: W. A. Benjamin, Inc.

Crossley, John N.
1975   Reminiscences of Logicians. In J.N. Crossley, ed., *Algebra and Logic. Papers from the 1974 Summer Research Institute of the Australian Mathematical Society, Monash University, Australia,* Lecture Notes in Mathematics no. 450, 1–62. Berlin: Springer.

Dauben, Joseph Warren
1979   *Georg Cantor, His Mathematics and Philosophy of the Infinite.* Cambridge, Mass.: Harvard University Press.

Davis, Martin
1965   *The Undecidable.* Hewlett, N.Y.: Raven Press.

1982   Why Gödel Didn't Have Church's Thesis. *Information and Control* 54:3–24.

1987   Mathematical Logic and the Origin of Modern Computers. In Esther R. Phillips, ed., *Studies in the History of Mathematics,* MAA Studies in Mathematics 26:137–165. Washington, D.C.: Mathematical Association of America.

Dawson, Jr., John W.
1984a   Discussion on the Foundation of Mathematics. *History and Philosophy of Logic* 5:111–129.

1984b   Kurt Gödel in Sharper Focus. *The Mathematical Intelligencer* 6(4):9–17.

1985a   The Reception of Gödel's Incompleteness Theorems. In *PSA 1984,* Proceedings of the 1984 Biennial Meeting of the Philosophy of Science Association 2:253–271. East Lansing, Mich.: Philosophy of Science Association. Reprinted in Thomas Drucker, ed., *Perspectives on the History of Mathematical Logic,* 84–100. Boston-Basel-Berlin: Birkhäuser.

1985b   Completing the Gödel-Zermelo Correspondence. *Historia Mathematica* 12:66–70.

1993   The Compactness of First-order Logic: From Gödel to Lindström. *History and Philosophy of Logic* 14:15-37.

Dowling, William F.
1989   There Are No Safe Virus Tests. *American Mathematical Monthly* 96:835-836.

Einhorn, Rudolf
1985   *Vertreter der Mathematik und Geometrie an den Wiener Hochschulen 1900-1940.* Vienna: Verband der wissenschaftlichen Gesellschaften Österreichs.

Feferman, Solomon
1960   Arithmetization of Metamathematics in a General Setting. *Fundamenta Mathematicae* 49:35-92.
1984   Kurt Gödel: Conviction and Caution. *Philosophia Naturalis* 21:546-562.
1986   Gödel's Life and Work. In *Gödel 1986-,* I:1-36.

Feigl, Herbert
1969   The Wiener Kreis in America. In Donald Fleming and Bernard Bailyn, eds., *The Intellectual Migration: Europe and America, 1930-1960,* 630-673. Cambridge, Mass.: Harvard University Press.

Fermi, Laura
1971   *Illustrious Immigrants,* 2d ed. Chicago: University of Chicago Press.

Ferreirós, José
1993   On the Relations Between Georg Cantor and Richard Dedekind. *Historia Mathematica* 20:343-363.

Finsler, Paul
1926   Formale Beweise und die Entscheidbarkeit. *Mathematische Zeitschrift* 25:676-682.
1944   Gibt es unentscheidbare Sätze? *Commentarii Mathematici Helvetici* 16:310-320.

Floyd, Juliet
1995   On Saying What You Really Want To Say: Wittgenstein, Gödel, and the Trisection of the Angle. In Jaakko Hintikka, ed., *From Dedekind to Gödel: Essays on the Development of the Foundations of Mathematics,* 373-425. Boston and Dordrecht: Kluwer.

Fraenkel, Abraham
1921   Über die Zermelosche Begründung der Mengenlehre. *Jahresbericht der Deutschen Mathematiker-Vereinigung (Angelegenheiten)* 30:97-98.

1922a   Zu den Grundlagen der Cantor-Zermeloschen Mengenlehre. *Mathematische Annalen* 86:230–237.

1922b   Der Begriff "definit" und die Unabhängigkeit des Auswahlaxioms. *Sitzungsberichte der Preussischen Akademie der Wissenschaften, Physikalisch-mathematische Klasse*, 253–257. English translation in *van Heijenoort 1967*, 284–289.

Frege, Gottlob
1879   *Begriffsschrift, eine der arithmetischen nachgebildete Formelsprache des reinen Denkens.* Halle: Nebert. English translation in *van Heijenoort 1967*, 1–82.

Fried, Yehuda, and Joseph Agassi
1976   *Paranoia: A Study in Diagnosis.* Boston Studies in the Philosophy of Science, 50. Dordrecht and Boston: D. Reidel.

Gandy, Robin
1980   Church's Thesis and Principles for Mechanisms. In Jon Barwise, H. J. Keisler, and K. Kunen, eds., *The Kleene Symposium, Proceedings of the Symposium Held June 18–24, 1978 at Madison, Wisconsin, U.S.A.*, 123–148. Amsterdam and New York: North-Holland Publishing Co.

1988   The Confluence of Ideas in 1936. In R. Herken, ed., *The Universal Turing Machine. A Half-Century Survey*, 55–111. New York and Oxford: Oxford University Press.

Givant, Steven R.
1991   A Portrait of Alfred Tarski. *The Mathematical Intelligencer* 13(3):16–32.

Gödel, Kurt
1929   Über die Vollständigkeit des Logikkalküls. Doctoral diss., University of Vienna, n.d. Reprinted and translated in *Gödel 1986–*, I:60–101.

1930a   Die Vollständigkeit der Axiome des logischen Funktionenkalküls. *Monatshefte für Mathematik und Physik* 37:349–360. Reprinted and translated in *Gödel 1986–*, I:102–123.

1930b   Einige metamathematische Resultate über Entscheidungsdefinitheit und Widerspruchsfreiheit. *Anzeiger der Akademie der Wissenschaften in Wien* 67:214–215. Reprinted and translated in *Gödel 1986–*, I:140–143.

1931a   Über formal unentscheidbare Sätze der *Principia Mathematica* und verwandter Systeme I. *Monatshefte für Mathematik und Physik* 38:173–198. Reprinted and translated in *Gödel 1986–*, I:144–195.

1931b   Review of *Hilbert 1931a. Zentralblatt für Mathematik und ihre Grenzgebiete* 1:260. Reprinted and translated in *Gödel 1986–*, I:212–215.

1932a   Zum intuitionistischen Aussagenkalkül. *Anzeiger der Akademie der Wissenschaften in Wien* 69:65–66. Reprinted and translated in *Gödel 1986–*, I:222–225.

1932b Ein Spezialfall des Entscheidungsproblems der theoretischen Logik. *Ergeb-nisse eines mathematischen Kolloquiums* 2:27–28. Reprinted and translated in *Gödel 1986–*, I:230–235.

1932c Über Vollständigkeit und Widerspruchsfreiheit. *Ergebnisse eines mathematischen Kolloquiums* 3:12–13. Reprinted and translated in *Gödel 1986–*, I:234–237.

1932d Eine Eigenschaft der Realisierungen des Aussagenkalküls. *Ergebnisse eines mathematischen Kolloquiums* 3:20–21. Reprinted and translated in *Gödel 1986–*, I:238–241.

1933a Zur intuitionistischen Arithmetik und Zahlentheorie. *Ergebnisse eines mathematischen Kolloquiums* 4:34–38. Reprinted and translated in *Gödel 1986–*, I:286–295.

1933b Eine Interpretation des intuitionistischen Aussagenkalküls. *Ergebnisse eines mathematischen Kolloquiums* 4:39–40. Reprinted and translated in *Gödel 1986–*, I:300–303.

1933c Zum Entscheidungsproblem des logischen Funktionenkalküls. *Monatshefte für Mathematik und Physik* 40:433–443. Reprinted and translated in *Gödel 1986–*, I:306–327.

1934a On Undecidable Propositions of Formal Mathematical Systems. In *Gödel 1986–*, I:346–371.

1934b Review of *Skolem 1933*. *Zentralblatt für Mathematik und ihre Grenzgebiete* 7:193–194. Reprinted and translated in *Gödel 1986–*, I:378–381.

1936a [Untitled discussion remark on mathematical economics]. *Ergebnisse eines mathematischen Kolloquiums* 7:6. Reprinted and translated in *Gödel 1986–*, I:392–393.

1936b Über die Länge von Beweisen. *Ergebisse eines mathematischen Kolloquiums* 7:23–24. Reprinted and translated in *Gödel 1986–*, I:396–399.

1938 The Consistency of the Axiom of Choice and of the Generalized Continuum Hypothesis. *Proceedings of the National Academy of Sciences, U.S.A.* 24:556–557. Reprinted in *Gödel 1986–*, II:26–27.

1939a The Consistency of the Generalized Continuum Hypothesis. *Bulletin of the American Mathematical Society* 45:93. Reprinted in *Gödel 1986–*, II:27.

1939b Consistency Proof for the Generalized Continuum Hypothesis. *Proceedings of the National Academy of Sciences, U.S.A.* 25:220–224; corrigenda in *Gödel 1947*, n. 23. Reprinted with corrections in *Gödel 1986–*, II: 28–32.

1940 *The Consistency of the Axiom of Choice and of the Generalized Continuum Hypothesis with the Axioms of Set Theory.* Princeton: Princeton University Press. Reprinted in *Gödel 1986–*, II:33–101.

1944 Russell's Mathematical Logic. In *Schilpp 1944*, 123–153. Reprinted in *Gödel 1986–*, II:119–141.

1946   Remarks Before the Princeton Bicentennial Conference on Problems of
            Mathematics. In *Davis 1965*, 84–88. Reprinted in *Gödel 1986–*, II:150–
            153.

1946/9   Some Observations About the Relationship Between Theory of Relativ-
            ity and Kantian Philosophy. In *Gödel 1986–*, III:230–259.

1947   What Is Cantor's Continuum Problem? *American Mathematical Monthly*
            54:515–525; errata, 55:151. Reprinted with corrections in *Gödel 1986–*,
            II:176–187.

1949a   An Example of a New Type of Cosmological Solutions of Einstein's
            Field Equations of Gravitation. *Reviews of Modern Physics* 21:447–450.
            Reprinted in *Gödel 1986–*, II:190–198.

1949b   A Remark About the Relationship Between Relativity Theory and Ide-
            alistic Philosophy. In *Schilpp 1949*, 555–561. Reprinted in *Gödel 1986–*,
            II:202–207.

1952   Rotating Universes in General Relativity Theory. *Proceedings of the Inter-
            national Congress of Mathematicians, Cambridge, Massachusetts, U.S.A., August
            30–September 6, 1950*, I:175–181. Reprinted in *Gödel 1986–*, II:208–216.

1958   Über eine bisher noch nicht benützte Erweiterung des finiten Stand-
            punktes. *Dialectica* 12:280–287. Reprinted in *Gödel 1986–*, II:240–251.

1964   Revised and expanded version of *Gödel 1947*. In *Benacerraf and Putnam
            1964*, 258–273. Reprinted in *Gödel 1986–*, II:254–270.

1972   Some Remarks on the Undecidability Results. In *Gödel 1986–*, II:305–
            306.

1986–   *Collected Works*, ed. Solomon Feferman et al. 3 vols. to date. New York
            and Oxford: Oxford University Press.

Gödel, Rudolf
1987   History of the Gödel Family. In *Weingartner and Schmetterer 1987*, 13–27.

Goldfarb, Warren
1979   Logic in the Twenties: The Nature of the Quantifier. *The Journal of
            Symbolic Logic* 44:351–368.

1984   The Unsolvability of the Gödel Class with Identity. *The Journal of Sym-
            bolic Logic* 49:1237–1252.

Goldstine, Herman H.
1972   *The Computer from Pascal to von Neumann.* Princeton: Princeton Univer-
            sity Press.

Grattan-Guinness, Ivor
1979   In Memoriam Kurt Gödel: His 1931 Correspondence with Zermelo on
            His Incompletability Theorem. *Historia Mathematica* 6:294–304.

Gulick, Charles A.
1948   *Austria from Habsburg to Hitler.* 2 vols. Berkeley: University of California
          Press.

Hahn, Hans
1980   *Empiricism, Logic and Mathematics: Philosophical Papers,* ed. Brian McGuin-
          ness. Dordrecht-Boston-London: D. Reidel.

Hartmanis, Juris
1989   Gödel, von Neumann, and the P = NP Problem. *Bulletin of the European
          Association for Theoretical Computer Science* 38:101–107.

Heck, Jr., Richard G.
1993   The Development of Arithmetic in Frege's *Grundgesetze der Arithmetik.*
          The Journal of Symbolic Logic 58:579–601.

Hempel, Carl G.
1979   Der Wiener Kreis: eine persönliche Perspektive. In H. Berghel, A.
          Hubner, and E. Koehler, eds., *Wittgenstein, the Vienna Circle, and Critical
          Rationalism,* 21–26. Vienna: Holder-Pichler-Tempsky.
1981   Der Wiener Kreis und die Metamorphosen seines Empirismus. In N.
          Leser, ed., *Das geistige Leben Wiens in der Zwischenkriegszeit,* 205–215. Vi-
          enna: Österreichischer Bundesverlag.

Herbrand, Jacques
1930   Recherches sur la théorie de la démonstration. Doctoral diss., University
          of Paris.
1931   Sur le problème fondamental de la logique mathématique. *Sprawozdania
          z posiedzen Towarzystwa Naukowego Warszawskiego Wydzial III,* 24:12–56.

Hilbert, David
1900   Mathematische Probleme. Vortrag, gehalten auf dem internationalen
          Mathematiker-Kongress zu Paris 1900. *Nachrichten von der Königlichen
          Gesellschaft der Wissenschaften zu Göttingen,* 253–297. English translation
          in *Browder 1976,* pt. 1:1–34.
1923   Die logischen Grundlagen der Mathematik. *Mathematische Annalen* 88:151–
          165. Reprinted in *Hilbert 1935,* 178–191.
1926   Über das Unendliche. *Mathematische Annalen* 95: 161–190. English trans-
          lation in *van Heijenoort 1967,* 367–392.
1929a   Probleme der Grundlegung der Mathematik. In *Atti del Congresso Inter-
          nazionale dei Matematici, Bologna 3–10 Settembre 1928,* 135–141.
1929b   Amended reprint of *Hilbert 1929. Mathematische Annalen* 102:1–9.
1931a   Die Grundlegung der elementaren Zahlenlehre. *Mathematische Annalen*
          104:485–494.

1931b   Beweis des tertium non datur. *Nachrichten von der Gesellschaft der Wissenschaften zu Göttingen, mathematisch-physikalische Klasse,* 120–125.

1935   *Gesammelte Abhandlungen.* Vol. 3. Berlin: Julius Springer Verlag.

Hilbert, David, and Wilhelm Ackermann

1928   *Grundzüge der theoretischen Logik.* Berlin: Julius Springer Verlag.

Hilbert, David, and Paul Bernays

1939   *Grundlagen der Mathematik.* Vol. II. Berlin: Springer.

Hoffmann, Joseph

n.d.   *Sanatorium Purkersdorf.* New York: Galerie Metropol.

Holton, Gerald, and Yehuda Elkana, eds.

1982   *Albert Einstein, Historical and Cultural Perspectives: The Centennial Symposium in Jerusalem.* Princeton: Princeton University Press.

Jackman, Jarrell C., and Carla M. Borden, eds.

1983   *The Muses Flee Hitler.* Washington: Smithsonian Institution Press.

Janik, Allan, and Stephen Toulmin

1973   *Wittgenstein's Vienna.* New York: Touchstone/Simon and Schuster.

Johnston, William M.

1972   *The Austrian Mind: An Intellectual and Social History, 1848–1938.* Berkeley: University of California Press.

1981   *Vienna, Vienna: The Golden Age, 1815–1914.* New York: Clarkson N. Potter, Inc.

Kennedy, Hubert C.

1980   *Peano: Life and Works of Giuseppe Peano.* Dordrecht-Boston-London: D. Reidel.

Kitcher, Philip

1983   *The Nature of Mathematical Knowledge.* New York and Oxford: Oxford University Press.

Kleene, Stephen C.

1981   Origins of Recursive Function Theory. *Annals of the History of Computing* 3:52–67. Corrigenda in *Davis 1982,* ns. 10, 12.

1987a   Gödel's Impression on Students of Logic in the 1930s. In *Weingartner and Schmetterer 1987,* 49–64.

1987b   Kurt Gödel, 1906–1978. *Biographical Memoirs of the National Academy of Sciences* 56:135–178.

Kneale, William, and Martha Kneale
1962   *The Development of Logic.* Oxford: Oxford University Press.

Kolmogorov, Andrei N.
1925   On the Principle of the Excluded Middle (in Russian). *Matematicheskii sbornik* 32:646–667.

Komjathy, Anthony, and Rebecca Stockwell
1980   *German Minorities and the Third Reich.* New York and London: Holmes & Meier.

Kreisel, Georg
1980   Kurt Gödel: 1906–1978. *Biographical Memoirs of Fellows of the Royal Society* 26:149–224; corrigenda, 27:697.

Lifton, Robert J.
1986   *The Nazi Doctors: Medical Killing and the Psychology of Genocide.* New York: Basic Books.

Lindenbaum, Adolf, and Alfred Tarski
1926   Communication sur les recherches de la théorie des ensembles. *Comptes Rendus des Séances de la Société des Sciences et des Lettres de Varsovie, Classe III,* 19:299–330.

Löwenheim, Leopold
1915   Über Möglichkeiten im Relativkalkul. *Mathematische Annalen* 76:447–470.

Maddy, Penelope
1990   *Realism in Mathematics.* Oxford: Clarendon Press.

Mayrhofer, Karl
1934   Hans Hahn. *Monatshefte für Mathematik und Physik* 41:221–238.

McAloon, Kenneth
1971   Consistency Results About Ordinal Definability. *Annals of Mathematical Logic* 2:449–467.

Menger, Karl
1934   Hans Hahn. *Ergebnisse eines mathematischen Kolloquiums* 6:40–44.
1981   Recollections of Kurt Gödel (English translation by Eckehart Köhler of the unpublished typescript "Erinnerungen an Kurt Gödel"). Published in revised form as "Memories of Kurt Gödel," in *Menger 1994,* 200–236.
1994   *Reminiscences of the Vienna Circle and the Mathematical Colloquium,* ed. Louise Golland, Brian McGuinness, and Abe Sklar. Dordrecht-Boston-London: Kluwer Academic Publishers.

Mirimanoff, Dmitry
  1917   Les antinomies de Russell et de Burali-Forti et le problème fondamental de la théorie des ensembles. *Enseignement Mathematique* 19:37–52.

Mitchell, Janet A., ed.
  1980   *A Community of Scholars: The Institute for Advanced Study, Faculty and Members 1930–1980.* Princeton: The Institute for Advanced Study.

Moore, Gregory H.
  1982   *Zermelo's Axiom of Choice, Its Origins, Development, and Influence.* Studies in the History of Mathematics and Physical Sciences. Vol. 8. New York: Springer-Verlag.
  1988a   The Emergence of First-order Logic. In William Aspray and Philip Kitcher, eds., *History and Philosophy of Modern Mathematics.* Minnesota Studies in the Philosophy of Science, XI:95–135. Minneapolis: University of Minnesota Press.
  1988b   The Origins of Forcing. In Frank R. Drake and John K. Truss, eds., *Logic Colloquium '86,* 143–173. Amsterdam: Elsevier Science Publishers.
  1989   Towards a History of Cantor's Continuum Problem. In David E. Rowe and John McCleary, eds., *The History of Modern Mathematics, I: Ideas and Their Reception,* 79–121. Boston: Academic Press.
  1990   Kurt Friedrich Gödel. In Frederic L. Holmes, ed., *The Dictionary of Scientific Biography* 17:348–357. New York: Charles Scribner's Sons.

Moschovakis, Yiannis
  1989   Commentary on Mathematical Logic. In Peter Duren, ed., *A Century of Mathematics in America,* 343–346. Providence: American Mathematical Society.

Myhill, John, and Dana Scott
  1971   Ordinal Definability. In D. Scott, ed., *Axiomatic Set Theory.* Proceedings of Symposia in Pure Mathematics 13, pt. 1:271–278. Providence: American Mathematical Society.

Pais, Abraham
  1982   *Subtle Is the Lord: The Science and the Life of Albert Einstein.* New York: Oxford University Press.

Paris, Jeff, and Leo Harrington
  1977   A Mathematical Incompleteness in Peano Arithmetic. In Jon Barwise, ed., *Handbook of Mathematical Logic,* 1133–1142. Amsterdam: North-Holland Publishing Co.

Parsons, Charles
  1990   Introductory Note to *Gödel 1944*. In *Gödel 1986–*, II:102–118.
  1995   Platonism and Mathematical Intuition in Kurt Gödel's Thought. *The Bulletin of Symbolic Logic* 1:44–74.

Pauley, Bruce F.
  1981   *Hitler and the Forgotten Nazis: A History of Austrian National Socialism.* Chapel Hill: University of North Carolina Press.

Peano, Giuseppe
  1888   *Calcolo geometrico secondo l'Ausdehnungslehre di H. Grassmann, preceduto dalle Operazione della logica deduttiva.* Turin.
  1889   *Arithmetices principia, nova methodo exposita.* Turin.

Peirce, Charles S.
  1885   On the Algebra of Logic: A Contribution to the Philosophy of Notation. *American Journal of Mathematics* 7: 180–202. Reprinted in Charles Hartshorne and Paul Weiss, eds., *The Collected Papers of Charles Sanders Peirce*, 3:104–157. Cambridge, Mass.: Harvard University Press, 1933.

Post, Emil L.
  1921   Introduction to a General Theory of Elementary Propositions. *American Journal of Mathematics* 43:169–173. Reprinted in *van Heijenoort 1967*, 265–283.
  1936   Finite Combinatory Processes. Formulation I. *The Journal of Symbolic Logic* 1:103–105. Reprinted in *Davis 1965*, 289–291.

Princeton University
  1947   *Problems of Mathematics.* Princeton University bicentennial conferences, ser. 2, conf. 2. Princeton: Princeton University.

Putnam, Hilary
  1982   Peirce the Logician. *Historia Mathematica* 9:290–301.

Quine, Willard Van Orman
  1987   Peano as Logician. *History and Philosophy of Logic* 8:15–24.

Russell, Bertrand
  1906   On Some Difficulties in the Theory of Transfinite Numbers and Order Types. *Proceedings of the London Mathematical Society*, 2d ser., 4:29–53.

Sachs, Judith, ed.
  1955   *The Institute for Advanced Study: Publications of Members, 1930–1954.* Princeton: The Institute for Advanced Study.

Schilpp, Paul A., ed.
1944   *The Philosophy of Bertrand Russell.* Evanston, Ill.: Northwestern University
          Press.
1949   *Albert Einstein, Philosopher-Scientist.* New York: Tudor Publishing Com-
          pany.

Schimanovich, Werner, and Peter Weibel
1986   *Kurt Gödel: Ein mathematisches Mythos.* Vienna. Videotape.

Shepherdson, John C.
1953   Inner Models of Set Theory. Part III. *The Journal of Symbolic Logic* 18:145–
          167.

Sieg, Wilfried
1994   Mechanical Procedures and Mathematical Experience. In Alexander
          George, ed., *Mathematics and Mind,* 71–117. New York and Oxford:
          Oxford University Press.

Siegert, Michael
1981   Mit dem Browning philosophiert. *Forum* (July/August), 18–26.

Simpson, Stephen G.
1988   Partial Realizations of Hilbert's Program. *The Journal of Symbolic Logic*
          53:349–363.

Skolem, Thoralf
1923a   Begründung der elementaren Arithmetik durch die rekurrierende Denkweise
          ohne Anwendung scheinbarer Veränderlichen mit unendlichem Aus-
          dehnungsbereich. *Skrifter utgit av Videnskapsselskapet i Kristiana, I. Matematisk-
          naturvidenskabelig klasse* 6:1–38.
1923b   Einige Bemerkungen zur axiomatischen Begründung der Mengenlehre.
          In *Matematiker kongressen i Helsingfors 4–7 Juli 1922, Den femte skandi-
          naviska matematikerkongressen, Redogörelse,* 217–232. Helsinki: Akademiska
          Bokhandlen.
1933   Über die Unmöglichkeit einer vollständigen Charakterisierung der Zahlen-
          reihe mittels eines endlichen Axiomensystems. *Norsk matematisk forenings
          skrifter,* ser. 2, 10:73-82.

Stein, Howard
1970   On the Paradoxical Time-structures of Gödel. *Philosophy of Science* 37:589–
          601.

Stern, Beatrice M.
1964   A History of the Institute for Advanced Study, 1930–1950. J. Robert Oppenheimer Papers, Library of Congress. Microfilm of unpublished typescript.

Stritch, Thomas
1981   The Foreign Legion of Father O'Hara. *Notre Dame Magazine* 10:23–27.

Sudan, Gabriel
1927   Sur le nombre ω^ω. *Bulletin Mathématique de la Société Roumaine des Sciences* 30:11–30.

Tarski, Alfred
1932   Der Wahrheitsbegriff in den Sprachen der deduktiven Disziplinen. *Anzeiger der Akademie der Wissenschaften in Wien* 69:23–25.
1933   Pojecie prawdy w jezykach nauk dedukcyjnych. *Prace Towarzystwa Naukowego Warszawskiego, Wydział III*, no. 34.
1935   Der Wahrheitsbegriff in den formalisierten Sprachen. *Studia philosophica (Lemberg)* 1:261–405.
1956   The Concept of Truth in Formalized Languages. Revised English translation of *Tarski 1935*, in J. H. Woodger, ed., *Logic, Semantics, Metamathematics: Papers from 1923 to 1928*, 152–278. Oxford: Clarendon Press.

Taussky-Todd, Olga
1987   Remembrances of Kurt Gödel. In *Weingartner and Schmetterer 1987*, 31–41.

Torretti, Roberto
1978   *Philosophy of Geometry from Riemann to Poincaré.* Dordrecht-Boston-Lancaster: D. Reidel.

Turing, Alan M.
1937   On Computable Numbers, with an Application to the Entscheidungsproblem. *Proceedings of the London Mathematical Society (2)* 42:230–265. Corrigenda, 43:544–546. Reprinted in *Davis 1965*, 116–154.

Ulam, Stanislaw M.
1976   *Adventures of a Mathematician.* New York: Charles Scribner's Sons.

van Heijenoort, Jean, ed.
1967   *From Frege to Gödel: A Source Book in Mathematical Logic, 1879–1931.* Cambridge, Mass.: Harvard University Press.
1985   *Selected Essays.* Naples: Bibliopolis.

von Neumann, John
1925 Eine Axiomatisierung der Mengenlehre. *Journal für die reine und angewandte Mathematik* 154:219–240.
1928a Über die Definition durch transfinite Induktion und verwandte Fragen der allgemeine Mengenlehre. *Mathematische Annalen* 99:373–391.
1928b Die Axiomatisierung der Mengenlehre. *Mathematische Zeitschrift* 27:669–752.

Wang, Hao
1974 *From Mathematics to Philosophy*. London: Routledge and Kegan Paul.
1981 Some Facts About Kurt Gödel. *The Journal of Symbolic Logic* 46:653–659.
1987 *Reflections on Kurt Gödel*. Cambridge, Mass.: MIT Press.
1996 *A Logical Journey: From Gödel to Philosophy*. Cambridge, Mass.: MIT Press.

Weingartner, Paul, and Leopold Schmetterer, eds.
1987 *Gödel Remembered*. Naples: Bibliopolis.

Weyl, Hermann
1946 Review of *The Philosophy of Bertrand Russell*. *American Mathematical Monthly* 53:599–605.
1953 Universities and Science in Germany. In K. Chandrasekharan, ed., *Hermann Weyl: Gesammelte Abhandlungen*, IV:537–562. Berlin: Springer, 1968.

Whitehead, Albert North, and Bertrand Russell
1910 *Principia Mathematica*. Cambridge: Cambridge University Press.

Wiedemann, Hans-Rudolf
1989 *Briefe grosser Naturforscher und Ärtzte in Handschriften*. Lübeck: Verlag Graphische Werkstätten.

Woolf, Harry, ed.
1980 *Some Strangeness in the Proportion: A Centennial Symposium to Celebrate the Achievements of Albert Einstein*. Reading, Mass.: Addison-Wesley.

Yourgrau, Palle
1991 *The Disappearance of Time: Kurt Gödel and the Idealistic Tradition in Philosophy*. Cambridge: Cambridge University Press.

Zermelo, Ernst
1908a Neuer Beweis für die Möglichkeit einer Wohlordnung. *Mathematische Annalen* 65:107–128.
1908b Untersuchungen über die Grundlagen der Mengenlehre. I. *Mathematische Annalen* 65:261–281.

1930   Über Grenzzahlen und Mengenbereiche: Neue Untersuchungen über die Grundlagen der Mengenlehre. *Fundamenta Mathematicae* 16:29–47.

1932   Über Stufen der Quantifikation und die Logik des Unendlichen. *Jahresbericht der Deutschen Mathematiker-Vereinigung* 41, pt. 2:85–88.

Zweig, Stefan

1964   *The World of Yesterday.* Lincoln: University of Nebraska Press.

Index

Numbers enclosed in brackets refer to endnotes (pages 271-312). The suffix '*n*' appended to a page number indicates a footnote. Dates in italics, with or without a letter suffix, refer to Gödel's publications, as listed in the Bibliography.